CONSTRUCTION ENGINEERING and MANAGEMENT

CONSTRUCTION ENGINEERING and MANAGEMENT

DR. S.V. DEODHAR
EX. DIRECTOR,
SSVPS COLLEGE OF ENGINEERING,
DHULE, MAHARASHTRA

S.C. SHARMA
FORMER GENERAL MANAGER
THDC INDIA LIMITED

KHANNA BOOK PUBLISHING CO. (P) LTD.
Publisher of Engineering and Computer Books
4C/4344, Ansari Road, Darya Ganj, New Delhi-110002
Phone : 011-23244447-48 **Mobile:** +91-9910909320
E-mail : contact@khannabooks.com
Website : www.khannabooks.com

Dear Readers,

To prevent the piracy, this book is secured with HIGH SECURITY HOLOGRAM on the front title cover. In case you don't find the hologram on the front cover title, please write us to at contact@khannabooks.com or whatsapp us at +91-99109 09320 and avail special gift voucher for yourself.

Specimen of Hologram on front Cover title:

Moreover, there is a SPECIAL DISCOUNT COUPON for you with EVERY HOLOGRAM.

How to avail this SPECIAL DISCOUNT:

Step 1: Scratch the hologram
Step 2: Under the scratch area, your "coupon code" is available
Step 3: Logon to www.khannabooks.com
Step 4: Use your "coupon code" in the shopping cart and get your copy at a special discount
Step 5: Enjoy your reading!

Price: ₹ 458.00

ISBN: 978-93-91505-35-6

Construction Engineering and Management
by S.V. Deodhar, S.C. Sharma

First Edition: 2018
Second Edition: 2022
Reprint: 2025

To view complete list of books, Please scan the QR Code:

Published by:
Khanna Book Publishing Co. (P) Ltd.
CIN: U22110DL1998PTC095547

Visit us at: **www.khannabooks.com**
Write us at: **contact@khannabooks.com**

Printed in Bharat.

PREFACE

Construction engineering is an inter-related complex of operations whereby building and installation are created, the end-product of which is finished building of other inter-structural works for its function use. Construction process may be mechanized or non-mechanized and in degree of complexity simple or complicated. The construction projects effects the economics, social and cultural activities and values of society. Most construction projects complex are complex both technically and commercially.

Construction covers various forms of activities ranging from houses to high rise buildings, industrial structures, road construction, express ways, bridges, major dams, barrages, runways, ports, canals, railways and other infrastructures. These high value projects involve management of material, equipment, human and financial resources, information system, control management and other interrelated aspects.

In major projects with modern technology there is need for detailed planning and management techniques, with the growing use of machinery it has become necessary for construction engineers to be thoroughly familiar with the working application and upkeep of the wide range of the modern equipment.

Since modern construction projects require a huge amount of capital, we have to adopt latest technology, modern equipment and modern management technology to achieve economy, quality, and quick results.

We have not given any new ideas or new theories, but have made an attempt to write and compile the subject matter in a systematic form and in simple language as per today's need, We thank Khanna Book Publishing company for bringing this book in market in shortest possible time.

The book has been divided in two parts namely, "Construction engineering and management" and "Construction equipment". The first part deals with management aspects different function and construction methods about major construction activities worried out in various construction projects. The second part of the book deals with basis of construction equipment, earth-morning equipment, hauling equipments, concerting equipments, pumping and de–watering equipment, and housing equipment.

The book cover the syllabii of different universities on the subject, any suggestion for further improvement of the book will be welcomed.

CONTENTS

Chapter

MANAGEMENT

1.1 MANAGEMENT

Management is described as the art of directing industrial activities i.e. the ability to accomplish a plan when others do the work. It is the art of getting work done efficiently, effectively, solving and removing all hurdles encountered in between. It therefore requires a ablest, daring, energetic person, who can take correct decision at correct time and that too immediately if required. It also needs qualities of leadership of self ideas and self responsibilities. One who throws responsibilities on others can never be a good manager on long term basis; though he become successful for a short duration.

A management technique may be defined as a recognized method of analysing or solving a recognized type of management problem in detailed systematic way A recognized method means or recognized technique means it must have a name must consist of precise step by step procedure that is unique and peculiar to it, and reliable to certain quite specific results recognizably peculiar to it, so that one can verify that it has been correctly used. This management method is required to be applied by a manager to management problem i.e. a situation in which a manger has to take a management decision. The management problem can be further defined as this means that a situation has occurred, or some one thinks it may occur, in which there are number of difficulties and a manager has to take action.

Characterstics of Management are:

1. Management is intangible *i.e.* an unseen process which can only be realised in production planing.
2. It is always goal oriented maintaining quality.
3. Management to different organizations will be different though basic objective may be same.
4. It deals with stack holders as such is social process and is a group work.
5. It is being goal oriented it is always dynamic nature as per then existing conditions and taking future aspects into considerations.
6. It is simple to complex in nature and is mostly multidisciplinary.

The **purpose of management** of any work are

1. Optimise the various resources like men, machinery, money and materials.
2. To have effective leadership and motivation
3. To establish sound leadership between management and their employees.
4. To achieve goals effectively and efficiently without compromising with quality.
5. To develop organizations
6. To improve standard of living of employees working in that organizations.

Managements are of different types depending upon their applications. Some of the common types of managements are :

1. Industrial Management
2. Business management
3. Construction management
4. Financial management
5. Labour management
6. Safety management
7. Production management
8. Maintenance management
9. Purchase management
10. Event management
11. Human Resource Management
12. Office management
13. Materials management
14. Inventory management
15. Quality control management
16. Equipment management
17. Project management
18. Contact management
19. Environment management
20. Resource management etc.

1.2 SYSTEM THINKING

Management involves deciding, doing and then eventually evaluating the things done and applying directions if needed. This is regular process of management system. These can be characterized as consideration of planning, implementation and control.

The planning needs to decide.

1. What is to be done.
2. When it is to done.
3. By whom it should be got done.
4. How it to be done.
5. Need of available resources to get it done.

There is always a need of feedback to control any system properly and efficiently. In industry or else every problem encountered is an unique problem in itself. There is very little likelyhood that two parallel or similar situations occurring. To combat the diverse situations a standard technique or atleast approach becomes imperative. This is provided by the pattern of disciplined and systematic thinking in the system approach.

The standard approach towards problem solving and decision making in incorporated system thinking is as follows:

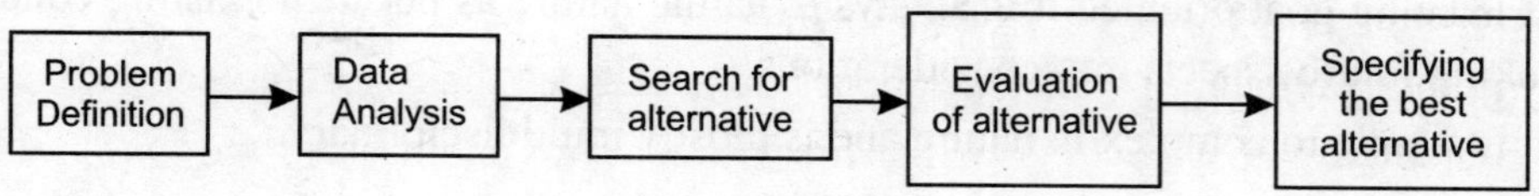

Fig. 1.1.

The search for alternative indicates that there could be many solutions to a problem. But there can be one and only one best alternative that is to be decided on the basis of merit. It is further to be noted that this solution is good enough according to those exciting conditions and hence can not be 'the best' solution.

1.3 HISTORY OF MANAGEMENT SCIENCE

Management is an age old science applicable to each and every day to day problem. Thousands of year ago also people were using management for managing their problems. Of course and probably it became science only after industrial revolution which led to the growth of workshops and development of factory system. People started its importance only later on and then scientific studies were conducted for proper and efficient management of the projects.

Probably the Taylor Frederic, a mechanical engineer, was the first person who applied management scientifically. He perhaps for the first time laid down definite principles and elaborated the management technique. His work was later on extended by Gantt, Emerson and Gilbert. In 1882, while working in Midvel Steel Company he carried out certain experiments for efficient working and development of efficient machine tools. In 1898, he joined Bethlehem Steel Company, did his analysis on design of shovels, wage payment systems etc. He endeavoured to find out the best way of doing a job so as to eliminate waste of effort, time and material. Scientific management may be defined as the system of management in which science replaces thumb-rule, harmony displaces discord, co-operation replaces individualism, maximum output replaces restricted output and there is development of each man to his greatest efficiency and prosperity. As it is clear and of course stressed by Taylor F.W. the principle object of management should be to secure maximum prosperity for the employee coupled with the maximum prosperity for each employer.

This fact of course known to everyone is not implemented in actual working. This is the basic reason of dispute between employer and employee. Scientific management believes in the interest of both whilst individual thinks of his own.

Taylor (1856-1915) stressed that efficient man should be paid reasonable wages, bonus and must have security of service. Machines must be properly maintained on maximum productivity of men and machine. Both should be maintained. At the same time it is duty of men to work such that the employer gets productivity. Employer must provide sufficient facilities to worker, assigning more responsibilities on labour, eliminate defects in working condition, maintain machineries properly etc. scientific and mechanized working improves productivity.

As such scientific management must adopt and follow the following important works:

1. Determining of task (avaluation).
2. Improving work methods: By standardization of tools, equipments, routing, scheduling and monitoring, doing time and motion studies ultimately assigning proper work to proper labour.
3. Performing proper responsibilities by manager.
4. By proper payment of wages to labour and improving their working conditions.

It is of course clear that Taylor was the first person who emphasized the importance of the systematic use of experience in management. He insisted on the separation of planning from the execution work recommending careful planning based on scientific analysis. He also insisted the wages to be based on quality as well as quantity of work performed.

F.W. Taylor performed various experiments on various industries like Symonds Rolling Machine Co., MIS Balhalehenry Steel Co. etc. and concluded that there are basic four elements which constitute the essence of scientific management. These are

1. The development of true science
2. The scientific selection of worker
3. Scientific education, training and development of worker and
4. Intimate friendly, cooperation between management and worker.

With his practical illustration, he proved that if friendly cooperation is maintained between management and worker mainly through sharing equally in everydays works, output of men and machine can be increased to such an extent that 30 to 100 percent increase in wages of workers can be increased, increasing the overall profit of that firm.

Inspite of all that there were certain defects in his theory, basically, teaching everyone to be efficient was little difficult. Nevertheless it is admitted that Taylor was a true pioneer of scientific management and did great work. Though thereafter a lot of suggestions and modifications have been made but the basic ides ideas are still the same. He also established many new concepts like 'rest period', standard method of working, wage incentive, standard task etc. He published his work in the form of books and technical papers. His popular books are 'Shop Management' and 'Principles of Scientific Management'.

Fayols (1841-1925) Contribution to Modern Management Theory. French industrialist Henry Fayol (1916) also contributed to the theory of management. His finding are based on his long experience as an industrialist and manager as such his findings are practical one. According to his theory, all the activities involved in industrial organization could be classified into six groups:

1. Technical related to production
2. Commercial i.e. buying, selling and exchange
3. Financial for optimum use of available capital
4. Security mainly for protection of property and person
5. Accounting
6. Managerial task related to planning, organising, command, coordination and control of project.

He listed fourteen principles of management based on his long term experience. These are :

1. Division of work
2. Authority and responsibilities
3. Discipline, sincerity and regularity
4. Unity of command
5. Unity of direction
6. Sub-ordination of individual to general interest
7. Fair remuneration to all workers
8. Centralization of authority and power

9. Scalar chain of supervisors
10. Order (material, arrangement of men and material)
11. Loyalty and devotion
12. Stability of tenure
13. Initiative
14. Espirit de corps i.e. union is strength and need of team work.

These principles if allowed there is no reason for a management not to be successful. In this theory only point eight centralization of authority, probably was not accepted by all. Even today also people think that there should de-centralization of authority. However his contribution to modern management task is very important and practical one, He probably was the first person to give set principles of management. His administrative aspects are very strong.

Other-Theories

Franks Bunker Gilbreth (1869-1924) was also an engineer. He did investigations on motion study, scientifically of works and workers. According to them (*i.e.* Gilbreth husband and wife); there is always a "*one best way to work*" for each work. If the work is carried out in a particular manner, it needs minimum/optimum time without causing any fatigue to worker. This couple Mr. F.B.I. Gilbreth and his wife Lillian M.Gilbreth a Ph.D. in Psychology contributed much to the human problems. One of their significant contribution was their definition of the elements of motion, which enabled individual motions to be studied and dealt with more effectively. In 1904, they studied motions of brick laying technique. They found that there exist 18 motion in brick-lying-technique which were not essentially required. They reduced the motion to 4 only increasing efficiency of labour without causing any fatigue or tireness. As such it made possible to each worker to handle 2700 bricks per day. In 1912, they also introduced science of micro-motion study. They also identified ultra-small basic elements of body movement known as "THERBLIGS". The research in this field led to elimination of unnecessary and unwanted movements and effective utilization of both the hands of the operator.

'*Urwick*' had also contributed almost on the same lines as of Henry Fayol and thus developed the concept of specialization, co-ordination, delegation of authority and responsibility, balance of work, span of control etc. '*Berger*' said that suitable incentive payment can motivate the workers. According to him, productivity increases when hostility, fear, suspicion etc. disappear. Moral motivation and mental peace influence productivity. Close supervision is dangerous. Hawthorne also emphasised that human psychological satisfaction is more important than merely providing physical facilities. Dr. Ralph M. Barness (1933) wrote his Ph.D. thesis in the form of book on '*Time and Motion*' study, micromotion study, work simplification. Elton Mayo theory dealt with human aspect of management. This theory gives more stress on human relations. Human emotions play an important role in scientific management. He described that no system of management can be successful if not based on human element in the organisation '*Douglas-Mc-Gragor*' late Professor of Industrial Management at Massachusetts Institute of Technology, had much offer supervision in his thorough work, "*The Human Side of Enterprise*" in 1960. He developed another view of science of human relation approach. He used theory of Psychology to study social behaviour of human and named them as Theory X. He modified it further and named it as Theory Y. McGrag, further suggested, that managers must start with the basic question of how managers see themselves in relation to others. This view point requires some thoughts on the perception of human nature, thus theory *X* and theory *Y* are two sets of assumptions about the nature of people. Basically theory *X* is the authority oriented, pessimistic, static and rigid. Control is primary external *i.e.* imposed on the sub-ordinate by the supervisor. Theory *Y* is goal oriented optimistic, dynamic and flexible with an emphasis on self direction and the integration of individual needs with organisational demand. Due to change in policies and behaviour in modern time. Theory *X* and *Y* became outdated.

Y. Hans Reynold in 1913 gave Principles of Management. He gave more stress on functional organisation. Halsey-Weir Rowan' had contributed scientific motivation to the subject of workers remuneration.

Most of the leaders of early work focused their activities on time and motion studies and related areas. However, there was another area developed in 1931, a concept on use of statistics in engineering problem of industry. Dr. Walter Shewhart of Bell Telephone Laboratories published economic control of the quality of the manufactured product based on sampling theory. Such a growth of scientific technique of industrial engineering was enhanced at a very high rate during second world war since most of the industries became busy in producing defense articles and to have higher productivity. But after the war was over, real competetion in industries started to cope-up the situation. Principles of management became more effective to economise the cost : that observed a tremendous growth in industrial engineering technique particularly after 1940. The decade of 1950's was a most active one in the transition from the prewar era and led to the development of more quantified methods. a-Operations Research Society and b-Institute of Management Science which led to two most significant development in the industrial setup : (*i*) Concept of diversification and (*ii*) Effective utilization of time available and resources available.

Industrial administration in twentieth century have started to be carried out by professional trained industrial engineers to fulfill the industrial objectives to reduce production cost. Development of computers have made it still competitive and efficient. Probably there is going to be a great revolution in management in 21st century.

Frederick Herzberg (1923-1940) was the fist to show that satisfaction and dissatisfaction at work place are based on various factors. He proposed two factor theories of human motivation at work place. There factors are

1. Satisfaction and psychological growth are result of motivation which may be of any form.
2. Lack of hygiene at work place causes irritation and as such dissatisfaction even if all other conditions are favourable.

Motivational factors are like promotion and growth responsibilty, recognition of organization, advancement work satisfaction etc. Factors cause dissatisfactions are relations with superiors, salary and status, security and personal life interference, policies of company, wages and other service conditions etc. He called these factors of satisfaction as hygiene factor. There factor either create positive thinking or negative feeling of organization and work place.

Abraham Maslow (April 01, 1908 June 08, 1970) an American psychologist contended that humans have a variety of needs. He described five motivational needs such as food, clothing and shelter, safety, social need of society, status and recognitions and self satisfaction.

Six Sigma and Lean construction are most modern theories of management used in construction industry. Six Sigma theory deals with management in construction taking into consideration of development and research developed where as lean construction deals with minimization of waste in construction and methods to reduce the cost of any project. Lean Construction has been dealt in detail in Clause 2.24.

1.4 DEFINITION AND OBJECTIVE OF MANAGEMENT

Management of any job means to get that job completed, effectively efficient and economically. For any tasks these are essential requirements. As such the main object of management of any task is proper execution of the work. The object of management in construction work is a planned manner in a specified period of time as per drawings and specifications laid down for that construction work at a minimum possible cost. These objective of management can be dealt in three phrases.

1. Planning phase 2. Scheduling phase 3. Control-monitor phase.

The planning phase consist of selection of project and its site, its planning and development of drawings design and estimate. Procurement of resources, store and recruitment of personnel agencies are also come under this. Scheduling phase consists of preparing time schedule, material schedule, finances and starting of construction work. Once the work starts control monitor phase consisting of control of all the items, their co-ordination etc. Complete execution of project should be undertaker as per the drawings and specification. Such as organisation doing all this is known as the management or project administration.

1.5 FUNCTIONS OF MANAGEMENT

Following functions are to be carried out for efficient project management.

1. **Planning.** The managerial function what to do, when to do and how to do is known as planning. This is the basic need to management. Its efficient completion depends up on sound judgment by planner based on his skill. Knowledge and experience use of scientific management theories is required to be made. This should be divided into various phases and each phase should be given due importance and thought, men, material and money required phase wise must be ascertained well in advance. Likely uncertainties must be considered and corrective measures likely to be taken must be taken care of. If planning part of any project is done thoroughly and completely the execution part becomes comparatively easy.
2. **Organisation.** Once planning is over, it is essential to setup an organization to execute the works planned. Different personnels their managerial level depend upon the volume and type of work. Various organization setups at various control and state government level, private level etc. are dealt in details in subsequent Chapters.
3. **Directing.** This is a job of a supervise manager at his level to motivate his sub-ordinate and guide them. Supervise them and lead them properly to get the work executed efficiently. Proper communication in the workers at different level and their co-ordinations is essential. The job of manager is to direct his sub-ordinates regarding works execution.

He should prepare bar-charts, networks to have the idea about progress of construction works. He is also to take steps in case the work is going behind the schedule.

4. **Controlling.** During execution, the job of manager is to monitor the progress achieved in comparison with the planned programme and to identify area of deficiency if any and to take remedial steps to bring further tasks on schedule. The jobs of controlling project constant of:
   (*a*) Keeping a watch on actual program of each activity of construction and to note activities if leading or lagging the schedule; further in case of activities going behined the schedule to take steps to bring remaining activities on schedule so that overall period of completion of project is not extended.
   (*b*) Controlling the quality of work as per specifications laid down earlier.
   (*c*) Controlling use of men, material and machine to avoid any wastage.
   (*d*) Controlling financial resources to keep the cost of project within estimated cost.
5. **Co-ordination.** Co-ordination of workers their views and finally completion of project are the necessary tasks of manager. He has to collect views of various persons at different level and then required to execute the job to be more correct, realistic and feasible. He is also required to co-ordinate various gangs, persons involved in construction. He has also to co-ordinate various activities in a project.
6. **Motivation.** This is very essential in modern management technique. The job of managerial person is to motivate the workers for effective working. This can be achieved by economic; psychological and sociological satisfaction of workers.

These six principles of managerial function or functions of management for the first time were given by Henry Fayol of France. As such planning needs forecasting, objectives, policies laying out, programming, scheduling, procedure finishing and budgeting, organising deals with activity determination and power delegation. Directing consists in initiating, guiding and supervision of subordinate in their activities. It involves issue of orders and instruction, guidance and training of sub- ordinates and their supervision. Controlling means achieving close correspondence between plan and performance. It includes establishment of standard of works, assessment of actual work, determination of deviation and taking corrective measures. Motivation deploy labours to satisfaction to get more work executed through them. Co-ordination means achieving team spirit and unity of action amongst the sub-ordinates. Co-ordination is required due to numerous person involved at work, complexity of work, delegation of authority and responsibility and chance of difference of opinion between executive etc. The co-ordination clarify the objective of each, maintain human relation and creates cooperation and compromising tendency.

1.6 LEVEL OF MANAGEMENT

Various levels of management right from Board of Directors to supervisor are illustrated in Fig. These levels are :

(*a*) Top level Management - Administration - determination of overall objectives and policies.
(*b*) Middle Management - Execution of policies and programmes.
(*c*) Lower Management - Supervision.

The inter relationships of information and instructions at different managerial level of management is illustrated in Fig. 1.2.

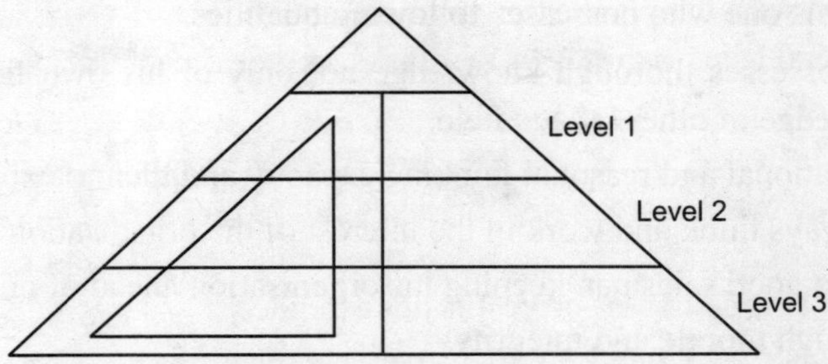

Fig. 1.2. Levels of management

At the activity level, the operator needs detailed operational instruction which may take the form of detailed work-sheet, memoranda, telephone calls or other verbal instructions, depending on works situation. The type of information passed up to the project leaders will be standard networking information, dealing with activity status, possibly including a cost element or data from which cost can be derived and possibly in the form of completed computer input document. The volume of information passing up and instruction passing down are maximum at this level of control.

At the project level, the project leader, will receive instructions from the next higher level of management. These vary considerably between organisations. If multiple-project working is performed and particularly if the projects are malleable, the instructions may include the request to stop or re-schedule certain activities in light of the overall-multi-project resources smoothing exercise, or possible the request to postpone or terminate the entire-project. Information flowing upto the project leader will be detailed status information on each activity, possibly in the form of computer output.

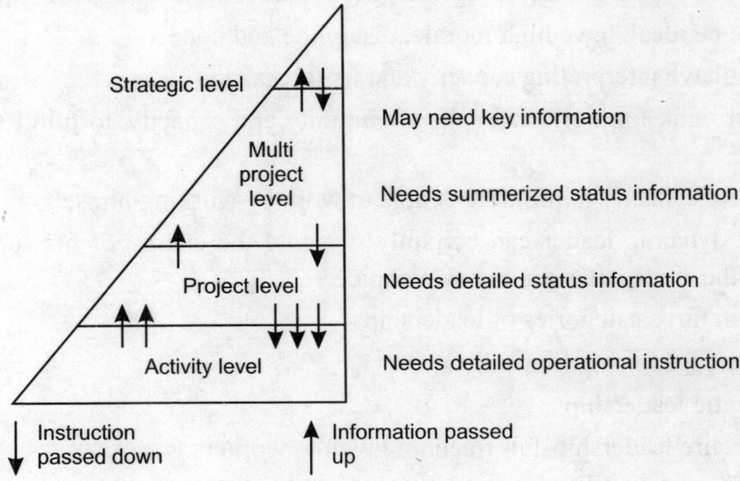

Fig. 1.3. Information for Management.

At top most level only summarised key information are passed while depending upon individual these instructions are passed down to individual project manager.

1.7 QUALITIES REQUIRED FOR MANAGER

A successful manager is one who possesses following qualities.

1. One who possesses thorough knowledge not only of his own field, but also possesses good knowledge of other related field.
2. Must have rational and reasonal thinking capacity and then to express his feeling clearly.
3. He must always think and work in the interest of the organisation as a whole.
4. He must be a good salesman keeping his organisation ahead of others.
5. Must have high morale and integrity.
6. Must be social, have skill to maintain human relation with theory.
7. Must have organisation capacity and leadership qualities, influencing others.
8. Must have capacity to inspire and motivate his-subordinates for best possible workmanship.
9. Must have capacity to fulfill the aims and objectives of organisation.

1.8 LEADERSHIP

Leadership may be defined as an art of human being which can bring together a group of people together and under his motivation. That person thereafter controls all activities of that group in an organised manner. A manager as a leader directs, guides and influence workers and other concerned with it in attaining a definite goal. Some of the important qualities required for an efficient leader to function well are :

1. Must have capacity to motivate people to follow him and capacity to guide them.
2. There must be working relations between him and his followers.
3. He must be capable to advise them as such must have thorough knowledge of his field.
4. He must be ideal, have high morale, discipline and honest.
5. He must have interpreting capacity and its forecasting.
6. He must think for welfare for the community and capacity, to fulfill their aspirations if possible.
7. Must have capacity to promote people to work by working himself.
8. A good dynamic leader can certainly generate the energy of his subordinate that can change the environment at his own choice.

There are main three categories of leaderships

1. Autocratic leadership like a dictator
2. Democratic leadership
3. Laissez faire leadership-full freedom given to workers.

In autocratic leaderships, leaders work as an dictator, hence dictates term. In this type of leadership, there is always a fear among sub-ordinates. In democratic leadership, leader gives due importance to his sub-ordinates. He takes their views and takes decision concerning all persons involved, as such he seeks confidence of his sub-ordinates of that organization. Such leaders advice is put to implementation by the people with their willing co-operation- In Laiss Fairs Leadership, leader gives full freedom to his sub-ordinates to work according to their will and wish. This type of leadership is rarely effective.

A leader should be an efficient, well behaved and must have knowledge that how to deal with different persons. He must have capacity to motivate and to control masses, further he should be acceptable to masses. The responsibilities of leader increases either during strike or during any accident. He is an essential link to the communication system.

There are various type of trade unions. Most of the industries, labourer are linked with one of them through their leader. Leader is the person who controls and guides the worker of that organization. The responsibilities of such leaders are many fold.

1.9 PROJECT PLANNING AND CONTROL TECHNIQUE

Everyone manages project but some people manage better than others. Management is thus can be regarded as 'a way of getting things done'. For effective management, it is necessary to achieve given targets and objectives with utmost efficiency, economy and in stipulated time.

During the planning of any construction project, it is essential to divide the entire project, primarily phased out sequentially and then dividing the operation of one phase into a number of jobs. Planning in granule can best be described as the function of selecting the enterprise objectives and establishing policies, procedure and programmes necessary for achieving them. Planning is of course, decision making since it involves choosing alternatives. There are four basis reasons for project planning.

1. To reduce uncertainty
2. To improve efficiency and effectiveness
3. To obtain better understanding of objectives
4. To provide basis for monitoring and controlling works.

This is a continuous process and must utilise the past experience in the field with sound judgement.

Following precautions are necessary to be observed during planning any project:

1. To consider eventualities and uncertainties such at weathering conditions, procurement of materials, training of man power, price fluctuations and labour absenteeism.
2. Correct judgement of time inception and complete the project in schedule time.
3. To procure correct type of machine material/men. If the project is large and complex, careful planning and analysis must be accomplished by the well designed organisational structure.

The ultimate objective of planning supervising a project to achieve maximum economy is the cost of construction of the project. The objectives of 'cost control' are :

1. Exercise overall financial control and relationship between value of works completed and valuation of works done.
2. Revision of output constant in respect to plant, equipment, machines due to improved techniques and
3. Revision of rates due to improvement in works.

The cost control can be exercised either during pre-construction or post-construction stages. In pre-construction stages, economy can be achieved in design and specification. During post construction stage, cost control can be achieved adopting suitable quantity buying policy and exercising proper supervision, continuous valuation of output, avoiding idle hours, keeping overheads and taking safety measure, etc. It is always useful to have three separate budget for each element, a- Direct Labour Budget, b-Support Service Budget, c-Budget for purchase items.

Quality control, schedule control, cost control and management control are to be made strictly as per the provision laid down. Quality control can be achieved by strict supervision and following the specifications. Schedule control assures that work is completed according to the planned schedule. The project manager has to be constantly vigilant to see that slippage is not accumulating at an unexpected amount. The slippage may be due to negligence on the part of organisation, lacks of credibility, insufficient and missing information in competence or on any of the reason beyond the control of organisation. Schedule slippage should be avoided as far as possible and prevented by assigning more responsibility, establishing checks points, reliable reporting system and supply of correct information. The cost control is based on task budget, achievements in work progress compared with budget expectation and alteration, additions and revision of plans. Careful management control must be established and management guidelines must be framed to achieve unity and coherence.

1.10 MANAGEMENT OF PROJECT-CO-ORDINATION

With rapid development in technology, the project management, is a wider process which includes systematic planning of all activities involved, efficient organisation, continuous co-ordination, skillful operations, periodic monitoring and effective control. All these diverse activities are interconnected and form a continuous process.

For effective co-ordination, recent trend in the U.S.A. Canada and West Germany is to form a team for professional constructional management in which the project planning, detailed designing and construction/erection are an integral task assigned to the team called Project Management Team (PMT). The typical structure of such a team is illustrated in Fig. 1.4.

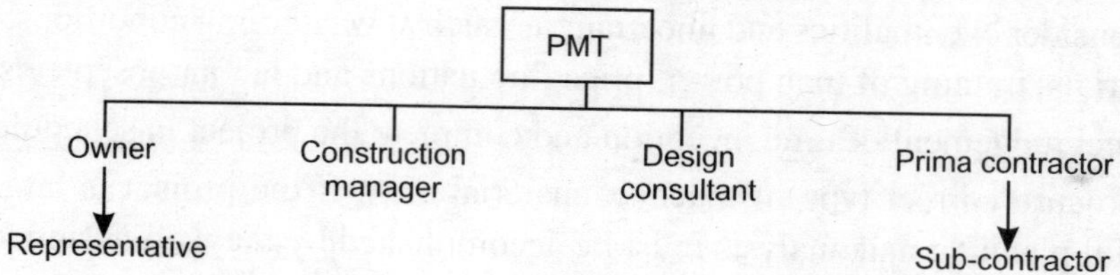

Fig. 1.4. Typical structure of PMT.

The team works together from the initial design stage to the project completion stage. The main objective is to serve owners interest.

If this concept could be implemented in India where the job completion of work lies with construction manger, he has to control consultant and contractor, co-ordinations of project can be done effectively. Construction manager himself can be a consultant engineer. This can become popular if owners are given advantage of it. Unless owner realises that he will get the project completed efficiently and effectively if under taken through PMT, construction manager being wholesale, there is no reason for failure of work undertaken through Project Management Team. It's conceptual form is illustrated in Fig. 1.5.

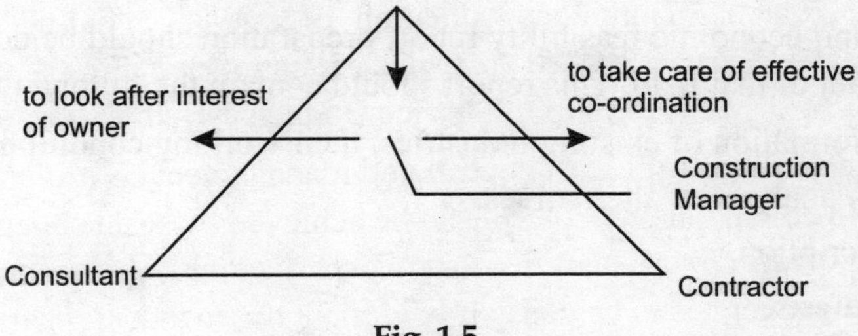

Fig. 1.5.

It can be seen that by implementing process involving several agencies in a time-bound programme of construction of any major development project and their effective co-ordination can play an important role in creating the right harmony for achieving a common goal-completing the project in a workman like manner within economic cost and time. Many industries in India have adopted this.

1.11 PROJECT MANAGEMENT

Project management in its widest sense includes both project formulation and its implementation. The project management would broadly cover such activities as preparation of feasibility report and obtaining sanctions, selection of proper technology, entering into foreign collaboration where necessary, construction of plant structures and utilities procuring and installation plant equipment and commissioning production. Construction of the factory structures has to be completed by the time plant and equipments arrive at site to avoid the problem of storage and delay in erection. The building of factory structure and utility services, therefore, forms a pre-requisite of plant installation and as such construction activity requires the highest priority. There is increasing awareness by the top management of the need for adopting the latest technical know how and sophisticated design and construction techniques to optimise construction costs and reduced consumption of scare building materials. There are examples of, that the old concept of steel factory structure has changed and has now given place to modern prefabricated, pre-stressed concrete construction, standardisation and modular co-ordination. Similarity the traditional methods of organising project are being replaced by modern project management techniques including network programming and monitoring progress and inventory control to reduce time and cost over runs.

The broad stages of a project prior to plant construction are:

1. Project formulation on the basis of reconnaissance ; funds available ; market and need of the country and its future planning.
2. Preparation of Techno-economic feasibility report.
3. Preparation of detailed project report.

The first stage of planning of project formulation begins usually with the sectorial studies for the various plans based on government policies and market trend and ends with the commissioning of one or more feasibility studies. The actual investment decision is taken only after detailed consideration of the feasibility report. As soon as a decision to commission the feasibility study has been taken, a suitable and competent person has to be appointed, who could be associated with the preparation of the feasibility report and should ensure that the project is processed with speed. This ensures efficiently of working and would help in better coordination and faster decision making.

The task of Techno-economic feasibility report preparation should be done by an experienced professional consultant of that field. This report should contain the following information.

1. General Information of existing industries, their working conditions etc.
2. Preliminary analysis of alternatives.
3. Project description.
4. Market Analysis.
5. Capital requirement and costs
6. Operating requirements and costs.
7. Financial Analysis
8. Economic Analysis.

This techno-economic feasibility report must contain all details discussed above. Adequate investigation, data collection and care at this stage can lead to significant economics.

The most important factor governs in this study is the site selection. The choice of location of the project has a great bearing on the economy of construction and viability of the proposal. The traditional considerations for site selection are :

1. Proximity of the sources of raw materials.
2. Availability of a suitable distribution system and transport facilities.
3. Nearness of the centre of demand.
4. Availability of adequate power and water.

It is a good practice to consider more sites and then to select one on their merits. In some cases particularly in government projects, sites are selected not only on Techno-economic considerations but on the basis of social cost-benefit analysis. Sometimes political considerations also become the basis of selection of site. Only minimum requirement of land including that needed for future expansion should be acquired at nominal cost. Given the choice, the most appropriate technology consistent with economy, profitability and the local conditions have to be selected. In some cases particularly in government projects or public undertaking projects, selection of technology is tied up with loans and foreign assistance available. If foreign technology is selected it should be modified if required to vital local conditions. Since process technology determines the design of the factory, the capital cost and the cost of production, the selection of technology are important for the success of the project.

It is always a good practice to prepare a masterplan of all the structures and utility services including future expansion and indicating phase wise construction. If number of projects are coming in the same vicinity it will always be economical to have integrated planning in collaboration with surround industries. This will reduce the cost of services like water supply, sanitary, electrification etc. The cost of plant and auxiliary structures form the main component of the cost of civil construction. A careful planning of these structures would lead to considerable economy and efficiency and better viability of a project. Adhoc provisions without systematic and detailed space planning often results in over sized structures and higher capital investment.

Advancement of technology has also brought in its wake problem of environmental pollution. It is essential to study the effect of project on environment and to take preventive measures. Parliament has also legislated for control of environmental pollution. The time schedule for the completion of the project should be accessed realistically.

The cost estimates and the profitability calculations in the feasibility report are based on present cost of the inputs and selling prices of the output. However, since the project implementation extends usually over a period of several years, the financial calculations at the time of preparation of feasibility report often do not hold good at a later date due to escalation in the price of materials and labour. It is therefore essential to make provision for escalation of costs. While doing all these studies, it is essential to do scrutinity of certain broad issues as:

(*a*) The contribution of the project to the economic and social objectives and adherence to the policies of the government.

(*b*) Availability of plan funds, desirability of diversion of plan funds to the new project from those already in hands.

(*c*) Plant capacity and the dimensions of the investments in the light of demand and supply balance including export possibilities of the product/services to be provided.

(*d*) Economic benefits of the project as distinct from financial return.

(*e*) Crucial assumptions in the feasibility report which are likely to affect the performance of the commissioned project in relation to the claims made thereon in the feasibility report.

(*f*) Major technological and structural aspects.

Once all these studies are made, preliminary reports are received and scrutinized, final and 'Detailed Project Report' is prepared, All these preliminary reports, civil works reports etc. are arranged in a sequential order with the analysis of each and overall view of the project. This report is then submitted, approved and sanctioned finally for execution. Funds are made available and civil work starts, orders for machineries placed, recruitment's are made. Here onward a manager has to administrate the whole work from this stage onward. He has to control all activities and has to see that the jobs are executed properly till final product comes to the market.

1.12 SYSTEM ENGINEERING AND MANAGEMENT

The society and the industries as on today are more complex then it was ever before. Higher education, technical skills, newer materials, more and more information data and so on are well understood by stockholders, it makes management more and more complex. Thumb rules existing earlier cannot be relied now. People have become more aware, they need specific information and database. As such use of system engineering, system concepts is becoming more important in management science.

System concepts give a systematic approach to get the task accomplished more efficiently, effectively and economically. System approach is an organized approach for simple to complex problems in day to day life and that to engineering problems like designs and management. It furnishes a frame of, how to manage the jobs under various adverse conditions. It solves the problems of business, human values, industries of all kinds and problems of science and technologies.

Systems of execution of any work were in existence long back. People were using some system to get work done as per their knowledge and then existing conditions. This was in existence right from development of mankind in one form or the other. However it became science and engineering only in 20th century. Each and every person was adopting some system to get work done effectively and efficiently, however they had no scientific approach.

System: It is a collection of interacting elements those operate to achieve a predetermined objective. It is a process to receive certain inputs, then to analyze them and to produce certain output with a purpose to optimize some function of inputs and outputs. To quote an example, is a concrete mixer. A concrete mixer is used to produce homogenous concrete. Raw materials like cement, sand, coarse aggregate and water are fed in concrete mixer and it is rotated for certain time at a certain designed speed. The mixed concrete is then taken out and put in the forms to get required shape. An engineer who had designed it must have collected data for its successful functioning. He must be having taken into consideration the shape of drum, its size, speed of rotation, motor type to rotate it with required speed etc. These all data are called as input. He might be having analyzed these data before reaching its design. A prototype might be have been developed and tested before final shape is given. There after based on problems faced during operation, mixer might have been modified to reach present shape and size.

A system is therefore is dynamic in nature. A few examples of systems are:

1. A human body is a system with various sub systems like nerves, blood circulation, bones etc.
2. An industry with various sub systems like purchase of raw material, processing, production, sales of finished product etc.
3. A transportation system of Rail, Road, Air, water etc.

   A system can be classified as

   (*a*) Mechanistic-A mechanistic system is one which is fully mechanized; however choice of system components is controlled by human beings. Ex. Lifting and placing of precast concrete panels by crane.

   (*b*) Quasi-Mechanistic-In this system, human carry out some of the mechanical functions.

   (*c*) Non-Mechanistic-Human elements are totally involved in this system. Truss fabrication,

System Analysis: System analysis is the study of each part of the system, both as an individual and in relation to the whole in order to design, modify or improve; it is immaterial whether it is an abstract system like inventory control or it is a physical system as power transmission in a vehicle.

It consists of study of system and then development of system. It can be for original wok or can be used for existing system to be modified as per problems encountered in that system. System analysis and rectification of existing system are continuous cycles and in general consists of following processes.

1. To define objective: It is a conceptual phase. In this objectives or problems to be resolved are well clarified and defined with a view to seledt a policy regarding action/process or decision making for solving problem.
2. Analysis of data: In this phase data available are analyzed and various possible alternatives are determined and one most feasible is finalized.

3. Model/Prototype Building: In this, based on final alternative in stage 2 , a prototype model is prepared and tested to study the problem encountered if any. This problem is then further studied and modified further if required. Depending on type of problem, there can be an overall process model, performance model, reliability model, time model, cost model or any other model type as decided based on available data and problem defined. Some times more than one model are prepared and tested to find suitability of each.
4. Then thereafter a final decision is made as per selected and verified model so that production process is started.

As such system design and engineering consists of

A- Define the problem

B- Find alternative solutions and to find one most feasible.

C- Develop prototype model and check the solution to ascertain that the selected alternative is performing well.

D- To start production planning.

System Engineering: It is an inter disciplinary field of engineering and engineering management on how to design and manage a system over its life cycle. System life cycle is a term used to refer the stepwise evolution of system from concept development, idea generation, through production planning, production and final disposal of finished product. In case of production life cycle consists of purchase of raw material to its processing and taking out a final finished product in market and to sell it out to purchase again raw material for its next life cycle.

System engineering is a process that is quite unlike a manufacturing process. A manufacturing process is focused on repetitive activities those achieve high quality output with minimum cost and time. The system engineering process begins by discovering the real problem that need to be resolved and identify the most probable or highest impact failures those can occur-system engineering involves finding solutions to these problems. It deals with the work process, optimization methods and risk management tool in that project under consideration. This can be applied to day to day problem of simple nature to very complex problems of engineering field.

History of System Engineering: The system concept might be existing right from the development of society and mankind, however is traced back to Bell Telephone Laboratory in the year 1940. When it was no longer possible to rely on design evolution to improve upon a system and the existing tools were not sufficient to meet growing demands, new methods began to develop those addressed the complexity directly. The continuing evolution of system engineering compresses the development and identification of new methods and modeling techniques.

In 1990, a professional society names as National Council on System Engineering (NCOSE) was founded by representatives from number of US corporations and organizations. This society was created to address the need for improvement in system engineering practices and education. In 1995 the name of this organization was changed as Inter National Council on System Engineering (INCOSE) due to its development and need at international level. Graduate program on system engineering have also started at many educational institutions.

The system engineering processes can be decomposed into:

1. System Engineering Technical process.

2. System Engineering Management process.
3. System engineering management and Technical process as a whole.

While the technical process includes assessing information, defining effectiveness measures, to create behavior model, create a structure model, perform trade-off analysis and creates sequential build and test plan, management process deal with sequential process and management of each. In third case both, technical as well as managerial, both are taken into consideration. That is to say whole process of system engineering consists of technical as well as management process both.

Else it can be defined as perspectives of system engineering are

System thinking related to process

System Engineering related to product and

Engineering system related to process and product both.

In almost all the cases one is required to find out what is need of development of system process or product or both. In case of an existing system the question arises why we are required to modify or revise existing system. What are the shortcomings of existing one and how theses shortcomings can be minimized without affecting the quality as a whole? The best way to resolve problem of system engineering is by dividing the system into sub-systems and then further to elements and sub-elements. This will make problem little easy to understand and to solve. Management of system is an important aspect of system engineering. If one could manage either production of any product or processes involved in production or both, he would be the best system engineer.

The need of development of any new system or modification in existing system can be determined by asking following questions.

1. Who are stakeholders?
2. What are system goals and objectives?
3. What is need of system under development?
4. What are external constraints on the system?
5. What is system architecture context?
6. What are interfaces to be satisfied by system?
7. What is key system attributes?
8. What are system functions that satisfy to key attributes?
9. What happens to success if system fails to meet above?

If these questionnaire is put either to new system to be developed or to existing system to be modified, problem solution becomes easy as the objectives are well clarified and defined and resolving of problem becomes very easy.

The process comprises in development of any system whether original/new or modification of existing one is:

1. To identify customer need. It means to define problem of stockholders.
2. To study system operation environment i.e. to find constraints and conditions of problem.
3. Planning of alternative methods of execution based on interfacing system.
4. Analysis of alternatives and to decide on one that suits most based on logistic support requirements. It also consists of capabilities of operating system.

5. To consider factors required to be considered in system requirement.
6. Modeling Prototype of final selected alternative and to decide modifications required if any.
7. Decide final one and to implement in actual execution/production.

Applications of System Engineering and Management: System engineering and management of system has various applications right from day to day life to any complex system of engineering, technology and management. Following are some of the applications.

1. Design and development of complex to highly complex industrial equipment system.
2. Design and development of military equipments and weapon system.
3. To decide tactical alternatives.
4. Management of operations of various natures.
5. In policy decisions.
6. And many more.

System engineering is a vast subject and beyond the scope of this book. As such only elementary aspects are dealt. Interested can refer any good book on this subject matter.

1.13 WORK BREAKDOWN STRUCTURE (WBS)

It is a name given to a technique in production management or project management in which a whole process of production is broken into various smaller component parts which when combined produce a whole production unit. A works break down structure is a key project deliverable that organizes the teams work into manageable sections. It is decomposition of project into phases. This process is very common in intestinal products however its use in civil engineering project has also becomes common This W.B.S is also called as Indent use level structure. In general, a product system is broken down in subsystem and then each sub system into sub-sub system and further till each sub-sub system reaches to a manageable unit as shown in Fig. 1.6

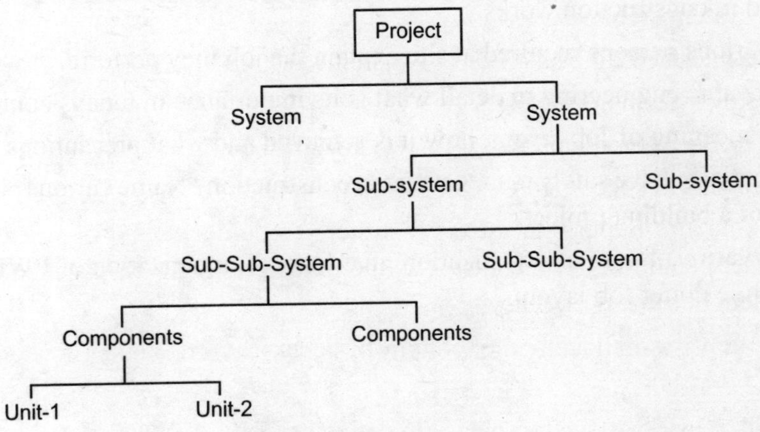

Fig. 1.6.

In case of civil engineering building a building can be divided into systems as various floors. These various floors can be further divided into section and each section into various items of construction like cement concrete, Masonry plastering colouring plumbing etc. The concreting can be further sub-divided into concreting of footings, concreting of columns of ground floor, concreting of beams of ground floor, concreting of first floor and so on. Similarity masonry can be divided into various units, section wise, floor wise. Completing each unit, section, floor, the final product building will be available finally to the user.

As such WBS represents a task oriented family tree of activities and displays the total work to be accomplished in order to achieve the final objectives of the project. It server as a common frame work for various exercises such as planning, scheduling, estimating, budgeting, configuring, monitoring, reporting, directing, controlling etc to entire project to get the project complete efficiently. This concept of WBS mainly came into existence due to network techniques like CPM/PERT. The whole project is divided into elements and units. Network is drawn for each element and units and they are linked to form a total network showing detailed process, time and cost involved in final product.

QUESTIONS

1. What do you mean by Management? Give History of management Science in detail.
2. Explain Taylor Frederic and Henry Fayols contribution in Management.
3. Explain various theories of Management. What do you mean by Levels of Management.
4. Explain Function of Manager and what do you mean by levels of management.
5. What qualities a good Manage and a Leader should have.
6. What do you mean by Project Management Team. What are essential requirements of a good Project Management Team?
7. Explain functions of Management. What do you mean by Levels of management.
8. What do you mean by construction economy .At what levels economy can be achieved to have most economical cost of any civil engineering Project.
9. Classify Various types of civil engineering construction works. Name various agencies involved in construction works.
10. Name various persons required at site explain the job they perform.
11. Explain value engineering in detail what is its importance in todays content.
12. Explain meaning of Job layout, how it is achieved and what precautions are
13. Is it essential to have safely management in construction? Name various possible accidents in case of a building project.
14. Explain various forms of organization give form of organization of PWD required to be taken while doing Job layout.

Chapter

CONSTRUCTION, ITS MANAGEMENT AND ORGANISATION

2.1 INTRODUCTION

Construction engineering is an inter-related complex of operation whereby buildings and installations are erected. The end-product of construction is finished building for its functional use. Construction operation may be classified according to the methods and purposes such as erection of form work , masonry construction etc. or by the type of material to be processed such as concerting, earth moving etc. According to field of applications construction operation may be general or specialised depending on their form such as concreting is a specialised operation whereas excavation of earth by manual labour is a general type of construction work.

Construction process in character of performance may be mechanized or non-mechanized and in degree of complexity-simple and complicated. For example levelling using a bull dozer is a simple mechanized work whilst excavation by power shovel is mechanized complex work.

Construction is vital for nation development. It may be defined as the creation, renovation, repair and demolition of immobile structures and the alternation of natural topography of the ground. Construction - both the activity and the resulting product are notably visible and tangible.

Construction reflects the economics, social and cultural activities and values of society. The building and engineering projects with which people come in daily contact affect, in some complex way, their conversely, the tradition, aspiration and vitality of people are reflected in their settlements and their constructed environments.

Most construction projects are complex, both technically and commercially. There is considerable specialization, both technically and commercially. There is considerable specialization, both among types of construction and among on site operations of a particular project. With the increasing, understanding of materials. improved designs, improved construction and management techniques and introduction of computers, modern construction is a multi-disciplinary activity. Technologically change in construction tends to be gradual and includes a wide variety of individual developments in materials, equipment, and methods.

Construction has been a part of the traditions in India, is also a nation heritage. Ancient construction in India such as the city of Mohenjodaro, the historic monuments, temples, and religious places forts and towers are a reflection of India's builders and craftsmen were responsible for a variety of civil works abroad. A large number of civil works, projects were completed in India as well. These includes the building of major cities, national highways, bridges, Railway networks; harbours and ports, dams and canals, various river valley projects etc. Soon after independence in 1947, India took to the planned economics growth. The successive five year plans have laid greater emphasis on infrastructure development, irrigation, energy; transportation and communication, health; housing and social welfare, rural development, civil amenities and other activities with the objective of improving the quality of life of its people. Consequently mammoth river valley projects were built and investment made in other spheres.

Table 2.1. Construction: Content of plans

Plan-period	*Construction content as percentage of total development plan outlay*
First Plan(1951-56)	50.4
Second Plan(1956-61)	42.6
Third Plan(1961-66)	42.3
Fourth Plan(1969-74)	43.2
Fifth Plan(1974-79)	43.4
Annual Plan(1979-80)	46.1
Sixth Plan(1980-85)	36.2
Seventh Plan(1985-90)	40.0
Eighth Plan(1992-97)	43.5
Ninth Plan(1997-2002)	Similar as above
Tenth Plan(2002-2007)	
Eleventh Plan(2007-2012)	
Twelveth Plan(2012-2017)	

The situation in many other developing countries is similar.

2.2 CONSTRUCTION INDUSTRY

The construction industry in India has grown and made significant along with many other developments. A large number of major, medium and small size construction firms have come into existence. The construction industry provides employment to over 25 lakh work force. Major irrigation projects, dams, roads and bridges, power plants and rural electrification projects, river valley projects, tall structures, offshore and ocean structure, ports and harbours,housing and other dwelling units, railways airports, nuclear structures and variety of other special and prestigious projects have been launched and completed. There are many challenges ahead as ocean and spaces programs, oil, and natural gas exploration, search for new mineral resources and deep sea mining programmes are increasing and also the challenging tasks for providing houses to millions are increasing.

Over the period of years a good amount of infrastructure built in India for technical manpower, generation, skill development, raw material production and organised construction agencies. In terms of man power generation and skill development, there are more than 3000 engineering colleges and institutes generating to about 15 lacs graduates including about 200000 as civil engineering graduates every year. There are many polytechnic institutions generating to lacs of diploma holders including about 20000 as civil diploma holders. Besides these; there are many Industrial Training Institute generating over 7.0 lakh trained craftsmen in related engineering trades in country.

In terms of construction equipment, more than 300 companies have been set up and a wide variety of construction equipment are manufactured indigenously and are available. A large number of contracting firms and some consulting organisation have also been established over the period of years.

As far as employment in construction industry is concerned, generation of large scale employment is another significant aspect of community polytechnics, advance training industry (ATS) the construction activity. It is estimated that every 20 lakh of rupees worth of construction generates 3100 man days of unskilled labour and 1300 man days of technical/managerial personal. According to some estimates, out of the 244.6 millions work force in 1983, there were as many as 11.3 million persons economically active in constructive including 90000 contractors estimated to have been registered with various government department and agencies all over India. There are skilled, semiskilled and non-skilled workers work with various contractors. There are some workers who do part work in construction and alterenate between construction and agricultural occupation.

Construction is the second largest economic activities in india; next only to agriculture. There is hardly any sector activity in india; next only to agriculture .There is hardly any sector which does not have construction component; it varies from 10 to 20 percent in scientific research and education 40%in transport and communication, 75% in power; 80% in irrigation and flood control and 100% in housing. Construction activities, but by itself creates both backward and forward linkage in the rest of the economy. In other words, it has an economic multiplier effects.

What surprise most be an economist is that despite its massive contribution to nation building activities; construction industry in India has not yet acquired the status which it deserves. It has not even been given weightage and importance to construction as industry in real sence.

The construction industry in India has certain setbacks. Probably this may be one of the reason of not getting it recognised as an industry. These set backs are:

1. There are no specified guidelines. They have not been framed; it has fragmental the construction sector.
2. There is no special scheme available to meet the financial requirements for large scale technological changes and modernization
3. Construction methods in India are still age old and traditional.
4. High construction cost.

5. Standardization and quality control is rather weak.
6. Many contracting firms run without adequate technically trained staff and without much appreciation.
7. Little attention is paid to appropriate choice of material and their quality.
8. Linkage between the R and D organisation and construction industry is weak. There is strong R and D base as far as cement industry is concerned however it is poor in construction industry.
9. Wages to the labour class in private sector not fixed. Mostly workers are employed on temporary basis.

Therefore it is essential to take suitable measures, technical, technological and managerial. There is a need to evolve and overcome setbacks and make the construction industry in India most modern to meet future challenges.

2.3 CONSTRUCTION TEAM

When new project comes up for consideration, the Civil Enginners will initially require some first hand knowledge of the locality in which the work is to be carried out, preliminary investigation and studies will then be made, and when these are approved, complete contract drawing and specification are prepared. Once financial arrangement is made , tenders are called for, a contract is awarded and construction begins. When the project is completed, it must be periodically inspected and maintained in good order.

Hence right form planning stage to completion stage and then to use it, number of different persons and specified firms are involved for its satisfactory completion. These number of specialist firms or personnels involved form (Building Industry).These different persons have their own special features and problem and a different approach in each case. Following are the persons who form a Building Industry.

1. The Owner : Owner is one who wants to get a work executed as economically and as quickly as possible and to know in advance his ultimate financial commitments. Owner may be a government, public body or some private enterprise or an individual one depending on type of work and financial involment.

2. The Architect : Architect is the first person involved in the planning stage of a building industry. He should be an artist, a technician and moreover a businessman. He has to satisfy owner, engineer, contractor and ultimately user. The architect after having obtained the instruction from the owner, design the function layout and provision of functional accommodation. He has to make structure beautiful and functional. He should give aesthetic effect to the structure. He is supposed to have knowledge of building bye-laws and regulations. During construction, he supervises the work as an agent to the owner.

3. Consulting Engineer : The consulting engineer is one who is wholesale responsible for execution of work. He develops the drawing, designs the structure, prepares the estimate, invites tender, gets work executed properly by contractor and gives it for the possession. He supervises the work and maintaine the record. He is responsible for the greater part of the technical design and construction which affects the cost. He advises owner as well as to the contractor whenever it is required. Government organisation employ their own engineers who look after all these works.

He also prepares the bills. In private constructions, he arranges labour, equipment, materials etc. for smooth execution of work. The consulting engineer, many times assisted by architect, town- planner and other specialised firms. In departmental works, he is assisted by accountants, administrative personnels and other junior staff. Some times he also needs legal assistance in case of dispute.

4. Specialist Firms : The term implied to those firms who are specialised in any one particular function within the industry; for examples electricians, decorators, accoustic engineers etc. These firms are sometimes nominated as sub-contractor or their services may be availed by general contractors at their own initiative. Now-a-days specialist of form erection, specialist of concreting etc. have also development. They can execute their work more efficiently than others. More than fifty per cent jobs are done now-a-days by these specialist firms.

5. Quantity Surveyor : He prepares the bills of quantities and values the work from time to time, during progress of the work. This job at present is mainly done by engineers in India.

6. Contractor : When all the plans are ready, complete details have been ascertained, and moreover tender notices have been issued, the contractor comes into pictures. He submits his tender for undertaking the work, along with financial risk. He employes the labour, arranges material and equipment required for the execution of work, control and co-ordinate the activities of the specialist firms, if required. This needs the careful planning, good organisation, a thorough knowledge of different technique of construction.

Contractor get the work executed under the supervision of engineer-in-charge. He is given all the drawings and detailed specifications required for the construction of work by the engineer in charge.

7. User : Strictly speaking he is not at all a part of building industry but whatever is planned and executed by building industry user is a judge to decide the quality of work and facilities provided. In building generally he decides whether house has been planned functionally well or not. He actually faces the difficulties of poor planning. Therefore, he also becomes a party of building industry.

In many countries, in big housing projects, in first stage, only a few houses are constructed. Theses houses are given to public for use to find the shortcomings of the houses, if any. The complete project is executed only after doing genuine modification, if any, as suggested by the users. This practice is not very common in India so far.

However, in India also, in case of apartment projects a sample flat is made and shown to would be purchase to give an idea about his flat. Some minor corrections are incorporated if found feasible.

8. Surety Bondsmen : It is a well established practice in engineering contracts for the owner to require a surety bond from each bidder, that is in the event of the contract being entered into with any particular bidder, the latter will fully and faithfully will make all its provisions. Similarly contractor also needs surety of payments of wages timely. Hence surety bondsmen give surety to the owner and to the contractor that he will be responsible for loss of wages, if any. Hence a surety bond creates a three concerned situation. First between the owner and contractor, secondly between owner and surety bondsmen and, thirdly between contractor and surety bondsmen. This practice of surety bondsman has not become very popular and common is India so far. Only for solving legal complication and to perform all official formalities between owner and purchaser or owner and supplier, Brokers are appointed/hired to do all this business. But they are not surety bondsman.

They do not assure the quality of work or payments of wages, they simply complete the formalities required for business transaction including official work, legal work

9. Material Suppliers : These are the persons who supply material to the contractor or owner on demand. 60 to 80 percentage cost of project is invested on materials. But whether they form a part of building industry or not is a matter of subject discussion.

10. Builder : Present trend in housing is through builder. Builders takes possession of certain land and build houses and then sold them to various persons willing and have capacity to purchase. Loan facilities are also arranged by these builders. As such indirectly, a builder plays a role of surety bond and becomes a part of building team. In big cities, persently more than 90 percent houses are built by these builders only.

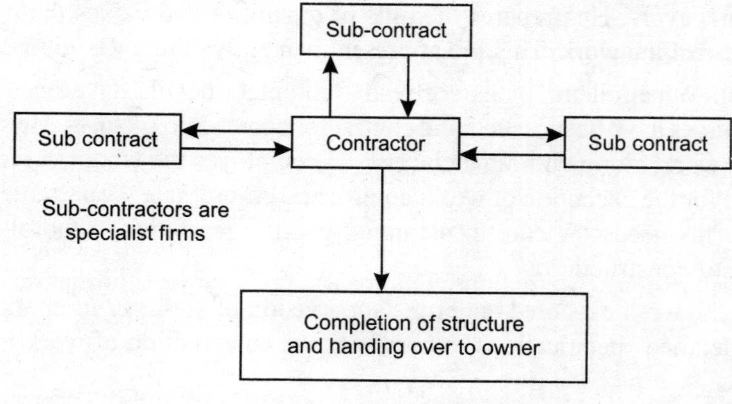

Fig. 2.1.

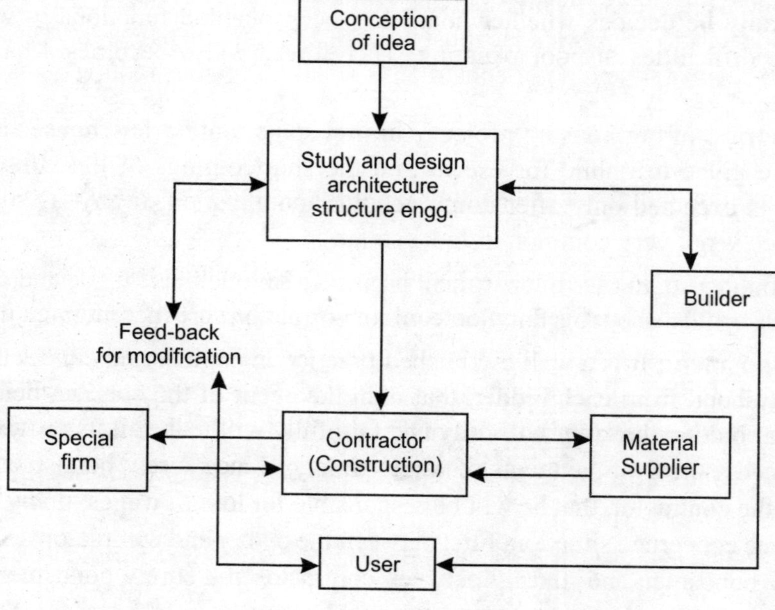

Fig. 2.2.

The object of all the persons involved in building industry is creation of some thing which will serve the purpose for which it is built. Construction is the ultimate objective.

As there is always a uncertainty in construction, a liason is essential between these persons. Figure 2.1 and figure 2.2, show how a few persons of construction team are inter related to each other.

2.4 CONSTRUCTION ECONOMY

The cost of project may be divided into the following items:

1. Materials cost
2. Labour cost
3. Tools, plants and machineries
4. Overhead and supervision
5. Profit

Economy can only be achieved if there is a control on almost all items either one way or the other. It is a prime job of an engineer to control and organise the work in such a way that considerable economy can be achieved. He should have latest information of new materials, new construction equipment and new methodology of construction. These information will enable him to modify design, construction methods, to economics the cost of projects and maintaining the quality control.

Contractors are also required to have latest knowledge of different developments which may reduce the cost. It has been found that due to less payment of wages to labour by contractor, work is held for a long time which consequently increases the cost of project. Hence minimum wages as per labour act must be paid to the labour. He must use modern equipment available which will have more efficiency thus reducing the time of construction to optimum thus reducing cost of project.

Economy by the Engineer

1. By incorporating components in the design which are either locally available or made within the country.
2. By avoiding special equipment and technique not available in the country.
3. Using standard or similar section in concrete structure so that same shuttering may be used over and over again (modular coordination techniques)
4. Giving full and detailed specification along with detailed working drawings to the contractor so as to eliminate chances of errors.
5. Giving the details of tests of each component where required so that the contractor is aware of the requirements.
6. Having a trained and experienced supervisory staff who can give suitable guidance to the field workers. In case of large projects special training cells are organised to make the supervisory staff acquainted with the job requirement.

   In order that the contractor fully knows the scope and requirement of the work so that he takes into consideration these factors at the time of building, it is desirable that the engineer-in-charge holds a conference with the bidding contractors, well before the date of receipt of tender, and acquaint them with the job requirements.

7. To provide all necessary documents to avoid any delay during execution of work (due to lack of decisions)or should not leave any work, decision to be done during the execution of project.

Economy by Contractor

The contractor can also plan his activities so that he also gets maximum output from the resources deployed by him. This may not affects the projects directly as he would have already given his rates but it would help in bringing down his subsequent rates and at the same time eliminate dispute since a satisfied contractor would not create disputes. Some of the important steps in his direction are:

1. Proper choice of plant and equipment. As a rule larger capacity plant are more economical.
2. Providing adequate spares so that plant and machinery are kept working order in arranging the stores/materials well in advance to avoid delay.
3. Employing trained and experienced staff.
4. Giving liberal incentive for greater output. For example, paying the labour or group of labour on the basis of production.
5. Following proper safety precautions and measures so that there are no accidents.
6. Initiating welfare scheme for the staff and the labour. This includes provision of medical facilities.

2.5 CONSTRUCTION AGENCIES

Various agencies involved in construction works of various types of civil engineering structures are:

1. The central government and its agencies such as central public works department, military engineering services, National highways; Central post and Telegraph office. Indian Railways, Central works departments etc.

2. State government and its agencies such as public works departments,Irrigation department public Health engineering department, various city development authorities, housing board, municipal corporation ;Block development offices etc.

3. Public undertaking department such as National building construction corporation, Indian Railways, National projects construction corporation, Civil Aviation authority of India, Oil and natural gas commission, National building organisation, Housing urban development corporation and various such public undertaking agencies.

4. Private companies and other enterprises such as Gammon India Ltd. Hindustan Construction corporation, contractors, societies, co-operatives etc.

5. Private individual construction owners.

Along with these there are a few more departments run by specific agencies such as-

(*a*) Environments Engineering wing-works under the secretory to Urban development.

(*b*) Ports wing-works under the secretory to home department.

(*c*) Energy wing-works under the secretory to Industrial, Energy and labour department.

(*d*) Command area development authority (C.A.D.A.) works under irragation department by an Administrator.

There are many other government, semi government and private agencies involved in construction as such the volume of construction work is therefore enormous.

2.6 CLASSIFICATION OF CONSTRUCTION WORKS

There are various ways of classification of construction works. It further varies from department to department. It is illustrated here.

1. Classification-I : The projects works includes scheme right from procurement of land and its survey; preparation of plans, estimates and construction to the final completion of the whole scheme. In project, number of construction works are carried out. These works are classified as major work or minor work depending upon their cost of construction. In general a works costs more than Rs. 2 lakh is called as major work otherwise a minor; but this amount various from department to department. Works having cost less than Rs. 2000/ are called pity work and executed by pity contractor. This amocent also varies from time to time.

2. Classification II : Permanent, semipermanent or temporary work, based on its nature of construction.

3. Classification III : Original work and repair work.

4. Classification IV : Light construction work and heavy construction work. Construction of heavy work involves use of machineries, Residential building, school building, ware house, light industrial shades come under light construction work. Heavy construction works are construction of tunnels, bridges, dames, thermal and hydroelectric power station, dock and harbour, airport building etc.

5. Classifiaction V : Industrial building, use of machineries may be required as per size and type of construction, oil refineries, steel plants, fertilizers and chemical plants; atomic reactors etc. come under this category

2.7 STAGES OF CONSTRUCTION

Various stages of construction are

(*a*) Pre-tender stage :

1. Initiation of proposal and preparation of schemes.
2. Technical sanction.
3. Acceptance of proposal in accordance to its feasibility and acquisition of land.
4. Survey of site.
5. Preparation of drawing, design estimates.
6. Accord of administrative approval and allotment of funds.

(*b*) Post-tender stage:

1. Preparation of drawing , design estimates.

2. Opening of tender, preparation of comparative statement and acceptance.
3. Allotment of work, and to issue work order.
4. Execution of work, its supervision and its completion.
5. To submit completion report.
6. Handing over of work to owner.

The stages of construction in P.W.D. comprises of

1. Planning stage -it consists of
   (*a*) Preparation of project report.
   (*b*) Administration approval and Technical sanction.
   (*c*) Provision in Budget.
2. Execution stage- it consists of
   (*a*) Preparation of detail drawing, design, estimate and working drawing.
   (*b*) Preparation of tender documents.
   (*c*) Publication of tender notice and inviting tender.
   (*d*) Scrutiny of tender.
   (*e*) Acceptance of tender.
   (*f*) Award of contract and issue of work order.
   (*g*) Supervision- quality control and execuation.
   (*h*) Completion report, it consists of two parts:
   Part I : Submitted on completion of the work/project
   Part II : Closing of account after disposal of surplus stores if any
   (*i*) Use of structure.

If required in case of land acquisition selection of site, finilization of alignment of work, acquisition of land, amount of compensation for land acquired are also required to be carried out in pretender stage.

If land acquired is of government, then site is selected by owner i.e. government. If land acquired is of some private person (owner) then though site is selected by government. The owner or his representative is taken into consideration. In case of government land, representatives of concerned government land owner is also required to be taken into consideration.

2.8 RESOURCES OF CONSTRUCTION

Any construction work either major or minor, large or small, industrial structure or else needs following resources for efficient and smooth execution of construction work.

1. Men skilled and unskilled, supervising staff, specialist personnels.
2. Material such as cement, steel, bricks, timber, aluminium, aggregate, fitting and fixtures, scaffolding and shoring, oil and lubricant etc.
3. Machineries trucks and dumpers, concrete mixer and vibrators, road rollers, scrappers, driving machines, earth moving machineries etc.

4. Money for almost all above items need financial resources without which nothing can be executed.

2.9 MANAGEMENT IN CONSTRUCTION

The most important factor involved in construction is financial resources which are in general always scares. Hence careful planning of construction, proper and scientific management of construction, proper use of men, material and machineries can reduce the cost of construction to a considerable extant, consequently saving the financial resources. This adds to national economy. Therefore there is always a need of management of construction works as a whole and construction activities in particular.

It needs proper planning of all the activities of construction, their directed organisation and efficient control. it also needs co-ordination of all the personal and activities involved in construction.

Stages in construction industry in a more simpler form can be classified as

1. Pre-design stage consist of initiation of proposal, preparation of schemes, technical sanction, acceptance of proposal according to feasibility report and acquisition of land if required.
2. Design stage consist of preparing counter plan, site survey, development of design and finalisation; structure design, and preparation of detailed drawing.
3. Estimates Preparation of details estimates;determination of quantities of construction materials.
4. Tendering preparation of tender documents; invitation of tender; preparation of comparative statement and award of contract.
5. Execution of work by any agency, departmental execution, contracting agency etc. procurement of labour and starting of construction, work site supervision, site inspection preparation of running bills; payment of wages, payment of bills, and final payment of bill, submission of completion report.
6. Maintenance daily maintenance; yearly maintenance; routine maintenance etc.

These various stage of construction are undertaken by appropriate persons assigned with that job and construction is completed for its final use.

Construction Scheduling

While planning the various events of any projects it is necessary that the sequence in which the project is to develop should ne known not only to the project administration, but also to all the links in the system namely the engineer, supervisor, contractor and other site workers. The requirement is met in a systematic manner by scheduling or programming not only the main construction activity but also the supporting elements such as labour, material, equipment and expenditure.

Once the scheduling for resources has been done it is comparatively easy to prepare time and progress chart also known as the Bar chart or gantt chart or the modified milestone chart and to prepare net work analysis in the form of Critical path method charts. The other advantages of preparing these schedules are:

1. Knowing the desired output of work along the various activities, the requirement of labour during different stages of construction can be worked out and phased recruitment retrenchment, inter-project transfer of labour can be prepared well ahead of the actual occurrence.
2. Useful deployment of plant equipment and machinery can be ensured so that these costly machines do not remain idle for want of skilled and unskilled labour or they are not off road for want of proper maintenance and repairs.
3. Cost control study of the project at different stages of construction can be periodically reviewed on the basis of expenditure, returns and progress of different items of work.
4. A phased programme of procurement of construction stores can be prepared.
5. Requirement of funds during the various stages of the project can be predicated fairly accurately.

The various schedules referred to above are given below:

1. **Construction Schedule:** The schedule is a time table showing proposed during of construction of various items of work such a table can be prepared when the following data has been collected.
   (*a*) Number of operation.
   (*b*) Output of labour and machinary.
   (*c*) Quantity of various items of work involved.
   (*d*) Number of field workers, and other resources (plant, equipment, machinary and construction stores)
   (*e*) Sequence in which the various activities are to be taken in hand.

Table 2.2. Construction Schedules (Weekly)

Project particulars............

Location

Administrative Approval.......... No. .... Data.........

Amount of administrative approval.

S. No.	***Item of work***	***Total quantity***	***Units***	***Month***								***Month***								
				Week								***Week***								
				1		2		3		4		1		2		3		4		
				Estim-ated prog-ress	A	E	A	E	A	E	A	E	A	E	A	E	A	E	A	

Note : E = Estimated progress

A = Actual progress

Alternately Construction Schedule may be prepared in the form of T and P chart or bar chart as shown in table 2.3.

Table 2.3. Construction schedule-1

Time and Progress Char/Bar Chart/Gantt Chart

Name particulars............

Location

Administrative No. .... Data.........

Amount of administrative approval.........

Progress as on.........

	Part 1st		*Stores*					
	Item	*Quantity required*	*1st month*		*2nd month*		*3rd month*	
1	Cement	500 T	1	15	1	15	1	15
2	Steel	50 T						
3	Timber	20 C.M.						
	Part 2nd		*Construction Equipment*					
1	Dolers	4						
2	Road Rollers	4						
3	Concrete	2						
4	Bitumen mixers etc.	4						

An improvement in the bar chart is achieved by having a time schedule of the completion of various activities.

The Part III of the chart illustrated in table 2.2. is reproduced in table 2.3 and put in the form in which events on completion are identified. The completion of each activity is numbered in the sequence in which it should be completed. The estimated completion of an event is shown as ▤ whereas the activities actually completed are shown as ▧ on the date of reporting if all the activities on the left hand side of the reporting date line are shown as ▧. It shows that the time schedule has been correctly followed. On the other hand, ends of activities denoted as ▤ on the left hand side indicate time lag in the completion of the target. Similarly, if the right hand side contains certain activities shown as ▧ it indicates that the item of work has been completed ahead of schedule.

Table 2.4. (Fortnightly Progress Report) ending 15th January

S.No.	*Item of work*	*Total quantity*	*1 January*	*February*	*March*
			1 15 31	1 15 28	1 15 31
1	Earth work in excavation	C M			
2	Foundation concrete	C M			
3	Masonry up to plinth level	C M			
4	Masonry up to lintel level	C M			
5	Masonry up to roof level	C M			
6	Rooting	S M			
7	Doors & windows	S M			
8	Flooring	S M			
9	Plastering	S M			
10	Finishing white washing painting distemper	S M			
11	Approach road	km			
12	Site clearance	S M			

Doors & windows

Reference: Actual progress; Estimated progress

2. **Labour Schedule :** The purpose of the labour schedule is to know well in advance, the type of labour and their numbers required from time to time so that steps required for their recruitment are taken in advance. A typical labour schedule for a schedule project is shown in 2.5 Table..

Table 2.5. Labour Schedule

Name of Project............

Location ....

Administrative Approval No. .... Data.........

S. No.	***Labour classification***	***1st month Feb weeks***				***2nd month Mar weeks***				***3rd month Apr weeks***						***4th month May weeks***			
		1	**2**	**3**	**4**	**5**	**6**	**7**	**8**	**9**	**10**	**11**	**12**	**13**	**14**	**15**	**16**	**17**	
	Mason																		
	Carpenter																		
	Mixer operator																		
	truck driver																		
	Supervisors																		
	Labourers																		
	etc.																		

Atypical expenditure return chart is given in table 2.6

Table 2.6. Expenditure Schedule

Name of the Project............

Location ....

Administrative Approval No. .... Data........

Expenditure Return for the months of ....

Nature of Expenditure (As per item of Adminis-trative Approval	***Expenditure during the month***	***Expenditure upto the end of the year (During the year only)***	***Expenditure upto the end of the year including expenditure of previous***	***Liability***	***Total expenditure***	***Remarks***
(1)	**(2)**	**(3)**	**(4)**	**(5)**	**(6)**	**(7)**

3. **Material Schedule :** The purpose of the material schedule is to know the requirement of various construction materials well ahead of the start of the project so that at no stage during the execution of the project, work is held up for want of stores. In the case of departmentally run project storage of construction stores would not only result in the delay in the completion of the project but would not only result in the delay in the completion of the project but would result in resources being undeployed or under deployed which may call revision of the delay in the procurement of store.
4. **Equipment Schedule :** The success of modern civil Enginner project depends to a large extant on the proper use of construction equipment. The use of heavy equipment in construction project is like having the allaudins jin in ones possession, who if constantly deployed on some mission or the other would do wonders but might cause considerable damage if enough work is not found for him. The heavy equipment plays a similar role in a Civil Engineering projects. If the machines are properly maintained and usefully deployed, they not only accelerate completion of the project but also bring down the expenditure on the project. If, however, their deployment is not properly planned, they can result in heavy financial burden on the project. The project management should therefore plan the use of construction equipment and machinery in such a way that they are usefully deployed all time taken out of the project at the earliest.
5. **Expenditure Schedule :** The engineer is primarily judged by his ability to provide most economical construction. It is, therefore, essential that the financial position of the work is reviewed at regular intervals to ensure the following
   (*a*) Expending during the financial year is kept within the allotment for the year.
   (*b*) Expenditure on each item of administrative approval does not exceed the administrative approved amount.

The return is submitted in the beginning of the month and the expenditure shown there in is verified with respect to the cash book entries.

2.10 FUNCTIONS OF CONSTRUCTION MANAGEMENT

The construction project administration/management carrier out the following function.

1. Phasing of project.
2. Exercising administration, financial and technical control.
3. Planning and design.
4. Procuring plant, machinery and equipment including workshop.
5. Procuring of construction stores.
6. Recruitment and training of staff and labour.
7. Arranging payment of salaries and wages.
8. Welfare of staff and labour.
9. Established round the clock communication between the administrate cell and field staff.

The above function are briefly detailed below:

1. Phasing of the project : This is based on the users requirements, availability of resources and availability of funds.

2. Exercising Administrative, Financial and technical control : The administrative control is exercised by having the work administratively approved so that the expenditure incurred on various items is as per sanctioned estimate of work and also the quantity of work done is not in excess of the quantity shown in the estimate for administrative approval of the work.

Financial control is exercised by provided budget allotment and obtaining the sanction for the expenditure for various projects and parts of projects and getting all the payment vouchers audited from time to time.

Technical control is exercised by ensuring that no item of any work is taken in hand without according technical sanction to avoid extra item/substituted items, and excesses in the quantities of items.

3. Planning and Designing : The planning of various activities and designing the sequence of operation of a project/work are very essential. A considerable economy can be achieved by planning and designing. The Construction period of project can be reduced which consequently reduces overhead overheads thereby economising the construction cost. Planning of each activity in advance, procuring construction material in advance, appointing subcontractor, labour etc, reduce construction time considerable. This all consist of planning and designing of various construction activities.

4. Procuring Plant, Machinery and Equipment including workshop cover : Before talking the execution of any construction project in hand it is necessary that the resources such as machinery, plant and equipment are made available and that suitable arrangements for the maintenance and repair as well as made. At the same time the project management has to see that the resources brought to the work site are not kept idle for want of oprators or for having not started the work. This requires a very carefully planning and coordination.

5. Procurement of Construction Stores : The construction administration or management has to ensure that the work is not held up at any stage for want of stores and construction material. A golden rule is that 2/3 of the construction stores should be at site (work site) and 1/3 in sight (procurement initiated)

6. Recruitment and training of staff and labour, arranging payment of salaries and wages and welfare of the staff. These are of special importance in case of departmentally run projects. when the execution of the work is through contractors, major portion of this problem is eliminated, as this is arranged by the contractor.

7. Established round the clock Communication : It is necessary that the project management is always in contact with site executives. This is best achieved by established telephone or wireless communications.

2.11 SCIENTIFIC METHODS IN CONSTRUCTION MANAGEMENT

This involved the use of modern technology, plant and equipment and scientific methods of exercising control on all the activities of the project. Main features of this are:

1. Proper planning, design and site investigation.
2. Use of scientific and mathematical instruments for quality control.
3. Training of skilled and supervisory field staff for specific jobs.
4. Use of CPM, PERT and bar charts for ensuring timely execution of different elements of the construction projects.
5. Use of work diaries for recording day to day progress and output.
6. Use of log Book to record consumption and output of plant, equipment, machinery and other vehicles.
7. Introduction of usage rate for purpose of debiting the project in respect of the usage charge of plant, equipment and machinery.
8. Cost control study to check the performance of the project as a whole.
9. Monthly Expenditure return to check the position of expenditure with reference to budgets allotment and administrative approval.
10. Arranging periodical audit of the expenditure incurred headwise(store, machines, labour, supervising staff).

2.12 FINANCIAL MANAGEMENT

The management of finance means control on available financial resources in a manner such that there should be maximum use of it with a minimum unnecessary expenditure. It is essential to control the expenditure on almost all the stages right from its inception through its development and design, till its execution and payment of final bill. This is essentially required as the available resources are always limited and it is primary job of an engineer to utitise them in the best possible manner by proper planning, designing and on all the items of construction to ensure that project cost is not exceeded from estimated and sanctioned cost. Cost control and financial management are important aspects of construction management to almost all the persons involved in construction industry as well as to national Economy.

The basic object of cost control and financial management are to minimise the cost of project and to minimize the efficient use of available resources.

Stages where cost control and financial management is needed

Theoretically speaking cost control and financial management are needed at all the levels. But there are two major stages where it has effected as at these steps only if proper control is made, considerable financial resources can be saved. These stages are:

1. Planning and Design stage.
2. Execution stage.

1. **Planning and design stage :** The planning of any project can control the cost of a project to a considerable extent. Studies have revealed that efficient planning of any project can reduce the project cost by about 10 percent. Similarly, specifications and design also affect a lot on project cost. If specification are duly specified, local available materials used, it is possible to control the cost to a considerable extant.

Accurate and realistic estimation of a project is also vital for its successful implementation.

2. Construction Stage : Construction of any project consist of:

(*a*) Material cost including transportation cost.

(*b*) Labour cost including wages, accommodation and travels.

(*c*) Plant, machinery and equipment including cost of fuel, and lubricants.

(*d*) Overhead cost such as guarantee, supervision, insure, liquidated damages agency charges, maintenance charges after completion, company taxes and engineers facilities, head quarter expenses.

There is a considerable scope to economise all these costs. Efficient use of material and efficient and effective use of machineries can control financial resources to a great extent. Wastage of material must be avoided and equipment must be maintained properly. Overhead cost also needs sufficient control.

Management of Estimation of Project

In the competitive environment, a construction company has to be able to excute the work at the least cost and for this, a systematic understanding of various cost components, which constitute the total cost is absolutely necessary. The accurate and realistic estimation of a project is very important for its successful implementation. The estimation of any project can be divided into two categories.

1. Preliminary Estimation : For inclusion of projects in the planning of the country and obtaining administrative sanction for budget purpose. There may be a large time gap between this stage and the actual implementation stage.

2. Detailed Estimated for project Implementation which precedes Execution of the Project: In respect of the project executed by the Government agencies such as public works department and Railways, as also the Government undertaking, the detailed estimate is basically done on unit rates of work, prepared and approved from time to time. These schedule of rates are prepared based on current market rates, mostly judges from tender of work executed at the time of preparation of the standard schedule of rates. This type of detailed estimates does not consider the time element of execution. This if speedy construction is executed either by better site management planning and technique or by better methodology, one gets advantage but not reflected in the estimate as such these estimates many times called as Execution Estimate. There should not be much time lag between time of preparation of this estimate and actual execution of work, otherwise there is always escalation in the cost of material, labour, machineries etc. These estimates generally remain valid for a period of four months in the international tenders. The execution estimate therefore must be accurate and should be prepared on actual basis not only on the basis of the unit rate of work.

The cost analysis is mainly based :

(*a*) Material

(*b*) Labour

(*c*) Equipment and machineries

(*d*) Overheads

The rate analysis must be carried out by an experienced person having thorough knowledge of various materials cost , labour cost and other items. Provision for tools and plant must be made in rate analysis. There is a real need to analyse the overhead expenses and their control. The overhead cost in the form of supervision includes the cost to higher level manager sites, their accommodation, travel, site office costs, vehicle for inspection, maintenance, office, stationary, telephones, telex, rest house facilities etc. These charges go on increasing if the work is delayed, consequently increase the project cost.

Various activities involved in proper estimation of cost of the project are:

(*a*) Study of tender document General and special condition of contract, specification, and drawings.

(*b*) Contradictions or ambiguous clauses in the condition of contract be avoided.

(*c*) Methodology for execution to achieve minimum cost.

(*d*) Quantities and scope of work, sequence of operation, bar charts.

(*e*) Assessment of quantum of materials of different items used in the work, their specification, cost delivery schedule etc.

(*f*) Study of field data with advance planning of items with which information is to be collected.

(*g*) Calculation of operating cost in detailed manner, avoiding approximation keeping in view the machine oriented or labour oriented technique of execution adopted.

(*h*) Study of alternative design to minimise costs.

(*i*) Compare economics of cost.

Indirect costs are those such as organisation overheads and management costs which are normally shared by various activities between the projects.

Time is an important element in some of those costs,particularly those relating to materials, services and sub-contracts. If the projects manager is concerned with the money which may be available to him to pressure alternative courses of action, he needs to known the combined value of the committed costs and the actual costs.

2.13 MANAGEMENT OF CONTRACTS

Projects involving civil engineering construction and more so the large projects ; are usually executed by engaging contractors. Even when turnkey contract for large project are awarded to big contractors; they are executed through sub-contractor; the prime contractor looking after only coordination activity and a part of the work on his own. As such management of contract is one of the important aspects of construction management. Since the agencies which deal with such contract are government department; corporation or other semi-government agencies and also as large number of contract are item rate percentage rate contract; it become still essential to have good management of contract.

Contract management should be laid early at the time of investigation of the project on analysis of rates for preparation of estimate. Natural situation are required to be taken into consideration while framing the contracts. Fairness or equity to both the parties of the contract is essential. Drafting of contract document; especially the specification; would need to be assigned by mature and experienced engineer. When works are of repetitive nature; adoption of standard specification would be desirable. Clarity or un-ambiguity is yet another attribute. The expression to the satisfaction of the engineer as well as the contractor;both should be taken into consideration. Avoidance of redundancy is desirable. It is a quality of good contract document. A standard model tender document should be used for all such purposes.

Every contractor is required to deposit some security money with the department as a guarantee of completion of work satisfactory. At the same time; contractor also need some seed money to start the work. If the seed money is not provided to him under contract; contractor has to borrow it at higher interest rate. He also loses interest on security at the same time, pays higher interest for money borrowed for starting the work. Hence instead of this; if a guarantee either in the form of bank guarantee or bonding companies bond are given; probably contractor as well as department; both may be be satisfied. The concept of Bonding-companies is not prevalent in India so far and bank guarantee from nationalised bank is a factor that must be considered. Extra items should be avoided and arbitrations must be minimised for smooth working of a contract deed.

2.14 MANAGEMENT OF PROJECT CONTROL SYSTEM (PCS)

The task of control is to ensure that plans succeed by detecting deviation from plans and furnishing a basis for taking timely action to correct potential or actual undesired deviation. The type of control system should be such that it measures in a verifiable way whether the plans under consideration are being accomplished or not. Control should be tailored to the specific needs of the projects under execution. More the control technique and information are understandable to individual manager, Who must utilise them for result, the more they will be actually used and the more they will result in effective control.

Essential Features required for proper and better functioning of project control system are:

1. A convenient framework to integrate time, cost and performance.
2. Ability to track progress according to defined parameters.
3. Ability to produce relevant, reliable and accurate data base to facilitate decision- making of various levels of managements.
4. Capacity to anticipate pitfalls.

The key of successful management is monitoring. Monitoring performs a distinct function in PCS. It renders possible, timely identification of problems and a realistic evaluation of alternative methods of corrective action.

2.15 PERSONS REQUIRED ON-SITE

Once a work is allotted to a contractor, before he starts construction, various categories of persons are required to be recruited by contractor for smooth working of the work. Though contractor has his office staff well equipped with facilities, still some other persons are required to be deputed. Many Contractor, have their own permanent workers, but in absence of these he appoints a few. These personnels are:

1. Site Engineers
2. Site Supervisor
3. Watchmen, store clerks
4. Mistries and labour

Each of these personnels have their specific duties. The job of site engineer is to maintain quality control and to get work executed as per drawing. He is also required to prepare measurement books, running bills, and to maintain stock of materials. The basic job of site supervisor is to supervise the construction work as per schedule. He also assists site engineer for preparation of bills, maintaining stock etc. It is his duty to report the contractor though site engineer the progress of work, requirement of material , labour problem if any, construction problem etc. The duty of watchmen is to have a close watch on the materials lying at site against any type of thefts. Unwanted persons should not allowed to enter the site. Registers are maintained to note the name of persons visited the site including their time of arrival and departure. Store clerk maintain the daily stock of receipt and issue of materials to the construction sites.

Contractor either appoints his own masons and labour paid either monthly or daily or sublets work to any known mistry at a reasonable rate as per that item. The later practice of appointing labour contractor sub-letting labour work to mistries is most common and maintain good progress of the work, moreover contractor gets relieved from dealing with labour class. The drawbacks with this system is that the quality of work is not maintained many times. Construction works as is the work is not approved by the engineer, his bill is likely to remain pending or unpaid.

Other personnels involved are electrician and mechanics. They are required on large sites if needed In case of construction involving use of heavy machineries and equipment, these are always required so that in case of any fault, repairs and corrections can be undertaken immediately and down time could be minimum. Many times the services of the persons are hired. Other personnels required depending upon type of construction and its extent are:

1. Plumber
2. Carpenter
3. Welder
4. Peons

2.16 VALUE-ENGINEERING

The engineers have always tried to reduce the cost of construction without affecting the quality and the functional utility,however their approach was based mainly on the past experienced. With the advancement of science and technology, it become comparatively easy to reduce the construction cost, but the concept of functional utility was not given due considerable. Reliability and durability were of little importance. Only in recent past. Engineers and Architects have started taking into consideration these important factors *i.e.* reliability and durability with functional utility to optimise the cost. This cost of the project is named as "functional cost"as the ultimate objective of project planning, construction, management and control is to have its functional utility with durability and reliability of its services at the optimum possible cost.

This subject has got emphasize in last few years whose object is to effect economy, in the cost of construction of project. '*Value engineering*' as such is a systematic application of recognised technique which identify the function of a product or service; established a monetary value for the function and provide the necessary function reliable at the lowest overall cost. In civil engineering it is related to minimum cost of project or construction work, without-affecting the quality. The projects are generally designed by engineers and constructed by contractor. The responsibility of engineer is to design project in a manner such that it must be economical in minimum cost and maximum output. The job of contractor thereafter is to apply his skill to construct the project in estimated cost or if possible even less than that.

Value engineering therefore may be applied to a project in a few stages. The stage being during planning of the plans, specification and design ie prior to release to the contractor. The purpose here is to study various alternatives in planning, designing, specification, use of local available materials to economise the construction cost wherever possible, of course without affecting the quality. The new saving if any shall be the total reduction in cost.

The second stage of value engineer study is after the award of the of the contract of construction works. Here the study is generally made either by the contractor or his value engineer. Some time contractor also awards this work to professional value engineer. Here value engineer after studying all details; informs if there is any scope for reduction in the cost of the project by adopting certain methods of construction or by changing or modifying design method or by any other legal method maintaining the quality if possible. In some cases contractor refers this to owner the finding of his value engineer. Also with his detailed statements describes the modification with estimates showing the anticipated reduction in the costs. Owner in such case after consulting his architects, engineer approves the modification with reduction in cost, the resulting net saving are being shared by contractor and the owner on a preagreed basis, generally an equal amount.

Value Engineering therefore may be defined as a systematic study to seek out the best functional balance between the cost, reliability and performance of a project. It can be defined as a systematic Inter-disciplinary examination of various factors those affect the cost of a project or service in order to device means of achieving the specified purpose most economically at the reduced standard of quality and reliability. The technique of value engineering was initially develop by Lawrence. D. miles a purchasing agent of General Electric Inc (USA) His studies concluded that

1. Conventional approach stifled imagination by restricting thinking to existing objects and design.
2. Concentrating on functional requirement produce greater mental freedom.

Arithmetical representation of value is

$$\text{Value} = \frac{\text{Function}}{\text{Cost}}$$

It is however a cyclic process as there is always a possibility in the modification of design due to continuous research and development. This develops new ideas of design and better utility of available resources.

Time Value of Money

It is well known fact that value of money changes with time. The time value is a relationship between the value of money today and its value at some future date, considering the interest charged for the use of money. Time value of money is essentially involved in the economic study to determine what is economical in the long run, over a period of considerable time. This gives an idea about the monetary loss occurred due to increase in the time of project. Hence it should also be taken into considerable while discussing construction economy.

The ultimate aim is to achieve construction economy using latest technique and know-how available for construction, of course maintaining quality of works.

Value-Analysis

A thing which is valuable to one may not be valuable to other. Hence it is essential to analyse the value. It is study of function. Value analysis is an organised approach for challenging unnecessary cost of any item of construction and procedure. It is the analysis technique designed to examine all the facts of cost and function of each element activity of a project in order to determine whether or not any item of cost can be reduced or eliminated while retaining all functional performance and quality requirement. The objective of the value analysis is to provide a means of total cost control anywhere within duration of construction of that project. It stresses only reduction or elimination of cost maintaining quality and reliability of function of the project to which the systematic approach is applied. The purpose is to seek out and eliminate un-necessary material cost, labour cost, indirect cost.

The value analysis sometimes referred as value engineering. Study should be made by one or more persons who are thoroughly experienced of such works, have thorough knowledge in methods of construction, materials and machineries required to be used in that project. Only such items which are repetitive and having potential saving is more than the cost of the studies should only be analysed.

2.17 JOB-LAYOUT

Proper job layout improves communication and unnecessary movement, increasing efficiency. It is one of the most important and essential job before starting any construction is to prepare a job layout indicating clearly the areas and site available for office; warehouse, storage of materials, equipment, excavated earth; formwork; reinforcing materials, fabrication etc. These are required to be arranged in such a manner that working should be easy and minimum time be consumed in carrying materials from storage areas to project site or construction site. Materials of similar use be stored close together where possible. The general office and ware houses should be located near the main entrance in order that person visiting site for one purpose or the other will have to walk minimum and will not have to travel around construction areas for no reason. This will reduce the danger of injuries and accidents. Cement godown should be located near entrance, first to avoid the time of collecting it form outside and second to have proper watch on it. It is desirable by a contractor to have sufficient area available with him near the site for storing the material as it reduces the labour cost indirectly. Layout of construction of a building is illustrated in Fig. 2.3.

As such due regard must be given to have complete implication of tideness accessibility and co-ordination of the following features.

1. Access and Traffic route : It depends on type of project; stage of construction and site condition. As far as possible two access if available; considered to have good access to the site. For very large projects involving use of heavy machineries, huge quantities of material at a congested site, access should be one way only. It should have proper traffic route. Permission from due authorised must be sought whenever or wherever required. Illegal use of public land must be avoided.

2. Materials storage and handling : The basic aim is to minimise wastage and losses of material due to careless handling, poor storage and theft. Stores are required to be maintained as per ABC analysis giving due importance to each material to be stored. The record of arrival of material, issue stock, present stock etc. must be maintained. Site plan must clearly indicate location of stores, their sizes, routes of access etc. in complete.

The use of machinery for storing and handling of stores makes efficient working of plant; consequently increasing the production output. Aggregates of various sizes should be stacked separately away from each taking precaution that these should not get mixed. Cement stores be such, not to have any atmospheric effect as cement gets deceriorated if comes in contact with environmental humidity. Similarly sanitary wares, Ceramic tiles being costly item must be stored in close rooms.

3. Administrative building and facilities : Administrative block at site must be given prime importance. It must give an overall view at site from the office so that contractor or his agent can have a complete watch on activities at site. Further it must be at calm and peaceful area so that if required official work can be undertaken without getting disturbed either due noise at site or due to other reason. Parking facilities may also be provided for all sorts of machineries and vehicles.

4. Plant- workshop and services : These are essential requirements for projects involving use of heavy machineries to avoid un- necessary downtime. Provision for temporary petrol pumps electrical connection water supply, telephone lines, compressed air and other essentials of workshop must be made.

5. Special problem : Many times depending upon site condition, special provision are required to be made. In case of dismantling of any structure , provision are required to be made for safeguard of neighbouring building. Tall building requires installation of temporary lift for lifting the materials.

Requirement of Site Layout

The basic requirement of any construction site are men, material machineries and equipment. As such while preparing layout, these should be placed and controlled in such a manner that:

1. Machines are placed in most advantageous position from security and use point of view.
2. Material should be stored on the basic of ABC analysis and must be easy to handle without creating problem to workers and circulation.
3. Working persons should have ample space for free working. They should not have any type of problem during working hours. It makes them free to work and creates a feeling of safety against any accident.
4. Adequate space must be available for storage of materials. Further materials like cement, lime, fixtures and fastening should be stored in close and be locked always.
5. The requirement of materials; their quantity go on changing as per the progress of the work, as such layout should have sufficient flexibility of planning.

Factors Affecting Job-layout

There are mainly three factors affect the job-layout for almost all civil engineering works and must be considered before any layout for any construction work is prepared. There are

1. Nature and type of work.
2. Site condition.
3. Method used for execution of work.

Every job has its peculiar method of construction depending upon its type; nature and size The layout required much more depends on these condition. Construction of building, bridge and canal will need altogether different layout. Use of machineries on big projects also affect a lot on its job-layout. The site condition also affect a lot on job-layout . In big cities, for building construction in narrow streets hardly spaces are available either to store material or for working of labour. Hence many times pump crete concrete is used for R.C. frame construction.

Use of precast wall panels, building blocks are also used under such condition. Construction of coffer-dams, river tinning work is found essential while construction bridge on perennial rivers. Similarly method used for construction of concrete deck bridge; steel bridge for railways need altogether different layout as their requirements of materials and machineries are different. Concrete bridge needs shades for storage of cement; their machinery requirement is concrete mixer, vibrator; etc. Whilst steel bridge needs mostly girders, their machinery consist of cranes; welding plant etc. construction of earthen dams need heavy machineries like bulldozers, shovels, etc.

Hence while preparing job layout for any civil engineering structure all these factors must be considered and then it should be prepared. No construction work should be started unless detailed job-layout is prepared and drawn. Much more thought is required to be given on entrance and exit, loading-unloading platform, electric supply line, water supply line etc.

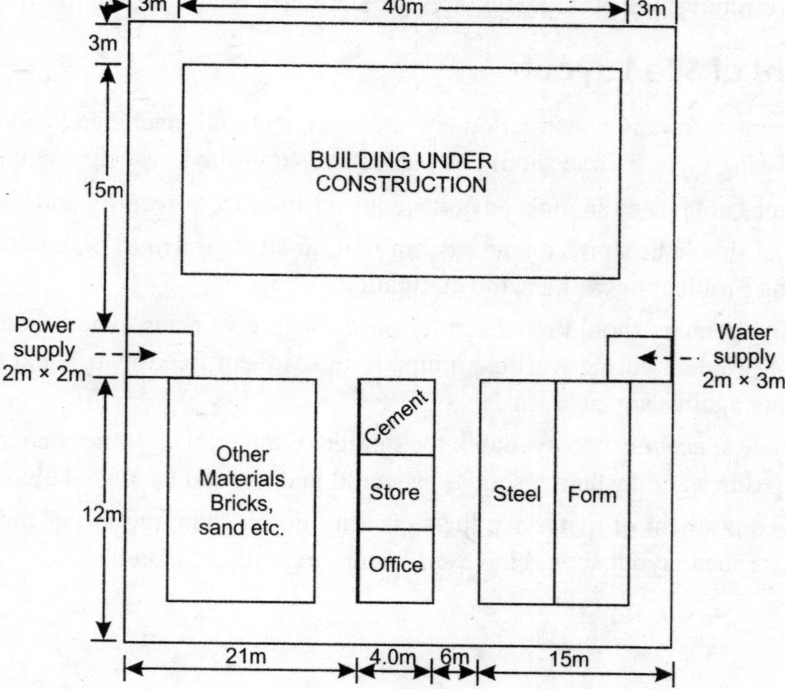

Fig. 2.3. Joblayout of a building project (not to scale)

2.18 MASS HOUSING

Industrialised approach to reduce the cost of housing construction are being explored and considerable attention has been given to technological measures in this regard. It is now increasingly realised that in modern housing projects, specially those involving construction on mass-scale, there is great need to adopt such operational measures while enable appreciable reduction on overhead charges.

"Development cycle approach" is one such method which can be applied to work to be executed on conventional contraction system. This system comprises of systematic development of the dwelling with the aid of investigation into needs, market research and analysis of technical and economic possibilities of construction , execution and use. Together with the building of proto type, a dwelling type is evolved which lend itself to large scale reproduction. During and after the finishing stages, analysis is made and the experience obtained in the construction and the use of the dwelling is fed back to the programming stage, where upon a new cycle may be started.

The fundamental point to be considered to this schemes are:

1. **Team work.** *i.e.* coordination of work of all the groups.
2. **Continuity.** Work should be continuous.
3. Integration of results of research and experimentation.
4. **Feed back.** Updating and revising the plans in light of new experience etc.

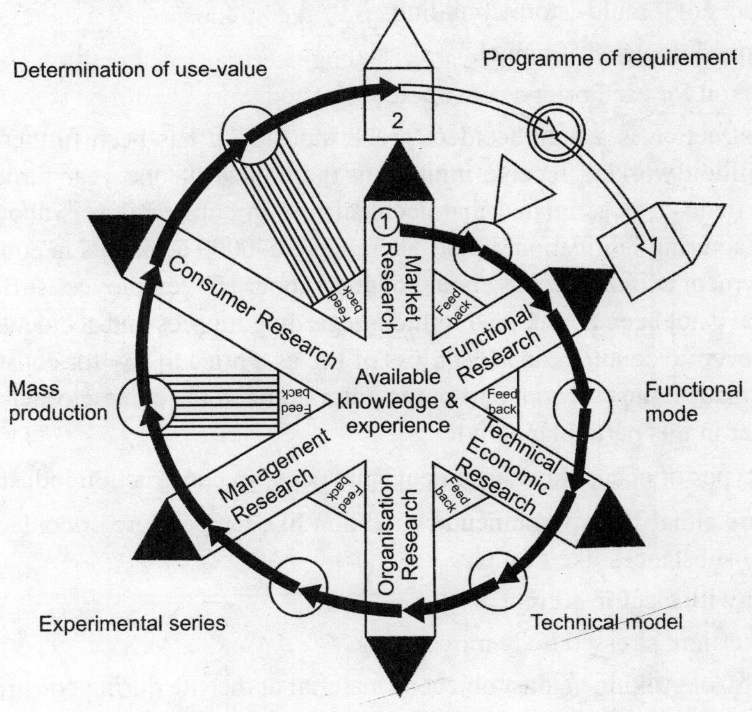

Fig. 2.4.

Possibility of improvement can be held only if there is a team work, there is continuity in the construction work.

The scheme of development cycle for obtaining optimum quality of building at low cost is illustrated in Fig. 2.4.

2.19 SAFETY MANAGEMENT IN CONSTRUCTION

Construction has been an important part of man activities since the men came out of cave. Construction has been considered as one of the biggest industry in India. Various construction made are either below the ground, in the form of basement etc. or above ground. Construction is also made are either below the ground, in the form of basements etc. or above ground. Construction is also made under water or in the space. Total investment in India since independence has been more than total invested during 150 years before independence. As per five years plan investment, in construction sector is of the order of about 40 to 45 percent of total outlay.

When such a quantum of works are executed; so much financial investment is made, accident are executed; but then it is prime-necessary to minimise it; to take safety precaution which are essential and need of today. It has been said of a certain section of a rail road that every tie laid represents the loss of one human life, the life of a worker on the gang that built the rail road. During earlier of construction it was common practice to presume that following constructions will cause life of one person either labour or other person involved in construction of site.

1. Double storey construction.
2. Two floors of a multi- storey building.
3. About one kilometer of tunnel.
4. One person for each bridge constructed.

As such construction is a high accident prone industry. It has been further investigated that more than 10 million workers receive injuries in the course of one year throughout the world. Based on world trend, it is estimated that accident rate in construction is about three times the rate in the manufacturing/fabrication sector. There were 640000 accidents in construction in 1983; When the employment in industry was about 3574000 about 16% employees suffered injury. So far proper records have not been maintained in India regarding injuries and accidents those take place every year. However no country can accept loss of life as a price to pay for construction. Accident prevention to be made as an essential part of the programme of the contractor who expects to enjoy a successful carrer in this particular activity.

The various types of occupational accidents that occur in construction industry are:

1. Exposure either for environmental condition like temperature or contact with potentially harmful substances like dust etc.
2. Contact with electric current
3. Over exertion; stressed body movements.
4. Struck by or striking against object or material at the site during construction.
5. Falling from height or at the same level including slipping.
6. Caught in machine or material.
7. Rubbed or abraded by object like machineries or materials or temporary and permanent structures.

8. Powered vehicle accidents.

These accidents cause following diseases.

1. Un-necessary mental stress.
2. Deafness, decompression sickness; skin diseases, hernea
3. Bursitis, muscle strain;
4. Cancer etc.

As per as the safety of workers is concerned, Factory act and Mines act are in prevailing in India; but they are legally not applicable to construction accidents as construction is not yet recognised as an industry. Further contract workers are not deemed to be governed under the factories act.

Therefore today there is need of accident prevention programme on every project in order to reduce the cost of construction measured in terms of

1. Human lives sacrificed.
2. Injuries to workers of any extent.
3. Loss of damage to equipment.
4. Loss of time during and after construction accident.
5. Compensation, insurance cost to worker if expire at work site, during construction period.

Indirect Accident Costs

Due to accidents at any construction site there are injuries to one or more workers which may lead to the death of one or more workers. No doubt the compensation in the form of insurance is paid to the workers; medical assistance is granted; still there are always certain indirect expenses in any form; the contractor has to bear. These losses can not be covered by insurance or compensation are called indirect cost and include the following:

1. Cost of lost time of injured employee and that of other employees who either stop the work due to accident or those who are employed for first aid and after assistance to that injured person.
2. Cost of supervisory staff in assisting that injured employee involved in investigation for cause of accident and preparing accident report.
3. Cost lost due to damaged equipment or damage of other property material etc.
4. Cost due to delayed progress of the work to accident.
5. Cost of payment of wages to the injured person, during the period of the injury; the labour or workers could not attend his duties.
6. Cost due to slowing down the work of other workers/employees due to accident , as there is always a tendency to take extra unnecessary precaution by each worker after the accident; hence works of other employees also slow down and suffers.

Studies have shown that loss due to indirect cost is about four times that of loss of direct cost. As such it is always necessary to take safety measures to avoid accidents.

Safety and Health (Modern Safety Management Practice)

Special features have been made by construction industry which adversely influence the safety and health and health of workers. Special problem that arise at construction sites are due to following reasons:

1. Short time duration of work sites to labourer.
2. Seasonal employment.
3. Extensive and exhaustive use of migrant labour.
4. Extensive sub-contracting practice.
5. Effect of weather.
6. Over-time working.
7. Competitive tendering.

Due to these reasons, workers change at every stage, safety measures are not adopted, further, no training is given to the workers hence accidents take place.

There is always a tendency in most of the construction sites to work unusual long hours during span of good weather to make up for time lost in bad weather. Sometimes due to need to finish certain operation as these can not be stopped such as concreting; workers are required to work for extra time ; even during night also. this also causes accidents, contractor also always tries to get more work done from migrant labours.

At present; bye laws framed by municipalities and corporation take care of safety in construction industry. Safety codes are framed by almost all municipalities and corporation which are applicable to that territory. Government department either central or state have also prepared safety codes for safety of workers involved in construction industry. But these provision are inadequate and there is no enforcing machinery to check whether safety codes are observed or not. Hence as such though the death rate probably is the highest in construction industry but no record is available rather it is also doubtful whether such authentic record has been made some where by some one else or not.

A balanced health and safety policy should include the following:

1. Safe and healthy working condition for workers at construction site.
2. Training of employees, workers regularly.
3. To comply with safety codes laws and regulations for the protection of heath, safety and environment.
4. Safety instruments and materials must by used during construction.
5. Ill workers should not be employed, children should also be not employed to at construction site.
6. Unnecessary persons should not be allowed at construction site.
7. Enforcing contractor, management to employ safety precaution, safety_codes before execution of any work is started.
8. Regular medical check up of workers.
9. Application of systematic safety training programme.

The objective behind all these safety measures, precaution, safety code are

1. To minimise incidents and accidents
2. Identification and elimination of risks before losses occur

3. Development confidence in workers.
4. Avoiding loss of time, life, machinery and of course work.

Accident Prevention Programme

An accident prevention programme must be started on every construction site in order to reduce the cost of construction in terms of human life and injuries; damage material and machineries and equipment and loss of time. It is therefore essential to collect useful information about the accident which have taken place at various construction sites and their reason. For this it is further necessary to evolve a proper accident monitoring system that can collect the relevant information from the persons concerned. The system should have positive thinking, that the aim is the prevention of accidents. An organisation chart of 'safety department' is given herewith in Fig. 2.5.

Next step in accident prevention programme is thorough examination of construction equipment and machineries and their maintenance property in safe condition. Many accident take place only due to faulty design and poor maintenance of construction equipment. Workers using machine must also be trained properly and be equipped with safety devices for safe procedure of works. Supervision by experience supervisor and proper training can go a long way in preventing most of the accidents. Management can play a very important and efficient role in accident prevention. Many times it has been seen that management itself do not pay much and proper attention to the safety programme for the sake of their own interest of increasing the production. This is not a good practice. Management must see that safety codes are followed and implemented.

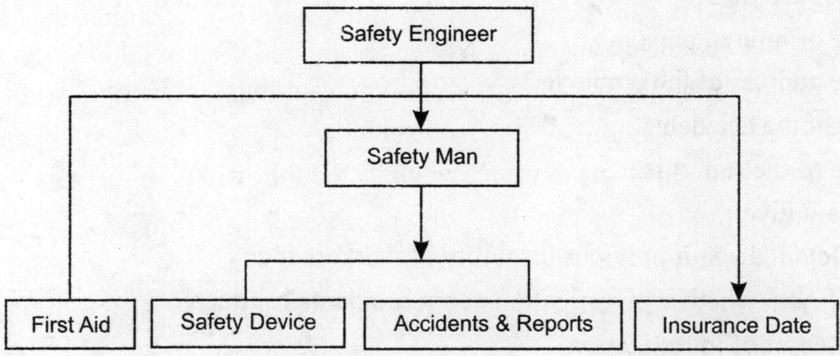

Fig. 2.5. Organisation chart of safety department

Every accident that results is a personals injury of any form must be reported and investigated and remedial measures must be taken. Investigation will also reveal that further efforts may be expected to achieve improvement in the safety performance.

Every site whether large or small must have adequate facilities for rendering first aid to injured employees. If an employee has suffered a serious temporary disabling injury or a permanent disability, it must be made compulsory to that employer to exercise every effort to rehabitate that employee.

Classification of Construction Accidents

Extensive studies carried out show that main causes of construction accidents may be grouped as follows:

1. Uncontrollable contact between men and equipment or between men and material such as cranes, trucks; materials storage etc.
2. Failure of temporary structure such as scaffolding, forms, ladders, cofferdams etc.
3. Inherent engineering hazards; such as the use of explosives, toxic gases; toxic dusts etc.
4. Unsafe practice of individual workers or personal hazards resulting from the carelessness of workers.

These can be avoided if proper precaution are taken. As per provision of Factories Act accident in general are classified into 5 categories.

1. Minor Accidents : Accident which make the worker disabled for less than 48 hours come under this category.

2. Reportable Accidents : If an accident causes worker disabled such that he can not work for more than 48 hrs. The accident is classified as reportable accident. Such accidents must be reported.

3. Fatal Accident : Result in a death of a worker.

4. Accidents which causes poisoning and

5. Dangerous occurrence.

All the accidents mentioned above except No. 1 must be reported and investigated. The report must include following details.

1. Name of organisation and department of working.
2. Date of accident.
3. Name of injured person.
4. Home address of the employee.
5. Time of the accident.
6. Name of the supervisor.
7. First aid given.
8. (*a*) Detailed about previous disability of workers if any
   (*b*) If yes, whether investigation about accidents conducted
   (*c*) Report of investigation.
9. Name of person who first reported about the accident along with witness.
10. Investigation report.
11. Certificate about disobedience to follow the safety rules on the part of management and worker.

This report prepared should be signed by appropriate authority.

List of Possible Accidents in Case of Building Works

1. Excavation : Earth sliding, falling of workers due to slippery area in case of water logged area , Blasted material may hit worker in case excavation by blasting and snacks/scorpion may bite if found during excavation.

2. Concreting : Framework may fall or fail, steel cage may collapse, slipping of formworks, plumb-bob may hit person at lower level, concrete bucket or metal pan may fall down, labourer or worker may slip from any position and fall below in case of slab particularity, ladder may slip erected for worker.

3. Masonry work and plastering : Brick may fall down, failure of scaffolding, props, bambooes may break, hammer may fall.

4. Plumbing works : Chiesel, hammer etc. may fall on persons working at lower level.

5. Painting works : Ladder may slip, ropes may get cut.

6. Truck working : While going reverse, may hit ladder on which a person may be working, vehicle may slip or overturn, brakes may fail.

Major causes of accidents. Though these are relative issue, there can be some other issue also, but chances of causing accidents are more in these nine cases.

1. **Trenching**
   - Excessive trench depth increases collapse hazard
   - Work must only proceed where collapse protection is in place
   - Collapses protection can be provided with proper sloping.
2. **Containers :** Gas cans must be clearly labelled as to contents and hazards.
3. **Rigging Slings :** Rigging must be taken out of service when heavily damaged or when wear detector are visible.
4. **Stairs :** Stairs need to be properly constructed with consistent rise and run, secured treads, stair rails in place.
5. **Office Trailer :** Unused access and entrance doors must be physically blocked from use to avoid falls.
6. **Loading platforms :** Railing, to prevent falls, must have mid and top rails in place. Under certain condition chains can be used.
7. **Flammable Liquid Storage :** A flammable storage cabinet must be installed and used in any building or trailer where over 25 gallons of inflammable liquids are stored.
8. **Fire Extinguishers :** Storage trailers containing flammable or combustible materials must have fire extinguishers within 15 meter proximity.
9. **Ladders :** Damaged ladders or ladder section must be tagged "Do Not Use" until removed from the job site and destroyed.

2.20 MANAGEMENT ORGANIZATION

Once an enterprise grows individual capacity to control; organisation becomes necessary. This means in general, assigning various tasks to different people and co-ordination their efforts. A detailed view would include continuously striving to create a tone, atmosphere and climate that is conducive to work, providing direction , leadership, inspiration to the worker through an effective working Organisation. Organisation theory begins by recognising that most employees are brought together by an employer so that this may help him to do, whatever he wants to be done *i.e.* to accomplishing the goal, employer wants to achieve.

It means organisation is an association of people. In terms of industry it means how manager, supervisors and workers relate with each other to accomplished certain objectives. It can also be termed as creation of a networks of relationship ; function to enable work to be done or the organisation to prosper. The basic object of organisation is to create and control working condition that will maximise its success in its mission.

For the most economical execution of any project it is necessary that it should be properly planned so that a planned development/progress of the same is ensured. In the initial/pre-contract stage of the project therefore only the minimum personnel required for planning phasing and designing should be established and other should be inducted into the work there on as required basis. An organised planning/design cell is therefore the first requirement of any project.

Once the project gets growing it has essentially to be divided into various section. Each incharge of a section deals with the special technique of the section concerned. For example, a senior civil engineer could be incharge of Civil Construction while a specialist in water supply and public Health Engineering would be looking after the works of his field. Similarly, Senior Electrical and Mechanical engineer could be entrusted with electrical and mechanical installation. Depending upon the type of project the overall direction may be given by the Civil Engineer incharge and comparatively junior electrical/mechanical staff may be employed under him. It would thus be seen there can be no hard and fast rule about the organisation in any construction project.

However, the project as a whole cannot be successfully executed unless each member of the team knows his work as well as the work of his associates, it is therefore necessary to have a format of organisation adopted giving duties, responsibilities and power of every unit in the organisation.

The term 'organisation' in project management may basically refer to the two thing:

1. Organisation of operation sequences. Covering the organisation of different project phases such as definition phase, concept phase, execution phase, test phase, and marketing phase.

2. Organisation of set up. Covering problem of personnel delegation both disciplinarily and technically as well as project management media.

If a systematic organisation of operation sequence render the execution of a project more transparent and thus easier to plan; setup organisation of project management is or decisive significance for the project's entire practical implementation.. There can be no doubt that the success or failure of project management is in many cases closely associated with the form of organisation adopted for a particular project. Project organisation is linked to the following objectives:

1. Completion of project according to schedule and in compliance with the terms of tolerance.
2. Completion of the project in accordance with quality specification at minimum expenditure.
3. Proper flow of informations and documents.
4. Optimum integration of any technical and policy and policy changes that may become necessary.
5. Observation of legal requirements and contractual condition.

In civil engineering works; the sequence of operation is more or less remains same. As such there is little scope of organisation in operation sequences. The only thing needs to be done here is that the different agencies involved for various operation must do their job properly. As such first type of organisation *i.e.* organisation of operation sequence needs little attention. But whatever set up is existing to perform these sequential operation are to be organised properly and scientifically. Thus there is a need and much more emphasis can be given on second form of organisation *i.e.* 'Organisation of set up'.

Principles of Organisation

The primary and basic test of any construction organisation is that it should get the job completed in the specified time (period of completion) in the most economical manner. To achieve this aim certain guidelines, known as principles of organisation, evolved on scientific lines and developed after years of experience have been formulated.

Some of the important principles are:

1. Unity of command
2. Cohegency
3. Job to be well defined
4. Effective control at all levels
5. Delegation of authority
6. Flexibility in job assignments
7. Interaction between executive and worker. These principles are briefly described below:

1. Unity of command A worker should not get order/ instruction from more than one individual. Dual nature of such instruction create confusion and delay. In some cases the worker is likely to exploit the situation.

2. cohegency The organisation should afford coordination and cohegency between the different branches of the organisation. The work of subsidiary units should be correlated with the progress of the mass unit so that the resources are not under deployed.

3. Job to be well defined In order that a job may be done well it should be well defined in respect of materials to be used, dimension, method of construction etc. There should not be ambiguity about the work to be executed. Otherwise there will be delay and may lead to dispute.

4. Effective control at all levels In order to have an effective control at all levels the ratio between the superior and supervisors should be properly arranged. For example, the top executive may have 3 to 5 advisors or assistants, whereeas a foreman may have 15 to 20 workers. The executive engineer for example may have three to four assistant engineer under him.

5. Delegation of authority Clear cut demarcation of work along with specific delegation of powers at various levels helps in fixing responsibility for any job and at the same time decisions can be taken without losing any time.

6. Flexibility in job assignment The organisation should be able to take care of unexpected work load. For this it is necessary that every worker should not be only thorough in his own work but he should also be in a position to take charge of other allotted jobs if required.

7. Interaction between executive and workers In order to create confidence of the workers it is necessary that the executive meets the workers and gets acquainted with their problem and try to help them as far as possible. In the army senior officers hold monthly 'DARBARS' where every person is free to speak out his grievances and the senior officers try to satisfy him.

Forms of Organisation

Based on the criteria, a variety of project specific forms of organisation have developed in practical work. A distribution may be made between 3 basic types:

1. Project Organisation Type A : The project specific single purpose venture "Line organisation"

2. Project Organisation Types B : The Project specific organisation within the enterprice with a demarcation line being drawn between project organisation and staff and line organisation called "Functional organisation"

3. Project Organisation Type C : The project specific organisation within the enterprice and within the staff and line organisation *i.e.,* 'Line and staff organisation'.

Project Organisation Type A

This type of organisation is chiefly employed in the implementation of large public projects, system or programme. Single purpose venture have first been adopted in space and air travel projects. This is one of the oldest and was earlier known as military type of organisation. A line organisation is an industrial system is characterised by a direct flow of authority from the top boss to the workers through the various executive. The one level officer or supervisor has full authority over his sub-ordinates for all their activities, though their legal position may differ. These forms of organisation are frequently characterised by the fact that they operate completely independently.

Due to clear distinction in the delegation of power of authorities; quick decisions are taken which leads to efficient control of the project. This further makes the system free from red tapism. The drawback with this system is lack of research and development work which many times lead to obsolescence cost. The authorities are more interested in getting the job executed irrespective of its cost and other factors which sometimes causes dictartarship in system.

The other drawback in this system is that no specialisation is present in the system;different procedures and crude method are followed by departments. Every activity of the department is carried out and controlled according to views and will of the executive head of various departments which leads to obsolence as these executives are busy mostly in planning and scheduling.

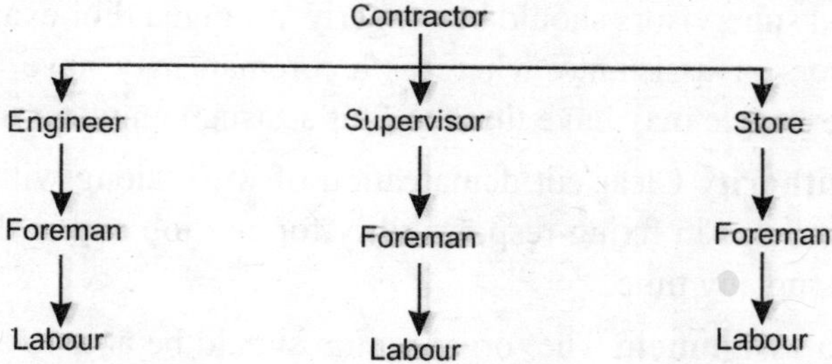

Fig. 2.6. Line organisation of a contractor

However this system is very effective in civil engineering projects. In this organisation, there is little scope for R and D in that small duration. Separate R/D cells are their which transfer their research and development work to the existing organisation for implementation in subrequest projects. The typical structure form of this organisation is illustrated in Fig. 2.6. This type of organisation is of course very easy: decision are in quick and complete responsibilities lies on one man hence effective coordination is formed due to direct chain and command.

Project Organisation Type B *i.e.* Functional Organisation

It is characterised by the fact that the responsibility for the project is taken out of the conventional line and transferred to a project manager or project team. This type of project specific form of organisation, may again be sub-divided into two groups *i.e.* 'matrix organisation, and 'task force organisation'. In practice work, these forms of organisation are mixed.

This was introduced by Fedrick W. Taylor with the view to bring the specialisation in the management and organisation. In this function organisation, the organisation is divided into different department headed by specialist, appointed to look after each department. Thus Taylor separated the specialist and routine functions as illustrated in Fig. 2.7.

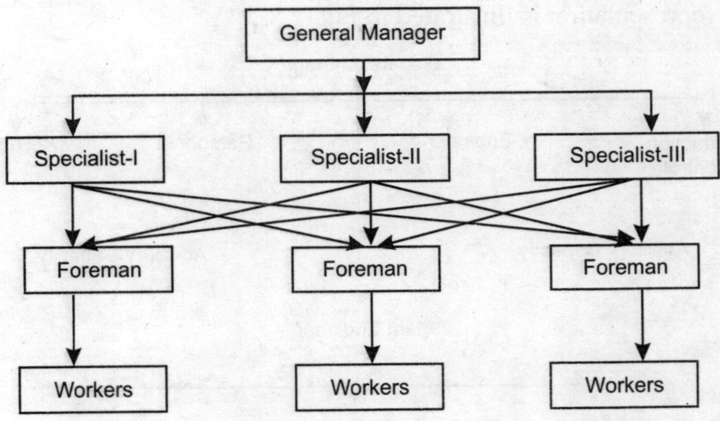

Fig. 2.7. Functional organisation by TAYLOR

In matrix organisation form, the manager largely uses line function for implementation of the project execution tasks while specialists use task force organisation. Project manager or General manger has no right to pass instruction to the line. As such function of each individual and as whole are well defined. Hence works successfully.

In this type of organisation each individual gets opportunity to show his ability in that field. It further relieves executives as most of the jobs are executed by specialist. As specialist are always available, work carried out in a most scientific manner giving more benefit to the company. It is possible for working head to improve the working condition, increase efficiency as they are relieved from routine works.

The drawback with the functional organisation is that every worker starts feeling that he is specialist, hence lacks in coordination and cooperation. As there exist several executives of the same rank; it sometimes leads to conflict amongst them; affecting the work. Further workers also get misguided or confused as having more than one boss. As there are more number of executive comparatively; involves more funds making this expensive type of organisation.

Line and staff organisation type C

The drawbacks of earlier two system have been taken care of in this type of organisation. The type C project organisation *i.e.* line and staff organisation is characterized by the fact that it is substantially more integrated in the conventional line or functional organisation, thus seeks to utilise the advantage afforded by project management. This type of organisation is frequently encountered in Germany. It is for instance adopted by firm specialising in the complementation of industrial projects such as erection of chemical plants, iron and steel works and production equipment. An outstanding feature of this form of organisation is that certain group are formed beneath the technical department level such as for sales, promotion, design and project implementation. Sometimes a project or contract executive is appointed within the technical department to assume the function of project manager.

It is basically a line organisation where some functional experts are introduced at different levels. The line portion carries the instruction and maintain stability and discipline; whilst functional portion serves to bring expert knowledge to hear on the problem; thus gets advantages of both the organisation. Line executive perform the job seeking advices from functional executives how to do job. An schematic representation is illustrated in Fig. 2.8.

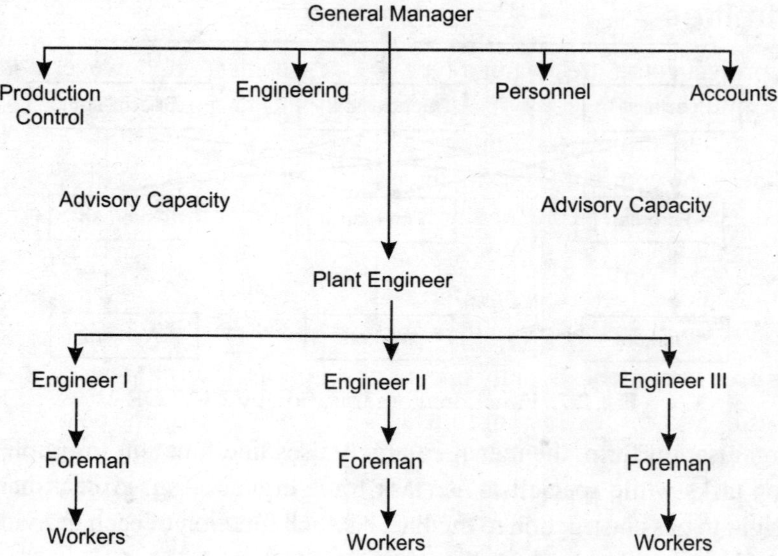

Fig. 2.8. Line and staff organisation

The functions carried by advisory committee are:

(*i*) Advises to working executives

(*ii*) Control and service and

(*iii*) Co-ordination

These specialists work with analysis ; research, standardisation, material management and solve problem of high technical standard whilst the line executive works according to advices of these specialist sitting together and discussing each and every aspect of it from practical consideration. These departmental managers are free to organise; control and supervise the productive activities if required.

Advantages

1. As the number of specialists are reduced; they remain only on advisory boards; hence this type of organisation become economical.

2. Line organisation are least concerned with planning, hence devote more on achieving greater efficiency and production.
3. Avoids confusion in general.
4. The technical department is present and co-operation is simple.

Disadvantages

1. Foremen are not given enough power; which leads him the feeling of lack of self respect.
2. Difficulties arise through enforcing action in competitive departments.

However there are little problem but it works most satisfactorily and efficiently. Many large project are working effectively and efficiently with this type of organisation.

Project Committee

Strictly speaking project committee is not a type of any organisation, however it is supplementary to the line organisation, functional organisation/line and staff organisation. In large projects to deal with complicated problem project committee are needed which work as the advisory committee of management and decision committee during the implementation of the entire project. The heads of the technical division participating are members of the project committee. Some times some experts not directly associated with the firms or project are also appointed as member of the committee.

Committee formed may be formal or informal in nature, advisory or managerial; permanent or temporary. In general four types of committee are formed to carry definite type of works.

1. The committee in control. Usually one man committee has full powers to act in the system.

2. Co-ordination and discussion committee. Formed to discuss the problem of management and then provide executive a wider base for making his decision. This committee is simply an advisory committee having no power to act. Co-ordination of action may also be discussed here and advises are given accordingly.

3. Investigation and advice committee. Formed to investigate specific problem of the system. This is also an advisory committee, Members of coordination and discussion committee forms this committee selecting a few members from themselves to deal specific problem.

4. Dissemination and education committee. Formed to collect information about system policies and projects, and to advise the top management about new projects and about existing policies. Also advice about the changes to be brought out in the existing policies.

When the committee are formed, it is essential to define their power and objective. The size of committee should be such that it can take decision quickly and members appointed should be appropriate. The members of the committee should be informed in advance and be acquainted with their authority and delegational powers.

2.21 METHODS OF EXECUTION OF WORKS

Almost all the civil engineering works are executed by either of the following system:

1. Employing daily labour *i.e.* Departmental execution
2. Contract system.

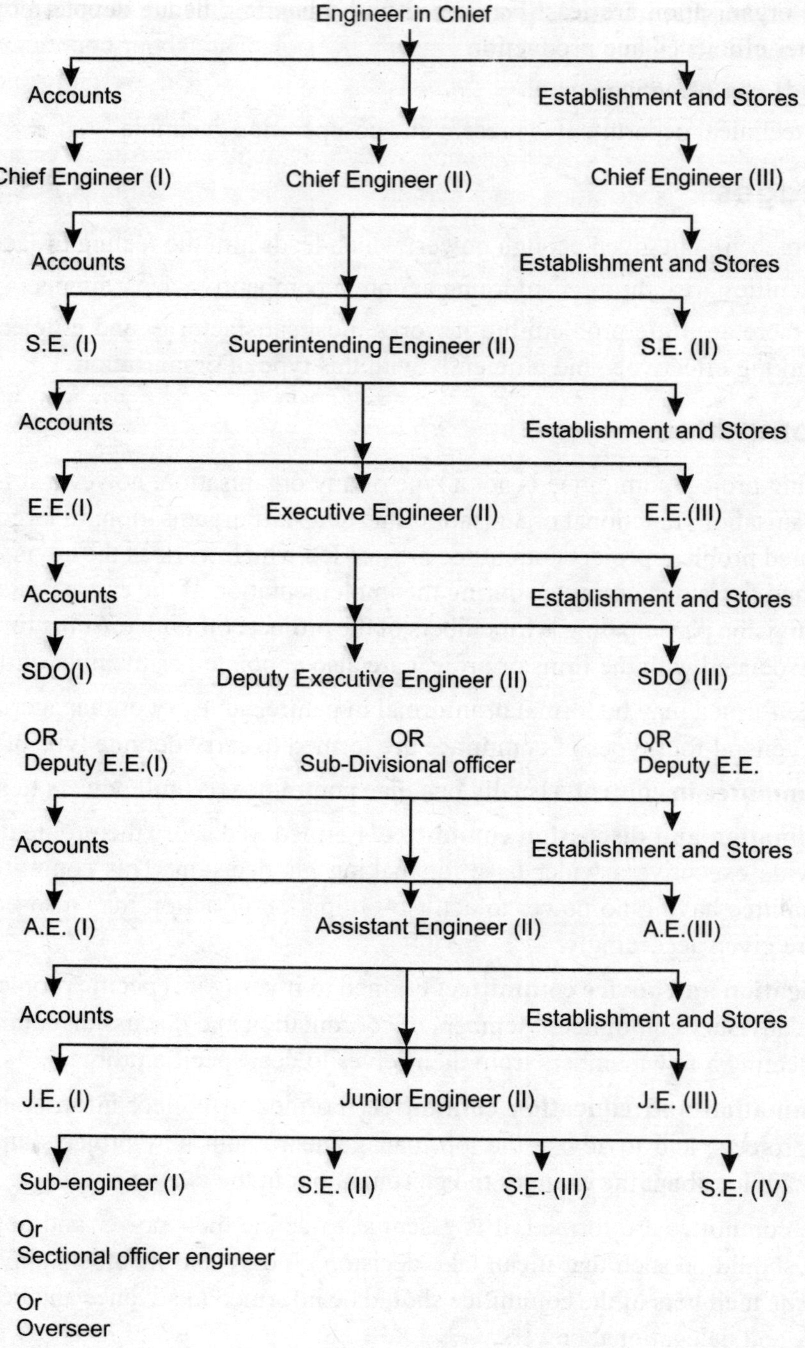

Fig. 2.9.

In first case, materials are supplied by the owner, and owner himself arranges labour on daily wages and gets the work executed through them. Payments either weekly or daily made by owner. The owner may be an individual person or any Government. Generally small works are executed by this system. In contract system the contract is given to some person or firm who undertakes the risks and completes that job. In labour contract all the materials are supplied by the owner and contractor arranges all labour required to execute that job. The labour contractor collects money from the owner and makes payments to labour on daily wages. This system is generally used for small works or in private works. In Government jobs generally contract is given for complete work. Contractor himself is required to arrange and purchase all materials required, he arranges all labour and executes the work on specified terms and condition and as per contract documents and tender deed. This is most common system adopted now-a-days either by Government or other bodies with little modification i.e. cement and steel are supplied by the owner and remaining materials are arranged by contractor.

An organisation in PWD/CPWD department consist of various levels of engineers having certain financial powers to spend.

Under each division, there is one establishment section and one accounts office to carry out all routine works. Drawing and design section is also provided under each division and there is one store also to store and issue various materials as per requisition. Establishment section is headed by office superintended and account section by Accounts officer or Accountant, under them various Senior, junior clercs work. All these persons work under divisional Engineer. Account office maintains the salary, TA/DA Bills and all other record pertaining to accounts.

Under each sub-engineer there are Mukaddam, Khalasi and labour those work on the job specified under that sub-engineer. The powers to sanction any work depends on the class of that engineer. Organisation of a contraction and railway organisation are show in Fig. 2.10 and 2.11

Organisation of a Contractor

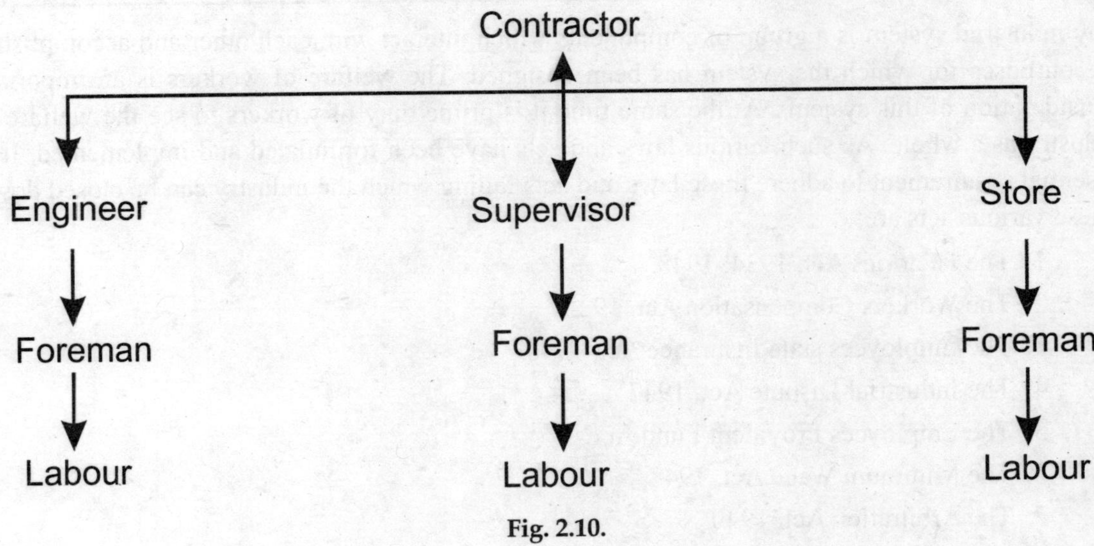

Fig. 2.10.

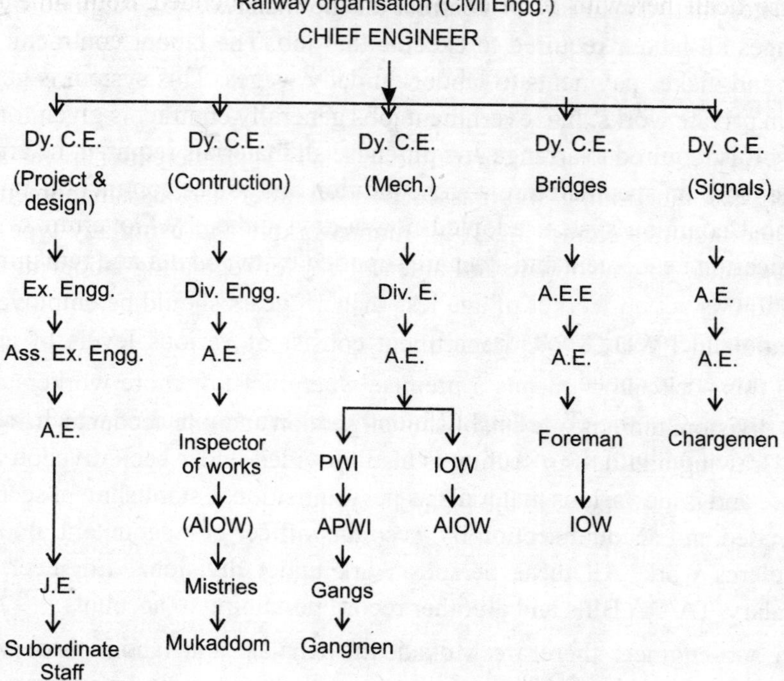

Fig. 2.11.

2.22 IMPORTANT ACTS & LAWS RELATED TO CONSTRUCTION INDUSTRY

Any industrial system is a group of components which interact with each other and accomplishes the purposes for which the system has been designed. The welfare of workers is an important consideration of this system . At the same time it is prime duty of workers to see the welfare of industry as a whole. As such various laws and acts have been formulated and implemented. It is essential requirement to adhere these laws and acts failing which the industry can be closed down. These various acts are;

1. The Factories Act, 1934, 1948.
2. The Workers Compensation Act, 1923.
3. The Employees state Insurance Act, 1948
4. The Industrial Dispute Act, 1947
5. The Employees Provident Fund Act.
6. The Minimum Wage Act, 1948
7. The Arbitration Act, 1940
8. Electricity Act, 1910 and Electricity Rule, 1956
9. The Boiler Act, 1923
10. The Indian Trade Union Act.
11. Apprentice Act, 1961
12. Mines Act, 1952

These acts are dealt herewith in brief. These acts are amended from time to time as per requirement.

1. The Factories Act 1948

The Indian factories act was passed first time in 1881. Then thereafter it was amended many times. The present act functioning was passed in 1948 after various amendments. This act is applicable to all the factories employing 10 or more workers and using any type of power. The main object of this act is to consolidate and amend laws for regulating labour in the factories in India. According to this act no worker of age less than 18 years should be employed. Further this act also classifies calender year; week, day.

According to this act Factory means a premise wherein 10 or more workers are working or were working on any day of the preceding 12 months, and in any part of which, a manufacturing process is being carried on with the aid of power.

Following are the main features of the factories Act:

1. The act prescribes that permission from the appropriate authorities must be sought before a factory is started.
2. Act also prescribes various provisions for maintaining health of workers. Cleanliness, ventilator, drinking water facilities, toilets etc. must be provided as per the norms. The temperature inside the factory where workers are working must be maintained within tolerance limit. provision for fresh air-circulation, humidification,must be made. Overcrowding should be avoided.
3. Factories act also dictate that safety provision for workers must be made from machineries and works. Safety measures must be made available against dangerous, toxic fumes and gases. Provisions for fire-fighting must be available for safety.
4. Welfare provisions like working facility, sitting facility, first-aid appliances, canteen, shelter, rest rooms, lunch rooms; creches etc. must be made available according to the factory act depending upon number of workers.
5. Hours of works, holidays, overtime wages, restriction on the double employment, employment of women etc. norms must be applied. Workers should not be allowed to work more than 48 hrs. in a week and 9 hours in a day.
6. It also restricts that no child, who has not completed his age 14th year, shall be allowed to work in a factory. Further fitness certificate must be granted by a doctor before a child is appointed. Further they should not be permitted to work during night hours. Further he should not be allowed to work for more than 4-5 hours a day.
7. Regular leave with wages shall be granted as per the provision and rules and regulation. Medical leave should also be granted if any worker suffers from any disease specified in the schedule which makes impossible to attend the work.
8. Those violate the provision of factory act should be penalised accordingly. Authorities reserve the right to shut down the factory under such circumstances.
9. Other provisions made are that the workers should work sincerely. Workers have no right to wilfully interfere with or misuse any appliance, convenience or any other thing

provided in a factory for the purpose of securing health, safety or welfare of the workers. Workers are required to follow the provisions of act, otherwise they are also liable to be punished.

2. The Workers Compensation Act

This act was implemented to make provision of compensation to the workers if injured during working in the factory due to accident. This protect worker from unnecessary hardship arising from accident. This act was made in 1923 and came into force from July 1,1924. This has been amended many times and finally drafted in 1962.

Accordingly to this act if any workers gets injured causing him partial or fully disabled to work; or dies during working in his working hours in the factory; he or his dependent as the case may be; must get compensation. This compensation is generally in terms of monetary given to him or his dependent or in the form of recruitment of one of his dependent. Along with this; the worker should also be provided the expenses incurred for medical treatment. The amount of compensation depends upon.

a – The pay of the worker concerned and

b – The extent of injury

The act classifies injury under four categories.

(*i*) Causing death

(*ii*) Causing permanent total disablement

(*iii*) Causing permanent partial disablement

(*iv*) Causing temporary disablement.

The amount of compensation is given in schedule 4. Further the percentage of loss of earning capacity for various injuries deemed to result in partial/permanent disablement is given as per clauses of safety engineering rules.

Schedule IV

Compensation to workers having monthly wages from ₹ 400 upto ₹ 1000 from May, 1976.

Monthly Wages workman		***Amount of compensation***		***Half monthly payment as compensation for temporary disablement ₹ (Half of the monthly wages)***
More than	***but not more than***	***Death***	***Permanent total disablement***	
400	500	21000	29400	118.75
501	600	21600	20240	135.00
601	700	23000	32340	148.75
701	800	24000	33600	160.00

801	900	27000	37800	168.75
901	1000	30000	24000	175.00

3. The Employees State Insurance Act, 1948

This act is brought into force; as many times it was found that worker's compensation act of fails to function. The examples are:

(*a*) It has been found that due to lack of finance available to the worker; he fails to file a case under court of law.

(*b*) The worker does not have knowledge of acts and other rules as such whatever compensation is given by employer; he accepts.

Though these deficiencies have been overcome by amendment; still contains certain loop holes like all diseases are not listed in compensation act; also during his sickness; worker does not get any amount for his medical treatment. This employees state insurance act 1948 takes care of all such items. According to this act; each and every worker working in a factory must be insured according to his work and working condition for any accident if occurs in factory. The workers as well as employer are required to contribute to this fund. The record is maintained by corporation as well as by employer. The corporation appoints Inspector for the inquiries and other details. The benefits that a person got insured under this act gets all like sickness benefit, Maternity benefit, disablement benefit, dependents benefit, medical benefit funeral benefit etc. as per the provisions of corporation.

4. The Industrial Dispute Act, 1947

This act was made inforceable for any sort industrial dispute between employers and workmen; and workmen connected with the employment and non employmentor with the condition of labour of any person. The basic reason for industrial dispute between labour and capital resulting in strikes and lockout. Low wages, non-payment of overtime, medical etc. favourism, exploitation, improper working condition etc. are various reasons for industrial dispute. The various type of strike, are go-slow strike, sitdown strike, tool down strike, sympathetic strike etc. Many times it also causes lock-out of factory, Gherao, picketing also take place which forces employer to seek police help. Various methods are adopted to resolve disputes are;

(*a*) Collective bargaining between two parties involved.

(*b*) Conciliation and mediation- a mediator is involved.

(*c*) Arbitration may be voluntary or compulsory.

As individual can not afford to fight for his right; collectively all the workers of that factory can ask employer for their benefits; if required can go to the court.

5. Indian Trade Union Act

As individual can not fight for his right; unions were made where all workers are members of that union and fight collectively for their right and betterment for themselves keeping interest of employer as well as of factory. The Indian trade unions are registered bodies of trade unions. Employers are required to recognise these unions, failing which matter is taken to labour court appointed by the government.

Unions collect membership fee from employees without any interference by the management. The collected amount is used for establishment and to meet recurring expenditure. The fund is also used for specific purposes as laid down by act. This amount is audited every year. It is also responsibility of unions that in case of strike, peace is required to be maintained so that there will not be any loss to the factory.

The first trade union in India was formed in Madras in 1918. The all India Trade Union congress (AITUC) came into existence in 1920 and The trade union act was passed in 1926 which gave a legal status and recognition to these trade unions. This act was further amended in 1947 and 1964. Various major trade unions working in India are AITUC; Indian National Trade Union congress (INTUC), Hind Mazdoor Sabha (HMS), United trade union congress (UTUC)and Bhartiya mazdoor sang (BMS) There are various small trade unions also. All these unions come under trade union act and has to follows the norms laid down by this Act.

6. The Employees Provident Fund Act

This act passed mainly to protect the welfare of the family members of that worker. Accordingly to this act; some fixed amount depending upon wages of the worker is deducated from salary of the worker monthly contributed to provident fund. This amount of provident fund is refunded to that workers after his retirement. This is basically to meet his expenses and his dependent during their old age. In certain cases equal amount is contributed by the employee and then it is called as contributory provided fund. According to this act; an interest on the total amount contributed in his account is also given and added to workers provident fund. In case of emergency worker can take loan from this amount or can have partly amount refunded after completion of 15 years of service. According to the act employer can not use this money for any purpose of the factory. Mostly this amount is credited to government account under the name of the worker *i.e.* the amount remains under the control of the government.

7. The Minimum Wages Act, 1948

This act was passed in 1948. This act dictates terms that no worker no employee should be paid an amount less than that required for survival of that worker. The worker must be in a position to fulfil his minimum needs of bread, shelter and clothing. This act is applicable to almost all the employee whether in industry or elsewhere. Further this act empowers the government central as well as state government to fix minimum rates. These rates are modified by time to time as per market index rate. This act also decides the wages to be given for overtime; piece work etc. The minimum wage consists of basic rate of wage and cost of living allowance and cash value of essential commodities. Provisions made in the act to appoint a competent authority to calculate the cost of living index. This act has brought uniformity in the standard of fixation of wages. A wage board for important industries to ensure equal pay for equal work has also been appointed by government of India. Accordingly to this act every employee as well employer are required to maintain the account of wages paid and received. Labour courts;industrial tribunals and conciliation courts made important contribution in their field.

8. The Arbitration Act, 1940

Arbitration is the settlement of dispute by the decision not of a regular and ordinary court of law out of one or more persons chosen by the parties themselves who are called arbitrators. 'Arbitration act' is an act to consolidate and amend the law relating to arbitration. This act is called as arbitration act 1940. It extends to the whole of India except the state of Jammu and Kashmir. The arbitrator should be an independent person or persons and must have knowledge of that subject matter. As per this act arbitrator has powers to give judgement or can refer the matter to court of law for opinion. He has right to correct clerical mistake or accidental slip or omission if any. According to this various types of arbitration are

(*a*) Arbitration without intervention of court

(*b*) Arbitration with intervention of court where there is no suit pending

(*c*) Arbitration in suits.

Arbitration clause 25 lays down that a contractor cannot have recourse to a count for redress of his grievances, unless he has exhausted the channel of arbitration as laid down in the Arbitration act and must be followed in case of disputes.

9. The Land Acquisition Act, 1984

To provide certain public purposes/uses such as Educational/Health/Transport Irrigation/ Housing/Buildings and communication etc. land would be needed for serving the public at large. All lands are not owned by the Government and under the Constitution, right to make property by the citizen is ensured. Then how do we get the land for above purpose when needed by the Govt/ Semi-Govt.

Then Land Acquisition Act of 1984 has been enacted for the only purpose of acquiring the private lands whether open or built-up by adopting the regular proposal in this regard.

Main provisions of this Act, stipulates that when any piece of land whether open or builtup (consisting with structure already built) is needed for any public purpose and if that public proposal is approved by the Govt. then Govt. would isue orders for acquiring the land describing the detailed schedule with the identify of land with its Survey No., Plot No., village and its area as recorded in the Revenue Record of the particular village.

After following the entire procedure stipulated under the Act in various section, the award has to be declared by the Special Land Acquisition officer appointed for this purposes by mentioning in the award the compensation payable as per market value as on notified date to the land owner/his bonafied representative. If the compensation is not acceptable to the owner, then he can accept the same under protest and file a claim to enhance compensation in the District court within prescribed time limit. If the Court is convinced about the claim filed by the owner, then court passes order to grant enhancement of the compensation and orders the Acquiring Body to pay the enhancement as per order.

Entire procedure under this Act is totally democratic and at every stage the owner has right to lodge his say in writing before the Special land acquisition officer. If the owner feels that the acquisition proposed is not justifiable for public purposes, then he can challenge the same before the high court and if court feels the same as mentioned by the owner, the acquisition stands cancelled.

The Act also provides or empower the Govt. to initiate the acquisition for the private institutions, which are registard under the Trust Act and Societies Act and are delivering public purposes whether it may be in the Educational or Health field.

The benefit of this act is that any land whether open or built can be acquired by the Government for public purposes. This act has been amended many times as per need.

10. The Urban Land Act 1976

This act has been introduced by the central Govt. all over India, with a view to have restrictions on the holding of land by each individual in the cities/agglomeration prescribed in the schedule of the act/under this act, there are three categories, which gives the quantum of owner of area in each agglomerations which can be held by the individual. These categories are (A) 500 sqm. (B) 1000 sqm. (C)1500 sqm. The Act clearly spells out that if the land determined is more than the prescribed limit of category A,B,C then the residual land stands excess.

This excess land under the law stands acquired without any choice of the owner. He would be entitled for payment of compensation as prescribed under the law. In this law the compensation is not payable immediately on taking over of the surplus land, but initially the amount as determined under the law is payable, while residual would be given in terms of the bonds which can be cashed after certain specified years.

The act also provided the exemption of surplus land to the owner on the ground of gift to the public purposes to the private institutions which are registered under the society Act and Trust Act.

The Act also provides the exemption under the pretext of the housing schemes sponsored by norms prescribed under the law by the Govt. from time to time.

Main intention of this law is to curb the speculation of land hoarding by the private individual and make available the surplus land to large public for housing purposes.

11. Environmental Protection Act 1986

After Bhopal disaster in 1984 claiming 300 lives or so, the Central Govt. enacted the government protection act in May 1986. It came force in whole of India on 19th November 1986.

Objects and Reasons. Concern over the state of environment has grown the world ever since the sixties. The decline in environmental quality has been evidenced by increasing pollution, loss of vegetal cover and biological diversity, excessive concentrations of harmful chemicals in the ambient atmosphere and in food chains, growing risks of environmental accidents and threats to life support system. The world community's resolve to protect and enhance the environmental quality found expression in the decisions taken at the United Nations Conference on the human environment held in Stockholm in June, 1972. Government of India participated in the conference and strongly voiced the environmental concerns. While several measures have been taken for environmental protection both before and after the Conference, the need for a general legislation further to implement the decisions of the conference has become increasingly evident.

Although there are existing laws dealing directly or indirectly with several Environment matters, it is necessary to have a general legislation for environmental protection. Existing laws generally focus on specific types of pollution or on specific categories of hazardous substances. Some major area of environment hazards are not covered. There exist uncovered gaps in area of major environmental hazards. There are inadequate linkages in handling matters of industrial and environmental safety. Control mechanisms to guard against slow, insidious build up of hazardous substance, especially new chemicals, for an authority which can assume the lead role for studying planning and implementing long term requirements of environmental safety and to give direction to and co-ordinate a system of speedy and adequate response to emergency situations threatining the environment.

In view what has been stated above, there is urgent need for the enactment of a general legislation on environmental protection which inter alia, should enable of an authority or authorities with adequate powers for environmental protection, regulation of discharge environmental pollutants and handling of hazardous substances, speedy response in the event of accidents, threatening environment and deterrent punishment to those who endanger human environment, safety and health.

The act seeks to achieve the above objectives.

Power to central government to make measures to protect and improve environment are

(*a*) The Central government should have the power to take all such measures as it deems necessary or expedient for the purpose of protecting and improving the quality of the environmental and preventing controlling and abating environmental pollution.

(*b*) In particular measures may include the following:

(*i*) Planning and executive if a nation-wide programme for the prevention, control and abetement of envirnmental pollution;

(*ii*) Laying down standard for the quality of environmental is its various aspects.

(*iii*) Laying down standards for the emission or discharge of environmental pollutants from various sources whatsoever;

(*iv*) Restriction in which any industry, operation or processes should not be carried out or should be carried out subject to certain safeguards.

(*v*) Laying down procedure and safeguards for the prevention of accidents which may cause environmental pollution and remedial measures for such accidents.

(*vi*) Laying down procedures and safeguards for handling of hazardous substances.

(*vii*) Examination of such manufacturing processes, materials as are likely to cause environmental pollution.

(*viii*) Carrying out and sponsoring investigations and research relating to problem of environmental pollution.

(*ix*) Inspection of any premise, plan equipment etc. to take steps for the prevention, control and abatement of environmental pollution.

(*x*) Establishment or recognition of environmental laboratories and institutes to carry out the function entrusted to such environmental laboratories and institutes under this Act.

(*xi*) Collection and dissemination of information in respect of matters relating to environmental pollution.

(*xii*) Preparation of manuals, codes or guides relating to the prevention, control and abatement of environmental pollution.

(*xiii*) Such other matters, as the central government deems necessary for the purpose of exercising and performing such of the powers and function(including the power to issue directions).

Power to give direction. The central government may issue direction in writing to any person, officer or any authority who shall be bound to comply with such direction.

(*a*) The closure, prohibition or regulation or any industry, operation or processor.

(*b*) Stoppage or regulation of the supply of electricity or water or any other service.

Rules to regulate environmental pollution. The government have made rules to regulate environmental pollutions.

(*a*) The standards of quality of air, water or soil for various areas and purposes;

(*b*) The maximum allowable limits of concentration of various environmental pollutants (including noise) for different areas;

(*c*) The procedures and safeguards for the handling of hazardous substances;

(*d*) The prohibition and restriction on the handling of hazardous substance in different areas; etc.

Prevention, Control and Abatement of Environmental Pollution

(*a*) No person carrying on any industry, operation or process should discharge or emit any environmental polluted in excess of standard prescribed.

(*b*) No person should handle or cause to be handled any hazardous substance execpt after complying with such safeguards as may be prescribed.

(*c*) Furnishing of information to authorities and agencies responsible for such discharge and the person incharge of the place at which any accident or other unforeseen act or event occur the person will forth with:

(*i*) Intimate the fact of such occurrence.

(*ii*) Be bound, if called upon, the render all assistance, to such authorities.

(*d*) On receipt of information should, as early as practicable cause such remedial measures to be taken as are necessary to prevent or mitigate the environmental pollution.

(*e*) The expenses, if any may be recovered as arrears of land revenue or of public demand.

Power of entry and inspection. Any person empowered by the central government in this behalf have a right to enter, at all reasonable times:

(*a*) For determining whether, and if, so in what manner, any such function are to be performed or whether provisions of this act or the permissions granted under this act is being or has been complied with

(*b*) For the purpose of examination and testing any equipment, industrial plant, record, register document or any other material object or for conducting a search of any building in which he has reasons to believe that an offence under this act of the rules made there under has been or is being or is about to be committed and for seizing any such equipment, industrial plant, record register, document or other material object it may furnish evidence of the commission of an offence punishable under this act.

(*c*) Every person carrying on any industry, operation or process or handling any hazardous substance shall be bound to render all assistance to the person empowered by central government and if he fails to do so shall be guilty of an offence under this act.

(*d*) If any person willfully delays or obstructs any person empowered by the central government he shall be guilty of an offence under this act.

Power to make sample and procedure to be followed in connection therewith. The central government or any office empowered by it in this behalf, have power take for the purpose of analysis, samples of air, water soil or other substance from any factory, premises of other place in such manner as may be prescribed.

Environmental Laboratories

(*a*) The central government may established/recognise one or more laboratories or institutes as environmental laboratories to carry out the functions entrusted to an environmental laboratory.

(*b*) Make rules specifying the function of the environmental laboratory.

(*c*) The procedure for the submission to the said laboratory of samples not air, water soil or other substance for analysis or tests, the form of the laboratory report thereon and the fees payable for such report.

Penalty for Contravention of the Provision of Act

(*a*) Whoever do not comply with provisions of this act is punishable with imprisonment for a term which may be upto five years or with fine which may extend to one lakh rupees, or with both and in case the failure or contravention continues, with additional fine which may extend to five thousand rupees for every day. These are amended as and when found necessary.

(*b*) If the failure or contravention referred continues beyond a period of one year and date of conviction, the offender shall be punishable with imprisonment for a term, which may extend to seven years or so as decided by act in court of law.

13. Indian Electricity Act and Electricity rules

The Indian electricity act was passed in 1910 to the supply of electricity to all consumers who apply for it. This, act was passed to supply electricity to all the consumers who apply for it and to control and regularise electric supply.

According to this all electric supply lines must be provided with safety devices used for this purpose. Accordingly on person except a license older shall be allowed to use or generate or transmit energy more than 250 watts and 100 volts. As per the act central electricity boards have been developed which consists of persons from government, boards, railways, mines and I.S.I. The jobs of these boards are to make rules and regulation to regulate generation, transmission, use and supply of electricity. According to this act any person interfering with meters; or misusing the powers of involved in the its of light and building is liable of the punishment as per norms of the act.

This act also classifies the voltage lines as low; medium, high or extra high depending upon the voltage. This act permits any electricity inspector to enter any premise and inspect the work. This act also dictates to rules of testing of installation and connections; switch board installation, variation in voltage, variation in frequency; reading of meters and cut-outs and connection with the earth. Electricity rules have been framed 1956 for strict implementation of act 1910. There have been various ammedments to it.

14. Apprentice Act 1961

This act was passed in 1961 to train the person for a period varying from 2 years to 3 years before they are put to actual production job. This training is exhaustive and sufficiently long. This is called as apprentices training and extracts are called as apprentices. As per this act apprentices must be taught guidance under the supervision of experts. The aim of this act is to overcome unemployment problem and to help young and promising technicians for entrepreneurship. Though this act was passed in 1961 but has been implemented since 1963 This act made it compulsory on the part of employer in 217 specified industries to engage apprentices in 138 trades. This act restricts that children below the age of 14 years should not be allowed to join the course. If a factory or concern is having more than 500 workers or employees then that concern must have separate training. Centre. Quota has been fixed for reserved categories for this apprentice training. Apprentices are also paid stipend as per the norms given in the act and modified time to time. The main advantages of this act are that it caters to the need of trained hand and increase production. It also improves the quality of work and reduces production cost. It also reduces wastages.

15. Mines Act 1952

This act has been passed mainly to safeguard the interest of mine workers. This act also lays certain norms like minimum wages, safety measures, working hours etc. to the mine workers.

16. Land Ceiling Act

This act has been introduced by the central Govt. all over India, of area in each agglomerations which can be held by the individual. These categories are (A) 500 sqm. (B) 1000 sqm. (C)1500 sqm. A category is applicable to Mumbai, while B is for Pune, Thane , kalyan and C is for Solapur, Sangli- miraj etc.

The Act clearly spells out that if the land determined is more than the prescribed limit of category A,B,C then the residual land stands excess.

This excess land under the law stands acquired without any choice of the owner. He would be entitled for payment of compensation as prescribed under the law. In this law the compensation is not payable immediately on taking over of the surplus land, but initially the amount as determined under the law is payable, while residual would be given in terms of the bonds which can be cashed after 20 years.

The act also provided the exemption of surplus land to the owner on the ground of gift to the public purposes to the private institutions which are registered under the society Act and Trust Act.

Main intention of this law is to curb the specution of land hoarding by the private individual and make available the surplus land to large public for housing purposes.

The uban land (Ceiling and Regulation)Act 1976, Known as act No. 33 of 1976 under the 252 of the constitution resolution come into force on 17 February 1976 with retrospective effect of applications for all the transaction or transfers on or after 17.02.1975. Since that the law is constantly getting amended and may continue to amend.

The objective of imposing this act for imposing ceiling on urban property as realised necessity by various state government. It was to control the disorderly development of urban areas with respective growth of population and increasing urbanisation. The urban land ceiling (ULC) Act has created an impact on value of properties and therefore on valuation also this act restricts and regulates transfer and use of urban property. The sellers and purchasers are required as per this act to file an application and affidavits under various sections, even if a seller has land lesser than the ceiling limits prescribed and the purchaser has more land than the ceiling limit.

Articles

1. Provision of village sites or the extension, planned development or improvement of existing village sites.

2. The provision of land for town or rural planning.

3. The provision of land for planned development of land from public funds in pursuance of any scheme or policy of government and subsequent disposal thereof in whole or in part by lease, assignment or outright sale with the object of securing further development as planned.

4. The provision of land for a corporation owned or controlled by the state. Tenant lands for old structures constructed before 1970.

5. Provisions of land for residential purposes to the poor or landless or to he persons residing in areas affected by natural calamities, or to persons displaced or affected by the implementation of any scheme undertaken by government, any local authority or a corporation or controlled by the state.

6. The provision of land for carrying out any educational, housing health or slum clearance scheme sponsored by government or any authority established by government for carrying out any such scheme or with a prior approval of appropriate government, by a local authority, or a society, registered under the society registration. Act 1860 or under the corresponding law for the time being in force in a state or a co-operative relating to co-pertain societies the time being in force in any state.

7. Provision of land for any other development sponsored government or with the prior approved of appropriate government by authority.

8. Provision of any premises or building for locating a public office.

There are state amendments to this clause and the base is becoming broader and broader to include several purposes in this definition.

Similarly all state governments have almost unanimously made amendments their respective state laws, improvement in Trust Act, Regional and Town planning Acts, Industrial Development Acts, and Highways acts, narrowing down the scope for differences. The various court ruling have sealed the probable loopholes and the Land Acquisition Act as on this comprehensive more than ever before.

Once a state government for competent authority has decided to acquire any land including built-up structures for any public purpose, the gazette notification is published to declare the intensions. This data of Gazette notification lays a pivotal role in valuation. There is a stipulated time limit of three years within which the authority has to take concrete steps otherwise the date is lapsed. However such authority has a power to publish fish gazette notification and in that case the pivotal date is extended to date if fresh notification.

There is also an obligation on authority to make proper surveys, to assess the fair market values of such properties to be acquired. The compensation payable is usefully decided on the basis of average rent of the property for five years preceding the date of notification and in absence, if found necessary, call for rent records from municipalities or Corporations for comparable rents of five years in adjoining areas. The survey is most important as it exactly notes the dimensions, the land use, tenancies, the unoccupied areas and even between tenant, and landlords. Thus use on the date of notification plays an important role. If the market value is specially high in consequence on the land being put to a use which is unlawful or contrary to public policy, that use is disregarded and the market value is deemed to be the market value of and if put to ordinary use.

If owner has erected or re-erected any structure other than mentioned in survey after the date of notification then value of that structure is disregarded. However if any improvements are made to the land which are as per the proposal for improvement in consequence to the lands under acquisition, such improvements are considered while adopting the market value.

The lands are acquired for various purposes such as road widening, slum improvement, any Town Planning scheme sewerage disposal, flood relief etc. However the existing land use on the date of notification(provided it is lawful and in line with public policy) is taken into consideration and the basis of the future land use is disregarded.

A valuer, while deterring the fair market value, of such properties, should consider the following:

1. The survey sheet mentioning dimensions, structures, names of tenants, and the prevailing land use.

2. The date of notification in official gazette, date of order of acquisition and award issued by competent authority. In case the date of acquisition is lapsed whether fresh notification is issued or not.

3. Any additions and alterations done after the date of notification should be disregarded. However any improvements made with the pricier permission of the competent authority within then notification period may be considered.

4. Rents of five years received or receivable as prevailing rents in the vicinity should be taken into account. In case of unoccupied areas or areas free of rent given to nearest relatives should be considered on the basis of prevailing rents in the vicinity.

5. An average or gross rents thus obtained from records for five years should be considered and 1/5th of such gross rent should be deducted in case of municipal taxes, insurance, collection charge and repairs and maintenance, as expenses are not known. However if proper accounts are maintained then actual expenses should only be considered to arrive at actual net rent.

6. In case of any fruit trees are standing then a fair assessment depending upon their life yield and further life may be added in the value of property.

7. Note down the actual land use and adopt proper years purchase to capitalise the value.

8. In case of rental valuation, reversionary value of land is disregarded. However in land and building method the land values should be added to the depreciated values of structures.

The purpose of valuation also plays an important role. If a valuer is asked to advice on the fair and reasonable market value of the properties to be acquired then his course of action should be as laid down above. However in the case where the property is already in process or acquired then the value should be determined from the award of compensation (received in receivable) after deducting the interest payable for the period from the date of notification and the date of payments of compensation. However 15% solarium (Whether awarded or not) on the market value as on the day of notification should be added. The awards issues normally take into account the various incidental charges etc. these charges should be deducted from the award. In absence, a reasonable percentage between 15% to 20% may be deducted from the compensation.

It is generally observed that the valuation for taxation purposes differs from valuation for acquisition purposes. The tax laws are constantly amended and issuing various directives and sometimes steps and methods. Such amendments have been made to simplify the procedures and avoid probable litigation and loss of time. Therefore several aspects could be ignored. A valuer should take a judicious decision in adopting the course while working out valuation for taxation purposes. If this observes that there is a difference (not more than 15%) between the two values adopted by different methods, then the average value should be certified as fair and reasonable value.

It is important to note here that a person aggrieved any ask for valuation for compensation purposes for lands in acquisition, he may need such valuation to be recorded in court matters. At the same time the same person may ask the valuer to produce a report indicating far less value asking advantage of the subjuris matter for above all and therefore valuer should be careful in finding out the market value since practice and the courts (if observation are recorded)may initiate lawful proceeding against the valuer.

17. Transfer of Property Act 1882 (Act No. of 1882) Amended upto 24th August 1984

The transfer of property act is extended to all states of India. Basic purposes of this act is to regulate the transfer of property whether movable or immovable. The transfer of property means as act by which a living associations or body of individuals, conveys property in present or in future to one or more other living persons, or to himself and one or more other living persons, and "To transfer Property" is to perform such act. We are primarily concerned with immovable properties.

The immovable property means the property which is "Attached to the earth" means

(*a*) Rooted in the earth as in case of trees and shrubs,

(*b*) Embedded in the earth as in the case of walls and building,

(*c*) Attached to what is so embedded for the permanent beneficial enjoyment of that to which it is attached.

The transfer of property is done by sale, mortgagee, gift, lease or any other negotiable instruments of transfer. The transfer of property by gift has religious sanction and it differs from religion to religion. All other means of transfer have common code of conduct. The rights of property are generally two types: 1) lease hold and 2) free hold.

A lease hold right may very from fraction of a year to perpetuity of 30 years and above. Whereas the free hold right vary from any number of years from 99 to 999 years. Similarly in agricultural lands the same is further classified into (1) Bhoomidhari means one who holds land and (2) Bhoomiswami means who owns the land. Like in rented properties the owners enjoy rights of benefits for a consideration whereas the tenant hold tenancy rights, it is beneficial enjoyment payable to the owner. There are various laws governing these rights and they differ from state to state depending upon the accepted practices in the respective societies.

The transfer of property act has been amended and it practically covers all the aspects of transfer. Wherever there is any other law having rules for transfer of properties such provisions are read in the line and with the same express meaning of the Transfer of property act and subject to certain condition ruling beneficial use in the public interest , all rules and regulation mentioned in the said acts. (for examples Urban land ceiling act) are to be read in the meaning and context of the transfer of property act.

The Indian registration act governing the procedure of registration and the stamp duty is also closely related with this act. Valuation is required for the purpose of assessing stamp duty and also assessing the value consideration to transferer or transferee. Values fourths two purposes is based on the broader principles of valuation. Various state government have made it mandatory to adopt the values proposed by their town planners in the form of indices for particular periods in different agglomerations. It is generally observed that actual sale values are considerably high or low in each transaction. However it is general tendency to mention less value for transfer of property purpose and adopt the value on the basis of indices for stamp duty purposes.

There are generally two parts for every transfer. One is an agreement or will and second is the document of transfer for registration purposes. Here also religion plays important role. For example in Hindu oral partition by a karta is acceptable whereas in other religious the cases are different.

It is vital for every transaction that the transferer hold goods and marketable title over the property to be transferred. The transferer hold such rights, as karta, as constituted power of attorney, as lessor, as assignor, as liquidate, as inheritor, etc. To established the transferer position in those regards, he has to establish certain documents such as sale deeds, gift deeds, documents should bear the same identity through earlier documentation. It is generally for more than a certain period and even if they are available, it becomes virtually counter such difficulties, there is a provision possession it is lawful, uninterrupted possession of the property for more than 12 years. This provision entitles the possessor to act as transfer. However there are certain exceptions like government lands which are excluded from this provision . For registration of deed or transfer, several no objection certificate and clearances are necessary. Various department such as income tax, city survey corporation or improvement trust etc. and competent authorities such as Town planning authorities under ULC act etc. have their own procedures to be adopted. Even though every such department and authority has income tax department, to control evasion of the improvement trust to control terms and conditions of leases and various building permits accorded, ULC act to control the concentration of urbanisable lands in the hand of few, city survey nazul department to regulate registration in their respective records etc. It is normally seen that the various laid down procedures of the departments have several conflicting thing, which creates chaos and harassment to the parties of the transfer apart from the money and time required.

It is ironical to note that a transferee has to satisfy himself about the title, rights and interest of the transferer and transferee in turn undertakes to indemnify the transfer to any defects found in later course time. It is invariably a slip shod method and by the time such defects are observed transferee is left with claiming the compensation through lengthy legal battles.

-The sections 53 A and 54 of transfer of property act deal with the transfer of immovable property.

-According to the section 54 of transfer of property act, the sale is a transfer of ownership in exchange for a price or promised or part paid and part promised. The transfer of immovable property of the value more than Rs. 100/- should be made only through a registered instrument. The sale is complete as soon as the deed is registered. The property does not pass or the ownership is not transferred until the registration is effected. but once the registration is effected the title relates to the date of execution.

-The section 53A of transfer of property act dealing with the part performance protects the buyers interest. Thus gives a shield to a buyer from the possible frauds of a seller. For the application of section 53A of the application of section 53A of transfer of property act the following conditions should be satisfied.

1. There should be contract to transfer for an immovable property for a consideration

2. There should be written agreement and its terms can be well ascertained with reasonable certainty.

3. The transferee in part performance of the contract has taken the possession of the property or any part thereof.

4. That the transferee has done some act in furtherance of the contract.

5. The transferee is ready and willing to performs his part of the contract.

-The section 53A while protecting the rights of the buyer does not give the buyer the title of flat. The transfer of property can take place only on completion of the registration as per section 54 of transfer of property Act.

Let us see the effect of these acts in respect of wealth tax liabilities of flat owner.

1. One of the methods to own a flat is to form a co-operative house building society having the persons interested in having a flat as a member. The society then constructs and allots the flats to the members on the land purchased by the society. By virtue of the being a member of the co-operative society, the flat allotted get the right to use the flats and sell their flats. In the case of sale of a flat by a member only a change of names are made in the records of the society and no transfer documents are made and registered with the registrar of assurance. This is done probably to avoid tax liabilities.

2. Another method adopted to own a flat is that a developer or a builder purchaser the land and construct the building comprising of flats and make agreements for sale of different types of flats to the intending buyers. On taking payments from various buyers he hands over the possession of the flats to actually these transfers of rights are to be registered under transfer of property act. But in order to avoid the stamp duty this is normally deferred indefinitely.

3. In an yet another method the developer constructs a building as per the agreement with the prospective owners of the flats. The undivided share of the land is transferred by the landowners only and registered. Whereas the building is concerned no restoration takes place as the building is constructed on a building contract with an individual owner. This is mainly done with a view to avoid stamp duty on the building cost. Generally it is the prospective owners are not identified at the start of the project. As such entering into contract for construction. This is done to avoid double stamp duty on the building portion. The undivided share of the land is also sold directly by owners (not by the developers) through the power of attorney received by the developers from the owners.

4. Main objective in delaying the registration and avoiding the registration is to avoid the stamp duty and tax liability. The people try to defer and avoid the transfer, by not registering the transfer as required under section 54 of transfer of property act. By not registering, the transferee does not become the legal owner, the transferee has been taking the advantage of the legal flaw to avoid the capital gain and the wealth tax liability.

5. When the documents of transfer is not registered, the legal owner is the promoter or the co-operative society. Though they are the legal owner, the user of the flats enjoy the maximum right. The value of the rights at the hands of the transferers as well as their tax liability is much less.

6. In order to make the real owners to pay the taxes by the flat owners who enjoy the maximum rights are made liable to pay the taxes by amending the sections 2 (*m*)(*iii*) of wealth tax act and section 2 (47) (*v*) & (*vi*) and 27 Income tax acts w.e.f. 01.04.1988

7. As per these amendments the transfer means to include the part performance of contract. In other works the non-registration does not come in the way of making the owner to pay the tax. The real users and owners have become liable to pay taxes.

18. Rent Control Act

All states in our country have there own rent control act with certain revisions and modification. This act is mainly to avoid by taking undue advantage by anti social elements and detrimental both to the landlord as well as to tenant. The Bombay rent control Act 1984 though his ended by March March 1988. Still enforce in Maharashtra.

19. Indian Contract Act, 1872

Important provisions and terms relating purchase and construction of contracts are:

(*a*) When one person signifies to another his willingness to do to abstain from doing anything, with a view to obtaining the aspect of that either to such act or abstinence, he is said to make a proposal.

(*b*) When a person to whom the proposal is made, signifies his there to, the proposal is said to be accepted. A proposal when accepted, becomes a promise.

(*c*) Every promise and every set of promises, forming the consideration for each other, is an agreement.

(*d*) An agreement not enforceable by law is said to be void.

(*e*) An agreement enforceable by law is correct.

(*f*) A contract which ceases to be enforceable by law becomes void when it ceases to be enforceable.

20. Arbitration Act

It is an act to consolidate and amend the law relating to Arbitration. This act called as Arbitration Act 1940 extends to whole India except Jammu and Kashmir. According to this act amended upto date; arbitrator has following five powers.

1. To administer oath to the parties and witness appearing.
2. To state a special case for the opinion of the court or any question of law involved, or state the award wholly or in part, in the form of a special case of such question for the opinion of the court.
3. To make the award conditional or in the alternative.
4. To correct in an award any clerical mistake or error arising from the accidental slip or omission.
5. To administer to any party to the arbitration interrogatories as may in the opinion of the arbitrator, be necessary.

According to Arbitation Act. 1940; various kinds of arbitrator are

1 Arbitration with no intervention of court of law.

2 Arbitration with intervention of court of law but with no suit under pending.

3 Arbitration in suits.

According the Arbitration Act. 1940, Clause 25; the procedure adopted for settlement of dispute in Civil engineering contract consist of following steps:

(1) Appointment of arbitrator by higher competent authority within 30 days after receiving complete detailed grievances and claim from contractor. Contractor is required to 'Submit his Statement of facts' within 90 days *i.e.* period of limitation.

(2) Preparation and submission of case for Arbitration. All documents are required to be submitted to Arbirator. These documents consist of agreement deed; drawing; specification; conditions of contract and the grievances and claims in the forms of 'statement of facts' and the bills; vouchers corresponding to dispute.

(3) Hearing of the case by arbitrator as per provisions in the Act. Both parties are permitted to plead their case with due arguments by other. Arbitrator requires to hear both the sides pleading and then to scrutinise and examine all documents. The total direction for this hearing should be completed within four months.

(4) Issue of Award: Arbitrator later on; going through all details give the decision. The award should be obtained by parties on non-judicial stamp paper. Once the award is issued by Arbitrator his duty and authority ceases. He can not make any second award.

(5) Finally is the filing of award *i.e.* If the decision is acceptable by both the parties; the award is written on stamp-paper and then duly signed by Arbitrator; both the parties involved and witnesses. This if required can be filed in court of law to its validity.

If the decision is not acceptable by either party; then the party has right to file a suit in court of law for further necessary action and then matter is resolved in court of law only by the regular judicially procedure.

As such Arbitration is an expensive way of settling the dispute between owner and contractor. As far as possible matter is solved at very first or second level only. In extreme cases matter is referred to court. The judiciary procedure and law takes very long time hence people avoid it.

Tribunal is the term applied to either the sole arbitrator or the panel of arbitrators conducting the arbitration and it is of comfort to the parties to be able to choose a tribunal in whom they have confidence. That is why it is important for the parties to endeavour in agree the sole arbitrator or, as the case may be, to appoint their party arbitrator otherwise they will have someone appointed leaving with essentially no scope for objection. In case of panel of arbitrators it consists of three persons- one appointed by each party and third is appointed as chairmen jointly. The chairmen has special right to cast his vote in the event of disagreement.

Arbitration has been around for a long time. It has been sugested that it was a method of dispute resolution during the Egypcian Civilization, but it was certainly used by the ancient Greeks and Roman. Traders at the medieval fairs in England and on the adopted arbitration to settle their difference regarding price or quality. Over a period of time arbitration became more wide-spread and is being used to settle an increasing variety of disputes and by the late 17th century it was considered necessary to put it on a more formal footing. The first English arbitration act was made in 1698.

The benefits of arbitration become widely acknowledged and by the end of the 19th century, international arbitration procedures were established for settlement of dispute between countries. International business had developed in magnitude and complexity during the 19th century, to an extend that in 1920, the international chamber of commerce was founded to promote regularise trade and investment. It is widely specified in International Construction contract. The intention of ICC was to provide a universally acceptable procedure to overcome the diversity in arbitration practice concorn regarding the disparities between and, often, the inadequacy of law on arbitration, gave rise to the formation of the United nations commission Arbitration which was adopted in 1985 and has been widely adopted by most of the countries. It can not be defined that the diversity worldwide can cause difficulties are certainly no greater than if the dispute is settled through litigation.

Other acts passed by governments are:

1. Dock workers regulation of employments act 1948
2. Indian Railways act 1930 and 1956
3. Indian Merchant Shipping act 1956
4. Air corporation Act 1953
5. Payment of Wages Act 1923
6. The tea district emigrant act 1932
7. The children (pleading) of labour Act 1933
8. India dock labourers act 1934
9. The first factories act 1881
10. The workmen's breach of contract act 1959
11. Employers and Workers Dispute Act 1860
12. The payment of bonus act 1965
13. Payment of gratuity act 1972

14. Coal mines labour welfare fund act 1947
15. Mica mines welfare fund act 1946
16. Conservation and safety act 1952
17. Industrial Employment(standing order) act 1946
18. Industrial dispute (appeal to Tribunal)act 1950
19. Mines act 1952
20. Boiler act 1923
21. Indian Electricity act 1910
22. Sale of goods act 1930

2.23 PROJECT REPORT

Whenever a project or construction work is required to be undertaken, it is essential to prepare report and to submit it to the appropriates authorities for its approval. Unless a project is approved by concerned authorities (related to that ministry) no budget provision is made. And unless budget provision is made, work can not be sanctioned. Therefore a project is very essential to prove its feasibility and necessity.

Preparation of Project Report

The project report consist of the following:

1. Introduction A summary from initiation of the proposal to its finalisation including brief described of various surveys, investigation and data collection carried out.

2. Design parameters and standards adopted, codes followed etc. For example, design speed, type and width of pavement shoulder, grade, curvatures and sight distance adopted are included. In the same way if the report is a proposal dam construction, the type of dam, structure features, area to be submerged, phases of construction and time schedule is to be indicated.

3. Salient feature of the project. In the case of a highway project, details of alignment starting from point of commencement to its termination, reasons for adopting the alignment and other features such as grades, curves and obligatory points are briefly described.

4. Foundation conditions including soil classification and various strata as well as particular of soil samples and marshy area are indicated.

5. Weather conditions prevailing. For example in the forward areas working period starts in the beginning of June and ends with the end of October. Arrangements for induction and deinduction as well as utilisation of sources has to be brought out.

6. Availability of resources. Phasewise requirement of men, material, machinery and equipment has to be given so that in the case of acceptance of the project procurement action for procurement of the resources required may be initiated. In case of machiner, available in the country be obtained.

7. Details of land to be acquired along with towns and villages affected. For example in the case of a dam the towns and villages to be shifted and compensation to be paid has to be spelt out in detail along with the environment effect of the project. Pollution control or environmental effect has now acquired priority over other consideration. The progress of Tehri Dam had been considerably retarded for environmental consideration. Similar is the case with Narmada sarovar Dam.

8. Specifications of the project. Under this heading may be included all such information from the point of view of the project which are important but may not have been included under any of the headings indicated earlier.

9. Annexures to the Project Report. The following annexures are normally supplied with the project report:

1. Index plan.
2. Ste plan
3. Longitudinal and cross-sections along with estimated quantities of earthwork.
4. Details of protective works.
5. Drainage details along with details of bridges, culverts etc.
6. Available construction material with details of resources available within the country and those to be imported.
7. Design details of structures.
8. Land acquisition and land compensation schedule.
9. Estimated cost of the project.

10. Conclusions. Under this the total estimated cost of the project specifying separately the amount earmarked for compensation of the property of various types to be acquired and contingencies is specified. A request for according administrative approval of the project along with specified demand for yearwise allotment of funds from the proposed date of starting of the construction is made.

The breakdown details and basis for arriving at the project is also specified. At the end the officer initiating the project signs the report giving his/her designation and date of signing.

2.24 LEAN CONSTRUCTION

Since 1940, lean production principles have been evolved and were successfully implemented by Toyota Motor Company strived to work towards the ideal of hundred percent value added work with zero or minimum waste. Then thereafter many production companies throughout the world implemented this technique to their production plants and found very successful. From 1992 construction industry started using this technique at construction sites. The researches also did some work on this technique and developed some fruitful finding. As a result, lean based tools have emerged and been successfully applied to simple and complex construction projects.

In general lean construction projects have been found easier to manage, safer, completed sooner at lower cost and are of better quality.

What is Lean?

Lean is a systematic, continuous improvement approach that focusses activities on reducing waste while aligning them to an overall growth strategy. Leaders of a Lean organization are dedicated to developing Lean thinkers and a continuous improvement culture.

A Lean Enterprise essentially eliminates waste throughout the business process. Waste costs you resources, but adds no value to the customers you serve. These non-value-added activities typically equal 95 percent of all manufacturing effort; a mere 5 percent of your manufacturing activities add value for your clients.

Lean will help you increase profits while:

- Reducing cycle time.
- Reducing inventory
- Reducing costs.
- Increasing capacity.
- Increasing productivity while improving quality.

Lean Enterprise is a business system for organizing and managing product development, suppliers and customer relations. Business and other organisation use lean principles, practice, and tools to create precise customer value- goods and services with higher quality and fewer defects- with less human effort, less space, less capital, and less time than the traditional system of mass production.

Many of the key principles were pioneered by Henry Ford, who was the first person to integrate an entire production system, under what he termed"flow production." Following World War II, The Toyota Motor adapted Ford's principles as a means of compensating for its challenge of limited human, financial, and material resources. The Toyota Production system (or TPS), Which evolved from this need, was one of the first managerial system using lean principles throughout the enterprise to produce a wide variety of products at lower volumes and many fewer defects than competitors.

Leaders today in a wide range of industry, nonprofit organisation, government agencies, healthcare, and other areas are finding ways to apply the principles of lean as a means of producing goods and delivering services that creates value for the customer with the minimum amount of waste and the maximum degree of quality.

Case Study: How Delay Increases Cost?

While constructing a building we normally keep a watchman at the site to take care of the things at the place. The watchman costs around Rs.6000/- per month.

If everything goes on time and construction completes on time in 6 months, the expenditure on watchman is Rs. 6000 × 6 (months) = 36000/-

Due to delay suppose the construction is delayed for 6 more months then, it adds 6000 × 6 = 36000/- more to the overall construction cost.

This is a simple example of how delay increase cost. Delay also adds more to the raw material costs, wastage etc.

What is Lean Construction?

It is a way to design production system to minimize waste of materials, time and effort in order to generate the maximum possible amount of value. Designing a production system to achieve the stated end is only possible through collaboration of all project participants *i.e.* owner, contractor if any, manager, supervisors, labour, end user etc. at early stages of the project. Lean construction aims to embody the benefits of the Master Builder Concept. Essentially, lean construction recognizes that desired ends affect the means to achieve these ends and that available means will affect realised ends.

Lean construction supplements traditional construction management approaches with:

1. Two critical and necessary dimensions for successful capital project delivery by requiring the deliberate consideration of material and information flow and value generation in a production system and
2. Different project and production management is planning-execution-control paradigms.

Lean Construction in Civil Engineering Project. Lean production was developed by Toyota led Engineer ohno. The term lean was coined by the research team working on International auto production to reduce the waste of Toyota production system and to increase the mass forms of production.

However, there was always a question mark that how this technique can be applied to Civil Engineering projects where the uncertainties are very large, having less planned construction system comparatively and many other reasons. Managing construction under lean is different from typical contemporary practice because it:

– has a clear set of objectives for delivery system.
– is aimed at maximizing performance for the customer at the project level.
– designs concurrently product and process and
– applies production control throughout the process and life of the project.

Typically the construction industry includes three parties an owner, a designer called as engineer/ architect and the builder or contractor. The material supplier is also one important party but is assumed to have link with builder or contractor. The owner and engineer may not come in contact directly with supplier. Traditionally there are two contracts between three parties: first between owner and engineer/ architect and second between owner and builder/contractor. Different system of project can have different parties also but the basic form more or less remains same. A third party relationship between engineer and contractor also exists indirectly but also equally importance for proper execution of work. However, in public sector owner as government department and engineer as employee of the department are one and the same. As such there are only two parties exists. These are government or government undertaking and their employees (engineers) as one party and contractor as other. However number of contractors may be many depending upon type and quantum of work.

Therefore an alternate contract or business model replaces from two traditional contracts with three party contract. The construction management company therefore becomes an additional party engaged in the project to act as advisor to all the concerned. The object of manager in charge of construction is therefore required to see that:

1. Optimum use of available resources(man,material,machinery and money)
2. Control of the scope of the work.
3. Project scheduling and monitoring.
4. Optimum use of design and construction firms; their skills and talents.
5. Avoidance of delays, un-necessary changes and disputes.

6. Maintaining construction quality with optimum flexibility in contracting and procurement.

A manager involved in construction project is required to have knowledge of engineering as well as that of management. He is also required to have knowledge of business procedures, economics and economy and human behaviour with marketing knowledge . He is also required to analyse the problem if any systematically. Use of cost dynamics and mathematics, science and technology, modern trends, latest known how etc. are also made as and when required as such the manager should have knowledge of all these facts involved in lean construction methodology.

Comprehensive management at every stage of the project, beginning with the original concept and project definition yields the greatest benefits to the owner in the form of avoiding waste, reducing time span of project, maintaining quality to satisfy end user from construction management.

Lean construction project gives more significance on reduction in project duration. This indirectly decreases total cost of project. The simple example is reduction in total amount paid to watchman/security guard appointed at site to avoid thefts. If project gets completed in less time, this reduces the amount equal to would have been paid to watchman if project had continued for longer period. As such indirect cost reduces considerably by reducing time period of project construction *i.e.,* construction period is reduced, delays are prevented and project is made economical.

With continuous decline in profit margin and increased competition in construction industry, contractors are in continuous search for ways to eliminate waste, reducing time period to increase profit. BOT technique is also one good example for lean construction. It tries to minimize or eliminate if possible the non-value-adding works like watchman. If all activities of construction are planned systematically, resources are made available as and when required; the project duration can be reduced considerably thereby increasing the profit by reducing indirect cost elements.

Lean construction is still in early stages of development and in due course of time is likely to become important time saving tool of construction industry.

Lean construction Principles include:

1. Established integrated teams of owners, architects, facility users, builder, contractor and suppliers.
2. Combining project design with process design, simultaneously designing the facility and its production process.
3. Stopping production rather than releasing a faculty assignments or product into the construction process.
4. Decentralizing decision making empowering project participants and making the process transparent so any team member can see state of the project phases.
5. Requiring a simple direct hand off between tasks in the work stream, with a clear way to request action and receive response to eliminate clogs between project phases.

In lean construction, collaboration beginning early in design is extremely important. Everyone is a stack holder in a project. It incorporates the ideas of relationship based contracting, meaning everyone on the team must keep their promises. The projects are actually networks of commitment. If one person fails to make the commitment he promised, its a domino effect on every stack holder in the project. Lean construction requires bringing contractor, sub contractor, supervisors, material suppliers and other stack holders to the planning table because their roles affect the overall project schedule.

Waste Reduction. In today's construction market most companies are cutting corners and margin in order to get work rather than seeking out waste in their processes. Reduction and elimination of waste (non-value-added) is a key part of the Lean project delivery process. Seeking out waste is a change in focus and will require behavioural change. Following 8 types of waste specific to construction process have been identified where waste can be reduced.

Table 2.6. Types of Waste

No.	Type of waste	Details
1.	Over-Production:	Producing over the customer requirements, producing unnecessary materials/products
2.	Inventory:	Holding or purchasing unnecessary raw supplies, work-in-progress inventory, finishing goods
3.	Transportation:	Multiple handling, delay in material handling, unnecessary handling
4.	Waiting:	Time delays, idle time
5.	Motion:	Actions of people or equipment that do not add value to the product
6.	Over Processing:	Unnecessary processing steps or work elements
7.	Correction:	Producing a part that is scrapped or requires rework/procedures
8.	Not Utilizing Human Resources :	Not following-up/implementing ideas/suggestions/manpower

Case Study:

1. Over-Production

Concrete- Planning should be made for making and utilizing the concrete. Its final setting is important. So it should be made just before its utilization and in a quantity as per the requirements. The over produced concrete will be a wastage.

2. Inventory/motion

Bricks- Normally a platform is prepared temporarily for collecting the concrete. The bricks are used for making this platform. These bricks are wasted after the use. Instead of these bricks a metal platform can be used. This metal platform is not wasted and can be reused many times.

Bricks/Tiles-Normally we waste 10% to 40% of bricks in just moving them around due to breakage or misplacement. Proper handling, storing, stacking of tiles and bricks will definitely avoid this wastage and will cut the overall costs. Many a times these brick bats are used in floor filling for no reason.

3,4. Transportation/Waiting-Most of the time the material needed comes in advance on the site. This material keeps on taking place well before its utilization. This results in improper handling and wastage. If material is made available on time only, then most of the transportation, stacking and handling wastage can be avoided.

5. Motion Movement of equipment from one place to other for no reason or due to poor management and placing. Trucks, dumpers or other machineries many a time are shifted from one place to other. It consumers diesel and manpower.

6. Over processing. In case of plastering, the habit of preparing of motor for a day work is common. It affects quality of mortar. To overcome this quality, adding cement in excess after every hour is real waste of cement. Moreover after the work is over, left mortar is mixed with sand for next day use. This means a motor was prepared more than required. This increase cost due to use of excess cement.

7. Correction. Constructing a part which do not match with drawing or of poor quality and require to be dismentled and reconstruction. This is most undesirable.

8. Many a time. Labour remain idle at site for want of job. Efficient use of labour must be made to reduce cost.

Lean Methodology: Lean manufacturing methodologies are not new. Many manufacturing in past have these methodologies in one form or the other. It is a line balancing technique used in conjunction with a series of material handling techniques. Mathematical models are created and iterated until an optimum utilization of manufacturing resources is identified for optimum solution. Lean methodologies are a serious of techniques those allow product to be produced one unit at a time, at a technique those allow product to be produced one unit at a time, at a formulated rate while eliminating non-value adding wait time queue time or other delays. It is a type of pipe line; once a liquid is made to flow; it come out at a specific time interval. This is a type where liquid flows without any stopping or delays. This metaphor of moving product through a serious of continuous operations. If during this operation, no stoppages occur, the product will take minimum time avoiding any delay, maintaining quality product to have optimum use of man, machinery and having optimum profit. A lean manufacturing line requires that a rate of flow through the pipe line be established.

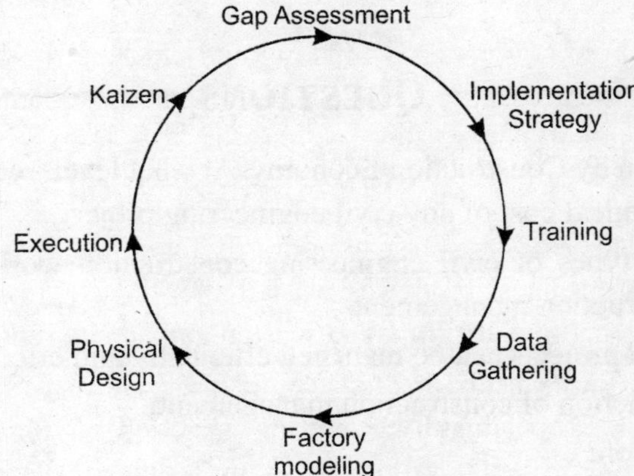

Fig. 2.12. A disciplined approach for implementing lean methodology

Tools used in Lean Construction. Reverse phase scheduling (RPS); Six Week Look Ahead (SWLA),Weekly Works Plan (WWP) Daily Huddle Meetings; first run studies (to redesign critical assignments)5 S process (visual work place) those are Serri (Sort), Seiton (Strengthen or set in order), Seiso (Shine *i.e.,* to clean), Seiketsu (standization) and Shitsuke (Sustain) etc. are a flew tools used in lean construction. In 5 S process, the works used are Japanese word their meanings are given in brackets. The purpose is to sort deficiencies and strengthen approach to implement lean methodology is shown in figure. This is of course for factories or industries but can be implement to civil engineering projects with slide modification.

Benefits of Utilizing Lean Technique in Construction

1. Improved response time to customer demand: Many a times, once promise is made to customer regarding occupation of flats in case of flat construction, it has been found that these are not made available at decided time. The reasons can be many. However, the reasons which are in the preview of engineer, contractor, owner or government can be resoved in advance. This makes reduction in time delay if unavoidable and customer satisfaction is achieved.

2. Reduces Inventories : At many project, unnecessary dumping of material is made. This creates problem to workers as most of the space is occupied by these inventories, causing delay in construction. Contractor thinks he is saving money because material is purchased in huge quantities at a low rate. However, he rarely thinks that money is blocked which looses interest. Moreover unnecessary shifting/transporting material from one place to other or from one site to other increases labour cost transportation cost and indirectly the cost of project. Moreover work is also get delayed increasing overhead expenses.

3. Reduces working capital requirements as un-necessary inventories are reduced : The money blocked can be used at other work site.

4. Simplicity and visual control by system MBWA *i.e.,* management by walking around is possible : It is very easy at floor shop *i.e.* at industries but little difficult in civil works However, if contractor walks around site; he can certainly find probWlem if any to resolve to matter timely.

5. There is certain improvement in productivity as well as quality and quantity : There are many operational benefits also those can be achieved implementing lean construction techniques in Civil Engineering.

QUESTIONS

1. What do you mean by Construction Economy. At what levels, economy can be achieved to have most economical cost of any civil engineering project.
2. Classify various types of civil engineering construction work. Name various agencies involved in construction management.
3. How construction projects can be managed efficiently and effectively.
4. Name various function of construction management.
5. Write short notes on:
   (i) Management of contracts.
   (ii) Financial Management.
   (iii) Project control System
6. Name various persons required at site and explain the job they perform at site.

7. Explain Value engineering in detail. What is its importance in todays context.
8. Explain various forms of Organization. Give organization of PWD.
9. Write short notes on:
   (i) Minimum Wage Act.
   (ii) Employees Provident Act
   (iii) Workers Compensation Act
   (iv) Land Acquisition Act
   (v) Environment Protection Act
10. Explain meaning of Job Layout, how it is achieved and what precautions are required to be taken while doing Job Layout.
11. Is it essential to have safety management in construction? Name various possible accidents in case of a Building Project.
12. Explain Arbitration act in detail.
13. Explain Lean Management in Construction. How wastages of construction materials at any site can be minimized using Lean Construction.
14. Explain Lean Methodology briefly.
15. What do you mean by Management? Give history of Management Science in detail.
16. Explain Taylor Fredrec theory of construction in Management.
17. Explain various theories of Management briefly
18. Explain functions of Management. What do you mean by Levels of management?
19. What do you mean by Project Management? Team what are is essential requirement of a good management team.
20. What do you mean by construction economy .At what levels economy can be achieved to have most economical cost of any civil engineering Project.
21. Classify various types of civil engineering construction works. Name various agencies involved in construction works.
22. How Construction projects can be managed efficiently and effectively.
23. Name various functions of construction management.
24. Write short notes on:
   (i) Management of contracts
   (ii) Financial management.
   (iii) Project control system
25. Name various persons required at site, explain the job they perform.
26. Explain value engineering in detail what is its importance in today's content.
27. Explain meaning of Job layout, how it is achieved and what precautions are required to be taken while doing Job layout.
28. Is it essential to have safely management in construction? Name various possible accidents in case of a building project.
29. Explain various forms of organization give form of organization of PWD.

30. Explain various precautions required to be taken while dismantling any building.
31. What is Lean construction.
32. Explain lean management in construction. How wastage of construction materials at any site can be minimized.
33. Explain lean methodologies briefly.
34. Explain meaning of project organisation ?
35. What are the functions of an effective organisation ?
36. Explain various forms of organisation mention merits and demerits of each ?
37. Explain line and staff organisation mention merits and demerits of each ?
38. What is difference between line organisation and line and staff organisation ?
39. Explain concepts of Modern Organisation approaches ?
40. What is project committee? What are various type of committees are formed. What are functions of each committee?
41. It is essential to train the workers selected? why ?
42. White a note on cost control record in management of works.
43. Sketch the job layout for a multi-storeyed building on a plot 30 m × 30 m.
44. (a) Define the term management, describe Henri Foyol's principles of management.
    (b) Describe the expression for economic order quantity (EOQ) for the material.
45. On a construction site cement is required in large quantities as follows:
    (i) Annual use–200 tons. price per bag at site Rs. 150/-
    (ii) Cost of ordering–Rs. 100 per order.
    (iii) Inventory carrying cost–20%
    (iv) Supplier's discount at order of quantity as follows:
        (a) Less than.... 20 tonnes.... No discount
        (b) 20 tonnes and less than 40 tonnes.... 5%
        (c) 40 tonnes and above 10% Suggest suitable order quantity.
46. "Planning construction of high rise structure is challenging". Justify it with suitable example.
47. Explain light, medium and heavy construction with suitable examples in each.
48. Explain advantages and disadvantages of mechanization of construction sector.
49. Explain following:
    (i) Construction schedules.
    (ii) Financial schedule.
    (iii) Draw a specimen tabulation of the above for a construction project.

Chapter

River Training Work for Construction Management

Training work and management during construction of any civil engineering project if water encounters, it is essential to make that area dry to execute construction work in general and concerting in particular. It happens mostly in construction of foundation.

In core of building foundation or others dewatering is done though pump as and process of concerting is executed adding 10% extra cement as cement is also likely to be thrown out with dewatering process. In case of construction in water bodies like rivers various river training works are used as per conditions existing and considering other factors. However management of construction activities under such conditions are pecular and essential. Different methods used to manage construction depends upon various factors like quantity of water, type of construction season during which construction is made, site and soil nature etc.

3.1 RIVER TRAINING WORK

For more than 3000 years, men have been trying to 'tame' some of the great rivers of the world. The first civil engineering work of any magnitude was probably the river training work. Today, the training of rivers is still an important branch of civil engineering, so diverce in its character, so wide spread in its practice and so specialized for each individual river that brief reference only can be made to it. We, despite the fact that the mendering to rivers, which river training works are designed to control is a part of the natural order, one phenomenon in the broad field of physical geology.

River-control work is essentially the control of a natural river flow within a certain well-defined course, usually that which the stream normally occupies. Since the deviation from this course will be. most likely to occur during periods of flood flow, control work is often a matter of flood regulation. In addition to this, the prevention of erosion of river banks during periods of normal flow constitutes another important branch of control work. Finally the improvement of rivers in connection with civil or industrial development such as the straightening of a river bed in connection with town planning and the canalization of a stretch of river to facilitate navigation are another type of river regulation that are often of importance.

For construction work river training work is important to preserve flooding of construction area and to provide the protection to approach embankment construction. Though cofferdam is not exactly a river training works but it serves that purpose also.

Four of the major river system of the world have well developed flood plains long occupied by human settlements. They are the Mississippi, the Nile, the Yellow and the Yangtz rivers of China. These rivers have caused tremendous loss of lives and property.

In India the river Brahmaputra had left its normal course spectaularly towards the end of the eighteenth century, seriously affecting the ancient city of Mymen Singh. The holy river Ganga also had left its banks causing tremendous loss of men and material. With the development of irrigation works of magnitude by British engineers in the nineteenth century, river training works became an established part of Indian engineering.

The importance of banks protection works must be emphasized since it enters into so many phases of river control work. Method of protection range from the construction of permanent retaining walls in the front of natural soil slopes for preventing the initial formation of soil gullies to the use of expedients such as mattresses of brush wood, dumped rock and even sand bags in cases of emergency. In the evolution of the many successful types of protection, due consideration has inevitably had to be given to the nature of the river action at the location. For example on sharp bends, use of wall of deep driven interlocking steel sheet piling will be proved successful as other method may fail due to continuous scouring of river bed until a condition of equilibrium is attained.

There are many methods of river training work as explained below:

Classification of River Terming work: River training refers to the structural measures which are taken to improve a rivers and its banks. It is also an important component in the prevention and mitigation of flash and general flood control as well as in other activities such as ensuring safe passage of a flood under bridge

River training is also needed to make some area of river water body dry to execute construction of foundation of bridge, dams etc in water bodies like river reservoirs etc. It is to make river course stable and to prevent erosion as river are dynamic and continuously change their position shape and other characteristics with variation in discharge and time

River training structures can be classified as

1. Installed perpendicular to the water course such as check dams, spurs sills screens etc
2. Installed on river banks parallel to river course such as embankments, sheet piles, rock rippers etc
3. Other protecting structures such as sandbagging channel lining. Bambs piles

River training works can also be classified as

1. Direct protection walks - one which are placed directly over the portion that needs to be protected such as Riprap over the erosion prone banks
2. Indirect protection works one which modify the flow conditions to protect erosion prone bank such as spurs, vanes etc.

Guide Banks

It was in 1887 that Mr. James Richard Bell proposed a method for training the Chenab River at Sher Shah, where a large new bridge was proposed. He suggested concentrating the low river flow in a relatively deep channel by means of low embankments soon known far beyond India as "Bull's bund". The same idea was further developed by Sir Francis Spring and has since been amended and improved. Their principle contribution was introduction of curved head at the upstream end of Bull's Bund. A typical example is shown in Fig 3.1. A section is shown in Fig. 3.2.

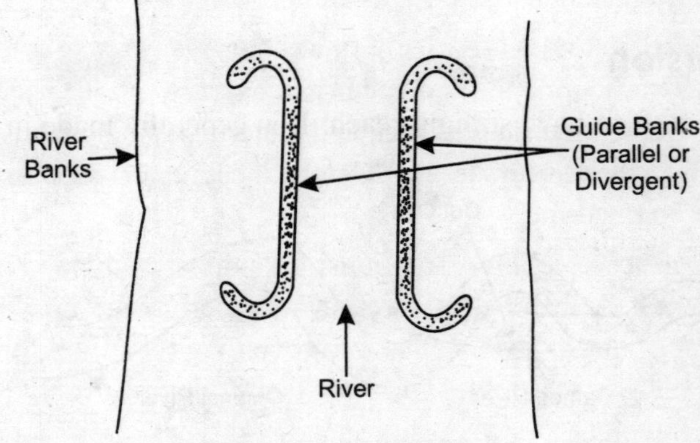

Fig. 3.1. Guide banks

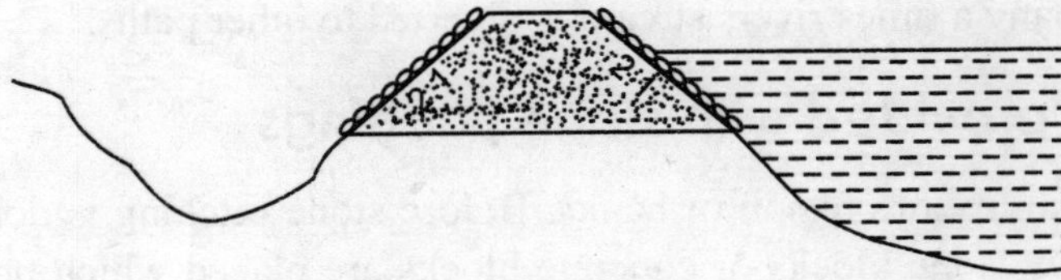

Fig. 3.2.

Spurs

These are the structures built transverse to river flow extending from the banks into the river. These can be constructed as per the requirements, i.e. to attract, deflect or to repel the flow along the desired course. These also create a pool of still water, variety of spurs are illustrated in Fig. 3.3 to 3.6.

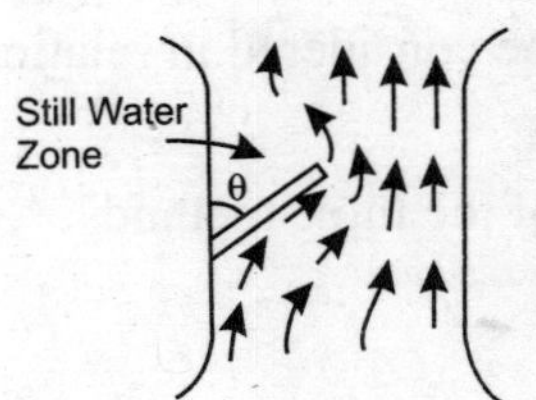

Fig. 3.3. Attracting spur θ = 40° to 60°

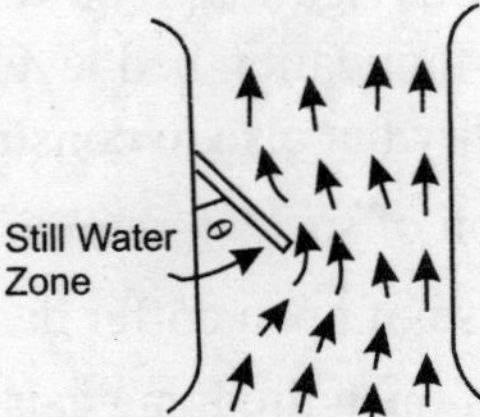

Fig. 3.4. Repelling spur θ = 60° to 75°

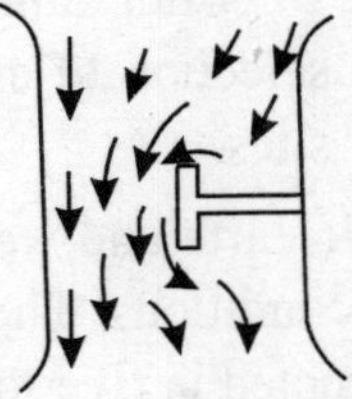

Fig. 3.5. T-spur

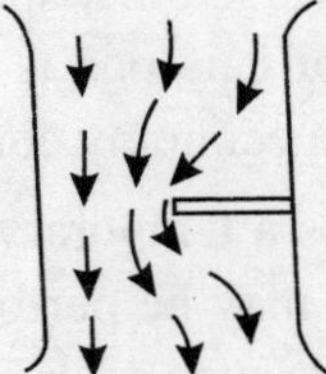

Fig. 3.6. Deflecting spur

These spurs may be made of permeable or impermeable type depending on material used for construction works. In impermeable spur a stone pitching or concrete blocks are provided on the surface to make spur impermeable.

Cut Offs and Diversion

Many a times river are cutoff to have straight reach. It is generally made in meandering rivers as shown in Fig. 3.7

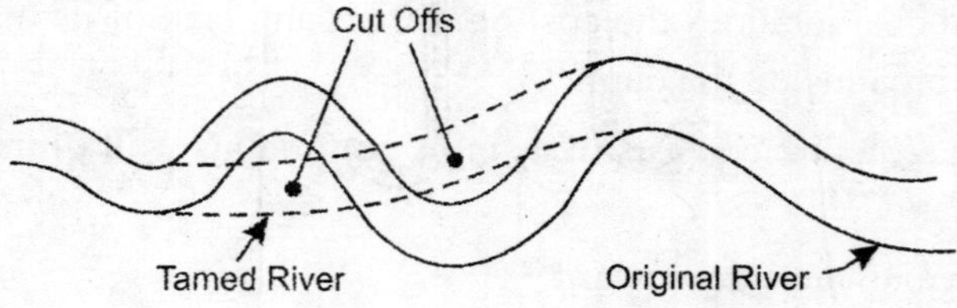

Fig. 3.7.

The earth cut-off is heaped to form embankment or placed in original rivers to have new straightened stream, many a times rivers stream is diverted to other paths.

River banks are provided with stone pitchings

This protects the possibility of erosion of banks. Before stone pitching is done, a stable slope is made over which heavy stone blocks or concrete blocks are placed which protect the banks and avoid the possibility of erosion.

Bushes and hedges are also planted many a times which protect the banks of river.

3.2 COFFER DAM

This design of bridge foundations, dams etc. can rarely be considered without specific relation to the problems of constructing them. Limitation in the restriction of a water course during construction requirements or navigation, requirements of traffic, depths of water and tidal range, and the depth below water to foundation bed level are all features that must be considered in relation not only to structural stability but also to construction methods.

Construction of such structures can be carried out by one of the three methods:

(*i*) The use of open coffer dams

(*ii*) Use of open dredging caissons

(*iii*) Use of compressed air caissons.

There are of course other special methods available but the use of coffer dams is very popular now-a-days. Use of caissons is also becaming very popular. The selection of any one of above method depends on economy considerations and their feasibility at sites.

A coffer dam is a temporary structure which is built to exclude earth and water from an area in order that work may be performed there under reasonable dry conditions. These are generally required for projects such as dams, bridge piers, etc. that are constructed in river or in other bodies of water. Many a times small bunds are also constructed to prevent the flow of water into foundation pits excavated.

Cofferdams are used for much general foundation works in addition to providing facilities for the construction of bridge piers and abutments. They are now so common a feature of construction work that they are generally accepted without much thought being given to their development.

It is not needed to be watertight because a water tight structure will be very costly. A cofferdam is a temporary structure, hence of little quantity of water seeps through it, the same can be dewatered by the help of pumps. The most satisfactory cofferdam is the one that has adequate strength to resist all destructive forces and that will permit the exclusion from the control of water inside the dam at the lowest cost. The total cost includes the cost of coffer dam, cost of its maintenance and cost of pumping water from the area inside the dam.

The usual cofferdam structure consists of a single continuous wall of inter-locking piling. The force acting on a cofferdam are:

(*i*) Weight of cofferdam acts downward

(*ii*) Water pressure acts horizontal

(*iii*) Scouring effect of moving water

(*iv*) Upward thrust

(*v*) Lateral pressure of earth, if any.

A cofferdam should be designed to resist the combined affect of all forces that may be acting on it. When the wall of a cofferdam is installed to resist the horizontal pressure of earth and ground water, it may be desirable to leave enough openings in the wall to permit the water to flow through it as a means of reducing the pressure against the wall.

Requirements of a Cofferdam

Following points should be taken into consideration while designing and constructing a cofferdam. These points are desirable but not essential. But if these points are observed a most economical and feasible cofferdam can be constructed.

1. The cofferdam should be reasonably water-tight. It should rest on an imprevious layer or otherwise a concrete bed should be provided at the bottom. It is to avoid the possibility of seepage of water from bottom. Pumping for dewatering should be done only after concrete has become sufficient hard in case of concrete bed is provided.
2. It is not needed to be water-tight because to construct a water-tight structure is costly. The design and layout should be such that the total cost of its construction, maintenance and dewatering is minimum.
3. While designing all the forces likely to affect the design should be taken into consideration. It should be checked against crushing, bursting, sliding, overturning. There should be no possibility of erosion due to water pressure. Scouring effect due to water pressure must be avoided.
4. This is being a temporary structure, it is dismantled generally after the main job is over. Hence it should be so planned as to facilitate easy dismantling and reuse of material, if possible.

Uses of Coffer dams

Following are the uses of coffer dam :

(*i*) To facilitate pile driving

(*ii*) To place foundations of buildings, bridge abutments and piers, dams, locks etc.

(*iii*) To enclose a space for the removal of sunken vessel.

Types of Coffer dam

Following are the most common types of coffer dams that are frequently used:

1. Dikes:
   (*i*) Earth fill cofferdam
   (*ii*) Rocks fill cofferdam
   (*iii*) Sand bag cofferdam
   (*iv*) Rods fill crib cofferdam.
2. Single wall cofferdam
3. Double wall cofferdam
4. Ohio type cofferdam
5. Cellular cofferdam
6. Concrete cofferdam
7. Suspended cofferdam.

1. Dikes. A dike is an embankment of some material. The material commonly used for the construction of a dike are earth, rock, sand bags etc. These are used for short period only. These dikes are designed on the basis of material used for their construction. Hence cofferdams are named with this material of construction.

(*i*) **Earth fill cofferdam or earth dikes.** This is the simplest form of cofferdam. Its use is limited to in the vicinity where impervious strata is available at the construction site and depth of water is low. Its height is generally 1.2 to 1.5 m. It is economical for low height only ; if height is increased it may prove to be more expensive then other types of cofferdams. A typical earth fill cofferdam is illustrated in Fig. 3.8.

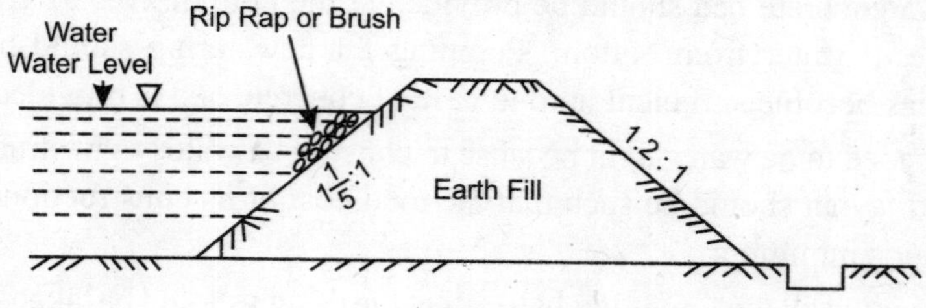

Fig. 3.8.

As the name indicates, the materials used for its construction are clay, murram, sand or gravl. A mixture of clay and sand or gravel is used. Many a time only earth is excavated from side trenches and dumped to form a heap maintaining slopes of 1 : 2 and $1 : 1\frac{1}{2}$ on exposed side and water side of it. The earth dikes are protected against wind and wave action by riprap, brushes, stones etc. A riprap is a piece of rock weighing about 10 kg. It should be maintained properly.

When the earth fill cofferdam is completed, water is pumped out from inside to outside of it and required construction work is started.

(*ii*) **Rock fill cofferdam or rock dikes.** This type of cofferdam is economical at places where plenty of rock is available. These are similar to earth fill cofferdams in construction and can be used for depths of water upto 3 meter and suitable even in swift water. It is illustrated as in Fig. 3.9.

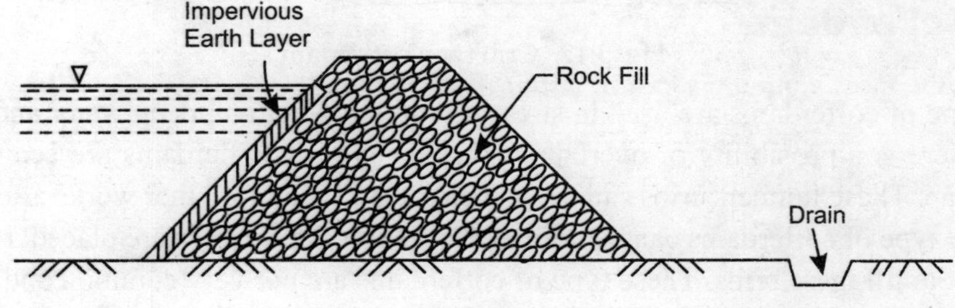

Fig. 3.9.

It is constructed by placing rock-pieces across a stream or around an area to be dewatered. It is imprevious hence an earth layer is provided on water side or on all the sides to make it imprevious. They can withstand the overtopping of water without and serious damage. Sometimes core wall either of clay or concrete or a steel sheet pipe is provided in the centre extending uptó the imprevious layer is provided as shown in Figs. 3.10 and 3.11.

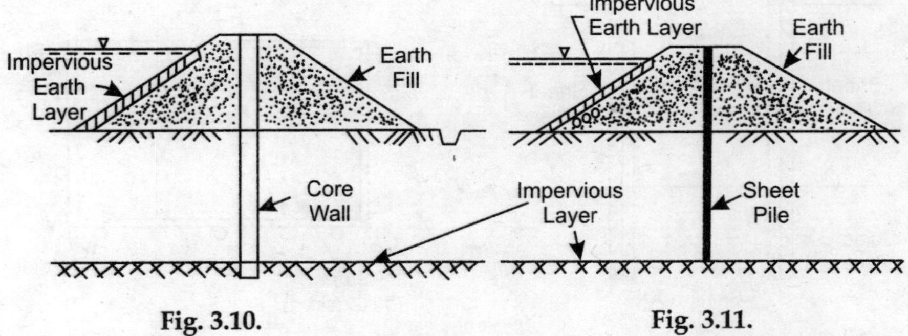

Fig. 3.10. **Fig. 3.11.**

(*iii*) **Sand bag cofferdam.** As the name indicates, a mixture of sand clay is filled in bags and then they are placed similar to that of earth fill or rock fill cofferdams. If cement bags are used to fill this mixture, small quantity of cement present helps to make these sand-clay filled up bags water tight and hence makes the cofferdam water tight to certain extent. These bags should be loosely filled so as to adjust themselves with the irregularities of any. These cofferdams are used for small heights upto 1.5 to 2 meter only.

(*iv*) **Rock fill crib cofferdam.** A crib is a framework of wooden horizontal, vertical and cross (if required) installed in such a manner to form pockets which can be filled with rock and earth. A crib type cofferdam is constructed by installing a series of connected cribs around or across an area to be deemed off.

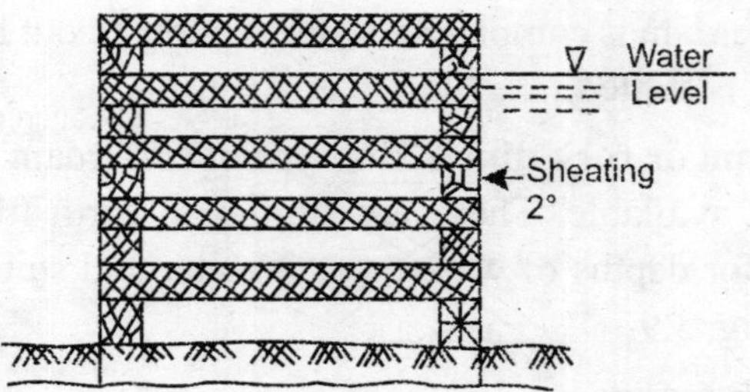

Fig. 3.12. Crib type cofferdam.

These type of cofferdams are used in streams with hand bottom, deep water and swift current and where there is a possibility of overtapping. These type of coflerdams are economical where timber is cheap. These lumbers cribs can be dismantled and used for other works after the main job is over. These type of cofferdams can be used in deep water. These cribs are placed, rocks and earth are fed into it to fill these cribs. These type of cofferdams are not very common and rarely used in construction, due to construction difficulty. Sheet piles are also required to be used to keep these cribs in position hence become costly.

2. **Single wall cofferdam.** This type of cofferdam is suitable when available working space is limited and area to be enclosed is small. Small area are enclosed by means of it and construction is done.

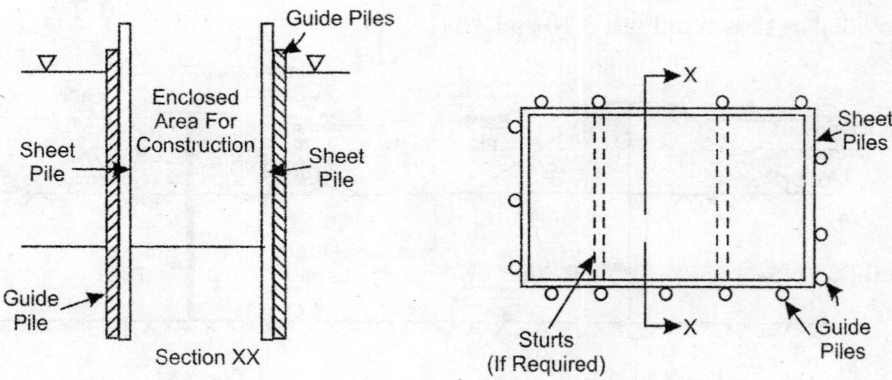

Fig. 3.13. Single wall cofferdam. **Fig 3.14.**

Single wall cofferdam can be used to a water depth of 25 m. The walls are normally made up of steel sheet files commercially available to a length of 22 m. Reinforced concrete and prestressed sheet piles can also be used. They are required to be braced. After erection of the wall, the portion inside is dewatered and construction work is carried inside it. Guide piles are also driven by the sides of sheet piles.

3. **Double wall cofferdam.** Double wall cofferdams are used to enclose large area. Double walls are constructed which gives stability to the cofferdam.

4. **Ohio type cofferdam.** This type of cofferdam was first employed on the river Ohio and hence derived its name as Ohio type cofferdam. It is a type of double wall cofferdam and consist of continuous rows of timber frames with cross-bracings and steel tie rods as illustrated in Fig. 3.15. It forms a large size crib. Vertical sheathings are provided on all the sides and space in between is filled with impervious material.

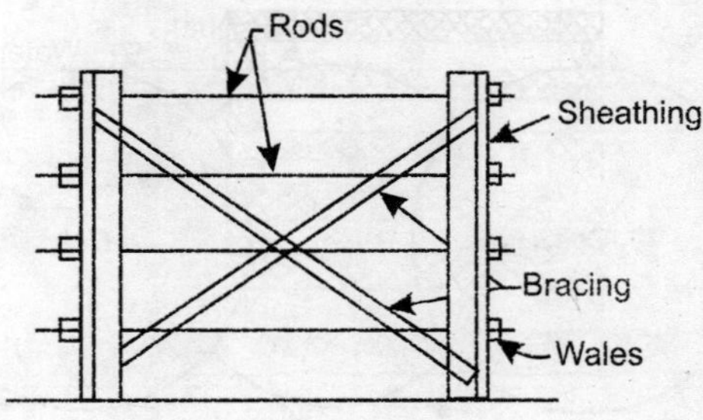

Fig. 3.15.

The main advantage of this type of cofferdam is that it is economical and can be built up rapidly. It is suitable for hard bed at the same time it can be used on soft strata if properly protected against erosion. This type of cofferdam should not be used where the velocity of the water is high or where there is possibility of overtopping hence it is unsuitable for deep water. Side elevation of this is as shown in Fig. 3.16.

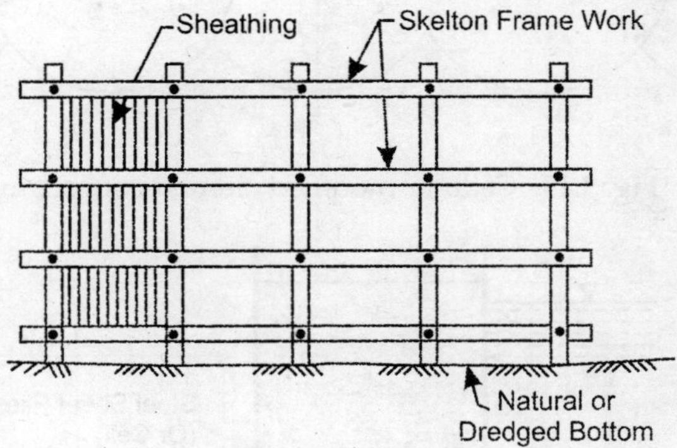

Fig. 3.16. Side-elevation.

5. Cellular cofferdam. Another type of cofferdam sometimes used for larger structures is that known as 'Cellular'; a steel sheet piling is driven in the form of large cell structures which are then filled with sand or some other material and which are stable as gravity structure One of the most extensive cofferdam operations utilizing steel cell unit was that carried out during the construction of the Grand Coulee Dam in Washington. This type of cofferdam has been proved successful for dewatering large areas of water bodies.

Plans of different types of cellular cofferdams are illustrated in Fig. 3.17. to 3.20.

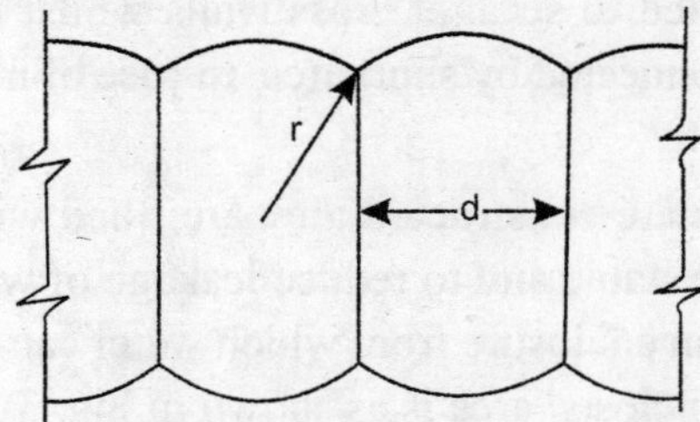

Fig. 3.17. Cellular cofferdam diaphragm cells

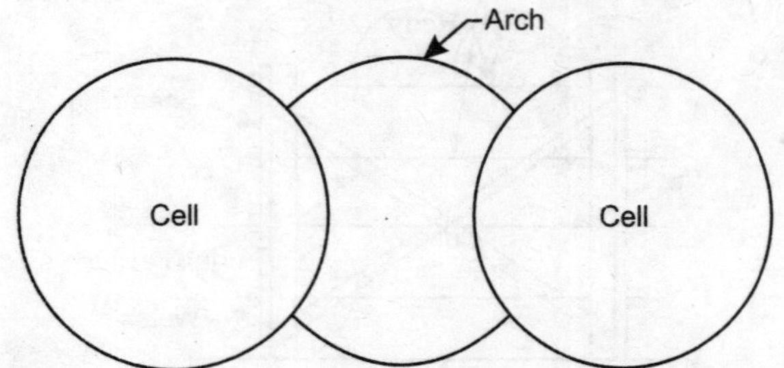

Fig. 3.18. Cellular-circular cofferdam cells.

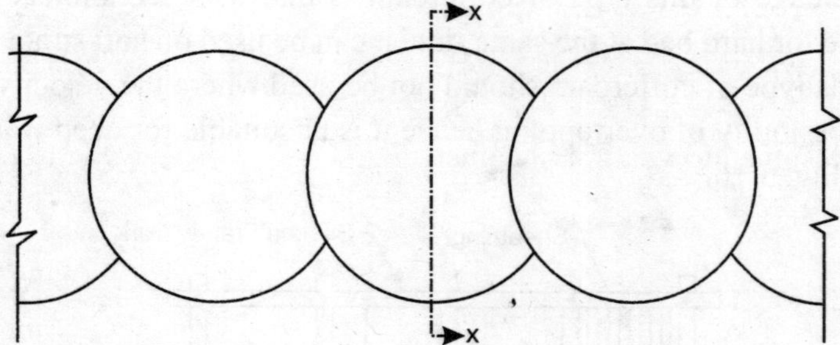

Fig. 3.19. Cellular modified circular cofferdam.

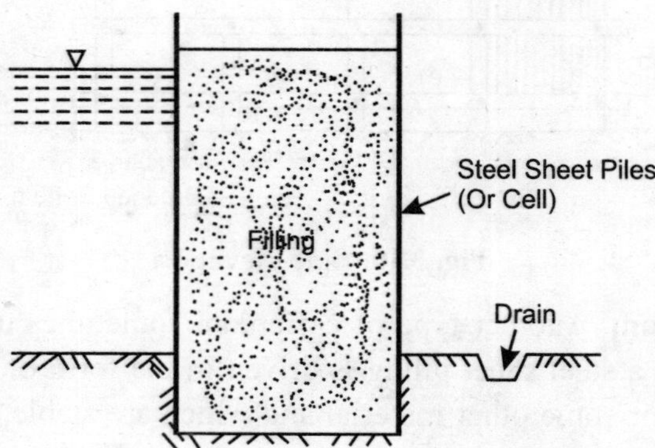

Fig. 3.20. Section of cofferdam at XX.

Cellular type cofferdams are comprised either of diaphragm cells or circular cells or modified circular cells as illustrated in Figs. 3.18 to 3.20. In case of cellular cofferdams with diaphragm cells, series of arcs are connected to straight cross walls whilst in circular cellular cofferdams a series of complete circles are connected by short arcs. In case of modified cellular circular cells the construction is slightly different.

After a number of the cells are constructed they are filled with earth, sand, gravel or rock to increase the weight to make the stable and to reduce leakage of water through them. The cells and arcs so constructed to produce an enclosure from which water can be excluded to form the working area inside the dam. A typical enclosed area is as shown in Fig. 3.21.

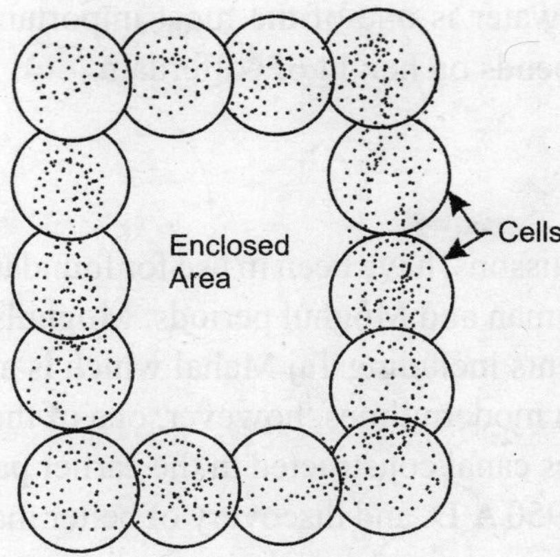

Fig. 3.21.

These cells are designed for stability against

(*i*) sliding

(*ii*) over turning

(*iii*) circumferential tension in the cell walls.

As each cell is a table unit, subjected to a tensile force in the walls when filled with some materials, all the cells can be filled up with these materials immediately. It has been claimed that of various ones, circular cofferdams are better to other.

The advantage of using cellular type cofferdam is its economy and little formworks requirement for its erection. Further construction being continuous each cell is used as a platform to other cell. These can be used on irregular river beds.

The drawback with cellular type of cofferdam is difficult in driving and pulling backs. This can be overcome by greasing them properly. Further these cells should be designed properly considering all probable forces likely to be exerted.

These types of cofferdams are becoming very popular now-a-days.

Concrete cofferdam. There are actually small concrete gravity dams used as cofferdam. Many times precast R.C.C. sheet piles are used and are driven in manner similar that of steel sheet piles. They are economical only if these piles form a part of main structure.

Suspended cofferdam. Sometimes a cofferdam is designed in such a way that a single unit of it is used several times. The cofferdam as such is lifted floated and placed in another position as soon as its purpose is served. Such cofferdams are known as suspended or movable cofferdam. These are generally constructed of timber.

Economy of cofferdam. When a cofferdam is constructed, it is necessary to determine height that will be most economical as well as feasible for required construction. The most economical height is one which will give the lowest total cost considering the cost of the dam and cost of risks that a flood may over top the dam. Hence it need not to be high enough.

The design of cofferdam depends on various factors such as hydrostatic head of water, area to be covered, soil conditions, fluctuation in water level, eddies current etc. It is not possible to consider all factors at the same time to achieve economy. Experience play an important role in its design. Hydrostatic pressure of water is one of the most important factor on which its economy depends which consequently depends on height of cofferdam.

3.3 CAISSONS

Wells, which are also known as caissons, have been in use for foundations of bridges, dams and other important structures since the Roman and Moghul periods. Moghuls in particular used caissons for the foundations of their monuments including Taj Mahal which is a standing testimony to the skill of mankind in the earlier days. In modern times, however, one of the earliest use in India is that for an aqueduct for the Upper Ganges canal constructed in the earlier part of the 19th century. With the advent of pneumatic sinking in 1950 A.D. and discovery of better materials like reinforced concrete and steel, use of wells as foundations gained popularity.

Materials commonly used are R.C., brick or stone masonry. Use of steel caissons has become very popular now also known as penstocks. This type of foundation is very common is U.S.A. and other

Western countries. A caisson of size 29.6 m × 60.1 m in section 74 m long was used in San Francisco Oakland Bridge.

Caisson are classified as

1. Open caissons
2. Pneumatic caissons.
3. Box caissons.

1. Open caisson. A caisson which is open from top and bottom during construction is called open caisson. They may have any shape in plan, *i.e.* circular, oblong, oval, rectangular but circular one of all is most common. The open caisson usually has a cutting edge. The cutting edge is first fabricated at the site and the first segment of the shaft is built on it. The soil inside the shaft is removed by grab buckets and segment is sunks vertically. When one segment has been sunk completely, another segment is added to the top and the process in continued. After the required depth is reached; concrete is placed under water on the open bottom as a seal to a depth that will contain the hydrostatic uplift pressure so as to avoid blowing in of the bottom when the water inside the caisson in pumped out. When concrete is completely cured, the caisson can be pumped out In present practice the caisson becomes a part of the foundation works.

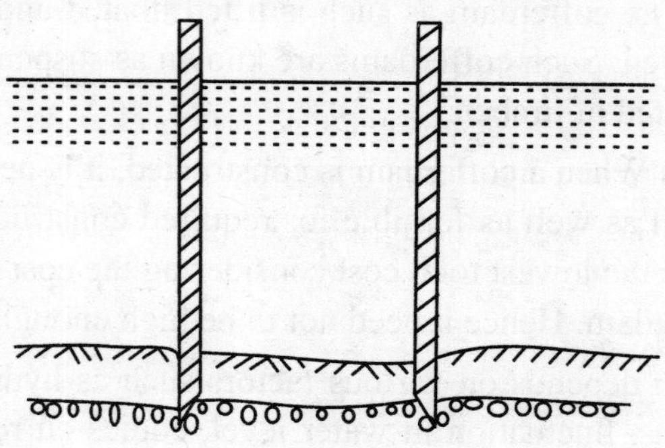

Fig. 3.22.

The advantages of open caissons are:

1. The caisson can be constructed to greater depth.
2. Construction cost is relatively low.

The disadvantages are :

(*i*) The clearing and inspection of bottom of the caisson can not be done.

(*ii*) If boulders are met the progress becomes considerably slow.

2. Pneumatic cassions. In this the working chamber at the bottom of the caisson is kept dry by forcing out water under air pressure. Air locks are provided at the top. The cassion is sunk as the excavation proceeds. Upon reaching its final depth, the working chamber is filled with concrete.

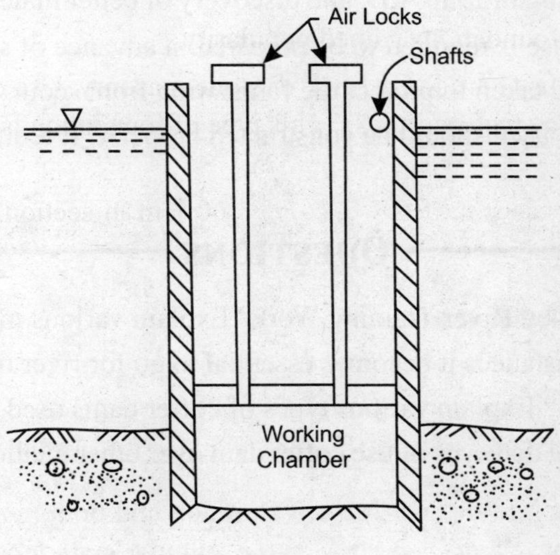

Fig. 3.23.

Advantages of pneumatic caissons are:

(*i*) The caisson is sunk vertically, further there is a proper control as work is done in dry condition.

(*ii*) The bottom of the chamber can be properly sealed with concrete as it is placed in dry condition.

Obstruction to sinking, such as boulders etc. can be removed easily.

Disadvantage is :

(*i*) Construction cost is quite high.

3. Box caisson. In the case of box caisson the bottom is closed. This type of caisson is first cast on land and then towed to the site and then sunks on to previously levelled foundation base.

It is sunk by filling inside with sand, gravel, concrete or water. It is also called as floating caisson.

Advantages of box caisson are :

1. The cost of construction is relatively low.
2. It is used when caisson of other type is not possible.

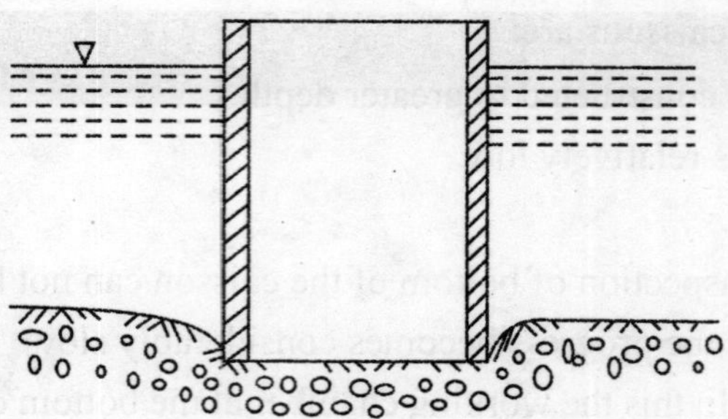

Fig. 3.24.

Disadvantages:

1. The foundation base is required to be prepared in advance of sinking.
2. Due care has to be taken to protect the foundation from scour.

The caisson foundation used should be constructed properly. It requires to have sufficient grip in the ground.

QUESTIONS

1. What do you mean by River Training Work? Explain various methods of it.
2. Under what circumstances it becomes essential to go for river training work?
3. What is coffer dam? Explain various types of coffer dams used.
4. What are merits and demerits of use coffer dam over other methods of river training works.

Chapter

Work Study

4.1 INTRODUCTION

Work study as per British standard is a genetic term for those techniques, particularly the method study and work measurement which are used in the examination of human work in all its contexts and which lead systematically to the investigations of all the factors which affect the efficiency and economy of situation being reviewed in order to effect improvement.

Work study can also be defined as the systematic study of an operation or process to ensure the best possible use of the human and material resources available. The prime aim is to improve productivity. The application of work study to a department or company or any other organisation is made to improve the existing methods of operations, as a result, change will occur which will affect all personnel irrespective of status. For successful achievements due regard will have to be given to all concerned.

4.2 OBJECTIVES OF WORK STUDY

The objective of work study is to obtain the optimum use of human and material resources which are available to it. The benefits may stem from improvements in one or more of the followings:

1. Increased production and productivity.
2. Reduced cost—labour, material, overheads.
3. Improvement of conditions which involve an element of excessive fatigue or danger.
4. Improved quality.
5. Better control of costs.

Benefits of Work Study. Work study can considerably increase productivity. People can benefit from less tiring work, better working conditions and because an efficient company can meet competition successfully, better employment prospects.

It raises efficiency by re-organisation of works. It therefore need no capital expenditure. It is systematic study, procedures are designed to ensure that no factors affecting the situations are overlooked. It is a best mean available for setting standards of performance and although often seen as the basis for incentives, the real value lies in the information provided for work scheduling, estimating deliveries and for accurate costing. Work study produces savings very quickly and these gains continue as long as the improved methods evolved are maintained. It is a universal tool not confined to workshops or industries but can be equally effective in the offices, warehouses, distribution systems etc. It is a most searching technique available because it takes every fact into consideration. It reveals weaknesses which are often overlooked in the day to day working.

4.3 COMPONENTS OF WORK STUDY

Work Study is the systematic examination of the methods of carrying out activities such as to improve the effective use of resources and to set up standards of performance for the activities carried out.

Method Study is carried out to find the best method of operation making full use of all available resources. Work measurement study find the time required to carry out the operations at a defined level of activity. These two in common usage are taken to mean work study.

One more important part of this work study is "Work Study Data" for the purpose of control, costing, estimating, planning and payment schemes. Work study is work nothing unless active use is made of available data and data it produces. It eliminates guess work and opinions by facts. These facts being obtained by careful observations using a trained work study, engineer working with staff of that organisation.

4.4 MOTION–METHOD STUDY PROCEDURE

Various steps involved in motion study consists are :

1. To breakdown work structure. A detailed list of all activities/ operations involved is prepared. All relevant informations pertaining to existing methods are recorded and studies in details and a chart is prepared to have clear idea of the same.
2. To carry out critical examination of recorded activities/operation and to find out probable places where rectification can be done, operations can be modified, sequences can be alter for better performance and output without causing fatigue to labour and without affecting output, efficiency etc.
3. Then to develop a new and better methodology to achieve same objectives. Elimination, combination, rearrangement etc. as and when required are carried out to simply by the procedure.
4. Once a new method is developed and found efficient and useful, the same is installed and feed back is collected for problems, convenience if any and finally implemented in operation.

4.5 TIME STUDY PROCEDURE

Time study is also conducted in a similar manner; however it is used for an average worker and average machines. Time study experts do this job. They measure time taken by labour to perform some activity and suggest measures to reduce the same without causing fatigue to labour. Each and every operation is observed, studied; time schedule is made and critical analysis is made. Method is standardized and efficiency is achieved. Stopwatch; time recording machines and motion picture cameras are used for rigorous analysis and implementation.

QUESTIONS

1. What is importance of work study how it is carried out?
2. What are the various components of work study?
3. Explain motion study procedure of work study.
4. What are objectives of work study and what are its advantages?

Chapter

Contracts and Tender

5.1 CONTRACT

In ancient Greek, when structures were built, Greek law laid down certain relationships between the contractor and the owner. They had certain rules and regulations. Such rules and regulations and other conditions of contracts are still prevailing today with certain modifications. At that period also contractors were required to hire sufficient skilled and unskilled persons, qualified worksmen. They were required to give the guarantee of the quality of works. Penalties were prescribed for failure or delay of the works. Payments were made in installments as per the progress of the works. These things are existing today also.

As per Indian Contract Act of 1872 "Contract" is an agreement between competent parties, enforceable at law whereby each acquires a right to what is promised by the other. Two persons can have any sort of agreement, they may promise to each other but unless such promises are enforceable at law they form no contract. Hence it should be noted that every contract is an agreement enforceable by law but every agreement is not necessarily a contract.

The term contractor means a person, syndicate .or firm that has made such an undertaking for execution or for services in connection with civil works.

5.2 ESSENTIALS OF CONTRACT

Following are the essential requirements for formation of a contract. These have been recognised universally.

1. Legally competent parties.
2. A lawful subject matter.
3. Proper and valid considerations.
4. Genuine agreement signed by both the parties.
5. Joint venture contract.

Engineering contract is a contract concerning to engineering project. It is also a type of business agreement. As per Indian Contract Act parties involved into a contract must be legally competent. Infants, lunatics, drunken people are not competent. There is no hard and fast rule that two persons doing engineering contract with each other should not have any close relations or blood relation but still as far as possible it should be avoided. In certain states "a wife cannot contract to her husband", such rules are existing, though they are legally competent parties. The contract should be on a lawful subject matter which is enforceable at law. These contacts must not violate state or union territory, *i.e.* if a contract has been made in one state it will be invalid for other state. There should be proper consideration between a contractor and owner whether Government Organisation or private party. In absence of these considerations it becomes very difficult to execute the work properly. In engineering contracts there are mutual promises. The contractor promises to execute the work properly, to perform the work faithfully. Similarly it is the duty of the owner to make the payments when due. Further a contract must be genuine. Things should be clear. There should not be any misrepresentations or fraud. The contract agreement is not valid unless until signed by both the parties.

For a contract to be valid there must be an offer from owner to get the work executed and similarly there must be an acceptance from the contractor to execute the work as per laid terms and conditions. Both the offer and acceptance must be definite and legal.

5.3 TYPE OF CONTRACTS

There should be a free consent by the free will of the parties involved. There should be no coercion (threat) duress (threat of bodily harm) undue influence, misrepresentation and mistakes.

There are many ways of getting a work executed in contract system; the work is got executed through an agency called contractor, who arranges all materials, labour, equipments etc. for the completion of the work in time. The main idea behind any contract system is

(*i*) To get lowest total cost of construction.

(*ii*) To achieve greatest speed of construction to have minimum indirect expenditure.

(*iii*) To have best workmanship and

(*iv*) To have future business between owner and contractor based upon previous experiences. These different system of contracts are listed below.

As on today contracts are classified as

1. Item Rate contract (B2 type)
2. Percentage rate contract (B1- type)
3. Labour contract
4. BOT system

1. **Item-Rate contract (B 2 - system).** In item rate contract the contractor undertakes the execution or construction of a work on the item rate basic. The amount the contractor receives depends upon quantities of various items of works executed. In this system; bill of quantities is prepared mentioning various items of construction and their quantities; contractors bidding contract are required to quote the rates of various item mentioned in bill of quantity with the stipulation that work will be subjected to measurement and payments are made at the unit price quoted. This particular system is called item rate contract. The difficulty with this system is; it is rather difficult to scrutinize the tenders

submitted by various bidders; hence modified. In modified system of item rate contract; the department generally quotes various items of construction along with their quantities and rates. The bidders i.e. contractors are required to quote various percentages at which he wants to execute that item. As such for each item a separate percentage is quoted. This system of contract is called as 'percent item rate contract' system. This system was very popular in P.W.D. and still used many times. This system is also called B 2 -system of contract in P.W.D.

Schedule B

Item No.	*Quantities estimated but may be more or less*	*Item of work*	*Estiamated Rates*		*Unit*	*Total amount according to estimated, quantities*
			In figures Rs. Ps.	*In words*		
		As Per Separate Sheet Attached				

Department Hand book

Note 1. All work should be carried out as per other specifications of the Division or as directed.

Note 2. Rates quoted include clearance of site (Prior to commencement of work and at its close) in all respects and hold good for work under all conditions, site moisture, weather etc.

Signature of contractor

Signature of $\frac{\text{Executive Engineer}}{\text{Assistant Engineer}}$

When the quantities of various items are not available; then there was practice to adopt item rate contract system. The drawback in this system, as the quantities of various items are not known. The contractor always gives higher rates for those item quantities which are likely to increase & low rates for those likely to be decreased. The same is found in percentage item rate contract. The contractor quotes higher percentage for items likely to increase and lower percentage for items likely to be decreased. This results in unbalanced contract. Therefore this type of contract system should be avoided as far as possible. Unless until detailed estimate is made; quantities of different items are evaluated correctly and got ascertained that there will not be more than five-percent variation in each quantity maintaining the overall cost variation less than 5 percent only; this contract system should not be used.

2. Percentage rate contract (B1 system). This system of contract is very popular and common now a days. Most of the works either in P.W.D. or private organisation are undertaken with this system only. In this system detailed estimate is prepared and contractors are required to quote single percentage either higher or lower at which he wants to execute this job. As such all item rates are changed with the same percentage; hence tender generally does remains balanced and scrutinity becomes comparatively very easy. Now a days materials like cement and steel are supplied by the department hence chances of manipulation becomes still less. Most of the engineering contracts are performed with this method only. The rate of recovery of material is given in schedule A. It helps the contractor to decide his rates to be quoted for item involved with use of these materials.

Schedule A

Schedule showing (approximately) the materials to be supplied from the Departmental stores for work contracted to be executed and preliminary and ancillary works and the rates at which they are to be charged for the construction of...........(work)

	Particulars	*Quantity*
1.	Cement in plastic bags	76.75 MT 1527 bgas
2.	M.S. reinforcing bars	2 MT

Rates at which the material will be charged to the contractor			*Place of delivery*
Unit	*In figure*	*In words*	
Per one MT	Rs. 2 lac	Rupee two laws	at departmental store
Per one MT	Rs. 50,000	rupee fifty thousand	at departmental store

Note 1. The person or firm submitting the tender should see that the rate in the above schedule are filled up by the Executive Engineer-in-charge on issue of the form prior to the submission of the tender.

Note 2. The contractor should furnish the account of all materials before placing demand for further material and furnish full account of all Department. If any material remained unused with the contractor, he should return the same to the Department. In the event of non return of such material recovery at penal rate of twice the issued rate with S.T. and G.T. will have to be effected from the contractor.

Note 3. Wastage of steel to be borne entirely by the contractor.

Signature of contractor   Signature of $\dfrac{\text{Executive Engineer}}{\text{Assistant Engineer}}$

3. **Labour contract.** This is the most common system of construction for private individual building; particularly in small cities and big towns. Under this system of contract; the contractor arranges all necessary labour; tools and plants and equipments required for. The construction and all the materials required are supplied by the owner. This type of system of contract is also particularly suitable for those cases where the owner or Government departments is in a privileged position to by large quantities of materials at cheap rates. This system is very common for small and private works. Most of the private dwellings are given on a labour contract only and the owner arranges all the materials required for construction. Generally owner also appoints some engineer to supervise the work to maintain the quality and to economise the cost of construction at a fixed fee or percentage of actual cost of construction as supervision charges.
4. **A BOT i.e. Built operate and transfer.** This is a system in which land is acquired by govt, and contractor is asked to built the civil works, operate it by collecting some fee from its users to collect the cost of construction and once the cost is collected, to transfer it to the owner *i.e.* Government. No investment is made by government under this system. Drawings, estimates are prepared by Government and contractor is required to execute the work as per designs and specifications given by Engineer. No payment is made to the contractor. All expenditures incurred are to be made by contractor. In return he is authorised to collect some charges as specified by government from the users of that civil work and once the cost is recovered by contractor, the site is trasferred from contractor to department *i.e.* department takes possession of that site. The security however is provided to the contractor by Government regarding law and order problems if any.

Mostly Bridges and Roads are constructed and developed by this particular system of contract.

5. **Joint venture contract.** This system of contracting has developed due to tremendous increase in the size of work or due to development of specialised persons for certain specialised job. In case of important and big projects, it is rather difficult to individual contractor to complete the work in specified period of construction. Further different specialist firms have developed who can undertake such work at a reasonable rate with less time for its construction. Hence it was thought and then decided to divide the whole work into different parts either on the basis of amount of work involved or on the basis of special type of work with which a specific firm may be familiar. Hence the work was divided into different contractors and hence joint venture system of contract developed. The first joint venture system involved six competing construction firms headed into line under one management at the construction of Hoover Dam. The system has become very popular and common now. In a building itself different works are given to different specialists firms under a main contractor and work is got done efficiently and early. Wiremen; plumber etc. are specialist persons for these jobs. Similarly gangs for concreting work are very popular in India. Their jobs is only to do concreting of various elements of reinforced concrete work.

These are the various systems of getting a work constructed and executed. Each method has its speciality in itself and can be used effectively as per the prevailing conditions.

5.4 LABOUR CO-OPERATIVE SOCIETY

In most of the states; labour co-operative societies have been "formed; and generally it is a practice to award contract for an amount less than Rs. 1.00 lakh to these labour co-operative societies without calling the tenders and at the estimated rates based on divisional schedule of rates, brought upto date. While awarding the works; class of society is taken into consideration and these classes are formed based on capacity of that society to undertake the work. Based on their financial status, managerial and technical capacity and staff; these labour co-operative societies in general as. Class A, Class B and Class C.

These labour co-operative societies are generally exempted from the production of solvency certificate for getting registered but are required to produce a certificate from the District Deputy Register/District federation regarding its capacity and financial resources. Chairman of that labour co-operative society is generally consulted for dealing with. The number of works allotted to one society is generally restricted to three only.

5.5 REGISTRATION OF CONTRACTOR

Each contractor is required to get himself registered to various departments where he is interested to work. In case of Government contract these contractors get themselves registered to Government and hence called as government contractor; however one should not get confused by the meaning Government contractor. These contractors are not serving the Government as an employee but are entitled to do Government jobs if awarded.

Class- Class I, Class II Class III class IV Class V Class VI and Class VII

The contracts are also classified by some departments as

Class- Class A1, Class A, Class B+, Class B, Class B1 Class B, Class B1, Class C, Class D and Class E

To get a contractor to be registered under certain class, he has to apply to the competent authority. He is further required to submit following documents along with the application.

1. Solvency certificate for an appropriate amount.
2. Attested copy of deed of partnership and power of attorney on stamp paper.
3. Current Income tax clearance certificate.
4. Works certificate for all works performed during last three years and for works in progress.
5. Undertaking (for class I and class II) for employment of the staff.
6. Application in duplicate with all documents.
7. Attested photo of all partners if any.

Pre-qualification of Contractor: Sometimes jobs are awarded on the basis of the quotations received from 'selected tenderer', which means tenders are invited from a group of known reputed contractors who are considered suitable for that job. This method is known as 'Pre-qualification' and saves a lot of time on the part of the employer, consultant and contractor. Sometimes, for special type of jobs, it is mentioned in the advertisement that specialist contractors having considerable experience alone apply. In order to judge the capability of contractor for that job, employer asks contractor to furnish relevant data and it is obligatory on the part of the contractor to supply all such informations required to the employer and obliqe. Sometimes department or consultants or owner have their own list of empanelled contractors those have already worked with them. The distribution of tender papers is then restricted to these contractors only. This, no doubt saves a considerable time but at the same time reprive other good contractors to tender the works, and show their capabilities. Sometimes for a very specialised job only one specialised firm is consulted and job is awarded by due negotiations. This is of course common in private works, however objectionable in government jobs. The possibility of malpractice can not be totally avoided in prequalification.

5.6 CONDITIONS OF CONTRACT

While preparing a contract agreement or agreement deed; certain conditions are laid down which are binding to the contractor as well as owner; it is mainly to avoid dispute during the progress of work if any. These conditions of contract though remain same for all works in general, but then changes in certain cases from person to person and also depends on type of work and amount of work to be executed. The certain basic principles have to be strictly adhered to while preparing the contract agreement. Variation in these conditions are liable to be challenged by either of the parties. If there are minor variations, they are generally accepted by both but for major decisions; the matter is solved jointly. If there is a wide variation in the cost; The matter is referred to higher authorities for approval.

There are two types of conditions that are generally made in tender document.

1. General conditions of contract.
2. Special conditions of contract. (Particular).

   The general conditions are those which do not affect the work. These are obligatory bindings on the contractor whilst special conditions are based on type of work and the conditions vary from work to work.

A few general conditions of contract are as listed below :

1. Work should be executed as per the drawings and specifications mentioned in agreement bond. There should be no modification in design; drawing etc. from those given in notice inviting tender.
2. Period allotted for that work should not change. Only when it may become essential to change the period due to unavoidable reasons or natural calamities; it may be changed but only with due permission from higher authorities.
3. It is obligatory that it is the responsibility of contractor to clear the site, to give medical aid to labour if required and to give free access to departmental engineers at site.
4. Contractor is bound to pay minimum wages to the labour as per Workmens Compensation Act.
5. There should be no flaw in agreement deed. It is usual practice in case of large works to get agreement scrutinised from law department so that legal aspect of the agreement are taken care of.
6. Contractor can not assign or sublet the work to other contractor. Only for special type of works this can be done but only with due permission of engineer in charge.

7. It is responsibility of engineer incharge to see that proper payments as per progress of works are made to contractor. He is also required to see that work progress is maintained by the contractor maintaining the quality as per specifications.
8. Dismantling of works not approved by the engineer if any and construction of new one, dismantling defective work is contractors job. No extra payment is made for dismantling and reconstruction of dismantled work to the contractor.
9. It is responsibility of contractor to provide potable water to the labours employed by him.
10. The contractor is required to make his own arrangement for electric power at his own cost and no claim can be made in this behalf. The layout and method of laying lines and wiring and the nature of connected leads must be approved by engineer incharge.
11. In case of errors, omissions and discrepancies; and doubts the matter should be referred to competent authorities and the decision of competent authorities shall be final and binding on the contractor.
12. All the materials excavated is of the government or of owner and should be disposed of as directed by the competent authorities as Deputy Engineer or Executive Engineer as the case may be.

All these conditions are binding on the parties involved in agreement deed. Along with these some other special condition can be laid down which are binding on both the parties. These special conditions depend on type of work and amount and quality of work. A few examples of these special conditions are:

1. The concrete used is of grade M 30 and contractor is required to get a mix designed from an approved laboratory before concreting is made.
2. Concrete should be compacted only be 60 mm diameter and 600 mm long needle vibrator giving 5000-6000 vibration per minute.
3. Through stone of size as mentioned in drawings should only be used at a spacing shown in drawings.

5.7 CONTRACT DOCUMENTS

The contract documents in general consist of following items.

1. Title page giving the detailed location of site; division/circle issuing tender, type of tender and type of work to be performed.
2. Index-page giving content of contract documents.
3. Tender notice.
4. Special conditions of contract including schedule A and B.

   Schedule A - is bill of quantities specifying the quantity of various items involved in that work and rates in case of Bl tender.

   Schedule B - is list of materials to be supplied by the department and their recovery rate.
5. Detailed specification all items of works.
6. Detailed Drawings.
7. Deed of pledge.

This contract documents called agreement deed is always in written/typed on required stamp paper duly signed by the owner as well as contractor. Each page of this deed is to be signed by both the parties. Generally three to four copies of the agreements are prepared and distributed as below.

1. One copy along with attested copy to higher authority. Higher authorities sometimes the returns original copy to engineer incharge and retain one attested copy for record.
2. One copy is kept in locker and is taken out only in case of dispute if required by a court of law.
3. One copy is issued to accountant general.
4. One copy is given to the contractor duplicate copies of contract documents can be held if required.

5.8 PERFORMANCE OF CONTRACT

Performance of contract means to deposit a sum equal to security deposit with the department and to sign performance bond for a value mentioned in tender notice. It is on a form approved by engineer, architect viz. government bond. Government security. It remains in the custody of architector department till the defect liability period is over.

5.9 BREACH OF CONTRACT

In every contract system; generally two parties are involved. These parties in civil engineering are contractor and owner where each acquires a right what is promised by other. If one of party fails to perform or fulfil his obligations, under the contract; it results into breach of contract which may lead to determination or termination of contract. It causes a loss to one party that gives an immediate cause of action and a right to damages as compensation for loss occurring from the breach.

As far as determination of contract by owner or employer ; this can be done under following conditions :

1. Default on behalf of contractor like
   (*a*) Contractor loses the job for any reason known to him only.
   (*b*) If contractor suspends the work and does not start again even after receiving notice.
   (*c*) Does not maintain the progress and quality inspite of several notices.
   (*d*) Does not dismantle and reconstruct the faulty work pointed out by department.
   (*e*) Sublets part of the work to other contractor without prior permission.

2. Bankruptcy: If contractor becomes bankrupt. Under such condition; owner takes the work in possession like materials and machineries lying at site. The owner however is required to make payment of materials and equipment at site. No payment however is made to contractor in any form.

Sometimes there is breach of contract from owner also. It is fundamental right of contractor also to recover the damages that have caused him due to breach of contract, not because of his fault. He will have to be put in the same position; where he could have been if had completed the work as per contract deed. Contractor is therefore entitle to recover the loss of profit on full job. He is also entitled to get interest on his capital invested for period of determination of the profit amount depending upon whether the breach is their before starting of work or part work has been executed.

In case contractor commits breach of contract; the employer is also entitled completing or correcting the contractors balance work.

Finally; it is to be noted that aggrieved party will have to prove that he has suffered loss on account of wrong action of other party.

5.10 PAYMENT OF BILLS TO CONTRACTOR

Like contractors responsibility to complete the work as per drawings; specifications and conditions of contract; it is responsibility of department to make the payments of the bills submitted by contractor as and when these are due. Generally two types of bill are submitted by the contractor.

1. According to progress of work contractor submits his bill called as running bill or Running account bill generally known as RA-1; RA-2; RA-3 i.e. running account bill No.1, running account bill no.2, running account bill No.3 and so on. Generally 5 to 6 running bills are submitted during the whole work.
2. After the completion of work the contractor submits the final bill for payment of remaining amount except S.D. which is paid only after one monsoon season or six months after completion of work which even is later. This is called defect liability period.

For the representation of running bills; measurements are taken in measurement books (called MB) for the works already completed till that particular date and market on drawings. Abstract sheet clearly indicating the quantity of each item completed. Its rate and amount of that item and total cost of work executed is also calculated. This amount is further increased or decreased depending upon percentage quoted in percent rate contract.

From this amount reached various deductions are made. These deduction are:

1. Deduction or advances given to contractor if any.
2. Security deposit money (part) if full S.D. is not deposited by contractor. If contractor has made payment S.D. in the beginning before the start the work no. deduction is made.
3. Cost of materials supplied by department to contractor as per schedule A.
4. Income tax @2% of the amount of running bill and this amount is directly remitted to income tax department through bank/treasury challan and a copy of counter foil is given to contractor for submission of his return to income tax department.

   Each running bill is then duly signed by contractor; engineer in-charge who has made the measurements before the bill is submitted to appropriate authority for payment. Each running bill must be given certain reference number and the same should be mentioned in next successive running bill.

5.11. ADVANCES TO CONTRACTOR

Once a work order is given to contractor after completing all necessary formalities like deposition of S.D. signing agreement deed etc; mostly contractors as involved in other contract work also may not have sufficient funds available with them to start the actual work. Hence Contractor's request to the owner for payment of some advance amount against the work which is further deducted from the running bill. These advances paid are called as mobilization advance or material advance or secured advance etc. These advances are paid to raise the initial finance needed by the contractor to mobilise the work smoothy and to have and maintain regular progress of the work.

'Secured Advance' is a term applied specifically to an advance made on the security of materials brought at the site of work; to a contractor whose contract is for the completed item of works. 'Mobilisation advance' is a term applied specifically to an advance made to the contractor to mobilise the work in the initial period as the funds required in initial period are comparatively large. This mobilization advance vary from 5 percent to 10 percent and in special cases upto 15 percent of estimated cost of work. Material advance against the material brought at site is generally given 50 percent to 70 percent of the cost of materials brought at the site. The practice of material advance now-a-days is always avoided as there is always possibility of thefts of the material from the site. These advances paid to the contractor are deducted from his first running bill only so that the S.D. remains with the department as guarantees of quality of work from the contractor.

Temporary advance is an advance given temporarily by the disbursing officer to a sub-ordinate officer to enable him to make payments of specific jobs like muster rolls or other vouchers while which have already been passed for payments. The account of this is also maintained in a proforma used for imprest. The account of temporary advance must be closed as early a possible.

5.12 TENDER

Tender is an invitation from the owner to the contractor to execute some work at specified cost in specified time and published in the form of tender notice in news papers; notice boards; gussets etc. according to the cost of woks.

These tenders may be invited by two classes viz.

1. Open tenders
2. Sealed tenders

Sealed tenders are invited for big and important works and they are in writing and hence requires wide publicity and no considerations for their validity. These tenders are opened on scheduled date and scheduled time in the presence of contractors who have submitted tender or their representatives. The presence of contractor or his representative during the opening of the tender is not essential but it is always desirable. An open tender is an oral talk between the engineer and contractor for certain small job to be performed. This may be in writing also; However this is also required to be published to have good competition so as to economise the cost of that work. It is generally used for small works.

For a contract to be valid; there must be an offer from owner in the form of tender notice to get some specified work to be executed and similarly there must be an acceptance from the contractor to execute the work. Both the offer and acceptance must be definite and legal.

5.13 TENDER NOTICE

It is an offer from the department engineer to various contractors to under take some specific job at some specified terms and conditions and to complete that job in specified period. Tender notice published must give sufficient informations to the contractor so that there should be no ambiguity to the contractor. It should contain following informations :

1. Estimated amount of the work and category of contractors entitled to tender for the work.
2. The place where and the time when the contract documents can be seen and blank forms of tender obtained; also the amount to be paid for such forms of tender.
3. The place where and date on which and time when tenders are to be submitted and are to be opened.
4. The amount of Earnest money required and amount and nature of security deposit to be paid on acceptance of tender.
5. The period during which the work is to be completed.
6. With whom or what authority the acceptance of the tender will rest.
7. The necessity of attaching the tender a list of documents accompanying the tender; signed by the contractor.
8. Name of work and its location.

A tender notices for a type of work is given below. For some special works where sufficient and capable agencies are not available to perform the work and where the cost of the work is very high. Global tenders (International level) are invited to undertake that job. Various terms used in tender notices are dealt later on.

Tender Notice

Sealed percentage Tenders in B-l form are invited on behalf of Chief Engineer for the work mentioned below from the contractors Registered with P.W.D. in Class III and above or Registered with C.P.W.D., Railway and M.E.S. in appropriate class.

Sr. No.	***Description***	***Details***
1.	Name of work	Industrial Area, Ph-11. Deepening and widening existing nalla including providing built up section and remodelling/ construction of addition CD. works.
2.	Estimates cost	Rs. 93,74,474/-
3.	Earnest money deposit (By demand draft)	Rs. 93,750/-
4.	Date of receipt of Pre-qualification form	2.6.98
5.	Period of issue of blank tender forms Upto 15.00 hrs. only on all working days	26.4.98 to 09.6.98
6.	Cost of tender forms	
	(*i*) In cash	Rs. 1500/- Non refundable

	(*ii*) By post	Rs. 1550/- Non refundable
7.	Last date of submission of sealed offers	16.6.93 upto 15.00 hrs.
8.	Place of submission of sealed tenders	Office of the Chief Engineer

Important Notes

1. Earnest Money be paid in the form of Demand Draft only drawn in favours of The Executive Engineer, Division on any scheduled/Nationalsed Bank.
2. Submission of offers in two envelopes system shall be followed as under.
   (*a*) First Envelope : Contained Demand Draft towards E.M.D.
   (*b*) Second Envelope : Technical/Price bid.
3. Any other information can be had from the office of the Executive Engineer, division on specific request during office hours on all working days.
4. Blank tender forms will be issued to the pre-qualified contractors only.
5. Right to reject any or all the offers without assigning any reason thereof is reserved by the competent authority.

SdA

Executive Engineer

Division

5.14 VARIOUS TERMS USED IN TENDER NOTICES

1. Estimated Cost

It is the cost of that work estimated on the basis of drawings and specifications. While inviting tender; it is essential to have detailed idea regarding cost of that work as earnest money deposit (EMD) and security deposit (SD) depend upon this estimated cost of construction. Further class of contractor who can bid the tender also depends upon this estimated cost only in general and in Government job in particular. The capacity of a contractor to undertake some work depends upon his current financial position and number of works in hand.

2. Earnest Money Deposit (EMD)

While submitting any tender, the contractor is required to deposit certain amount with tender document to the department as a guarantee that if his tender is accepted; he may not refuse to work. In case of refusal this amount is forfeited. This amount of deposit is called Earnest Money Deposit (EMD). Earnest Money of the contractor whose tenders have not been accepted is refundable. The amount of earnest money to be deposited with the tender should be sufficient large to be a security against loss in case of contractor is failing to complete the contract documents and furnish the required security within the appointed time, after the acceptance of his tender as the case may be. The amount of earnest money in general is 1 percent to 2 percent of estimated cost of works.

3. Security Deposit

On acceptance of the tender; contractor has to deposit some amount with the department before work order is issued. It is to have a check on the contractor so that he fulfills all the terms and conditions laid down in the contract and maintains quality and work satisfactorily. If he fails to do so; part or whole money deposited is forfeited by the department. This amount deposited as security against quality is called Security Deposit *i.e.* S.D. It is generally 5 to 10 percent of estimated cost of work and is inclusive of E.M.D. which is deposited by contractor along with the tender documents. In case of some faulty construction; if contractor refuses to demolish and reconstruct; the same can be done by the department from security deposit money. No security deposit is made in case of contract is for supply of material as supplied material itself becomes a security. This security deposit is refundable to the contractor after a specific period of time after completion of the project. This period is generally six month or one monsoon season after completion of the project which ever is more. No interest is paid on S.D.

Many a time when the cost of the project work is large; it is deducted in installments from the running bill of the contractor, generally 50 percent amount of S.D., contractor is required to deposit in advance and remaining 50 percent is deducted from 2-4 running account bills as the case may be.

4. Defect Liability Period

The S.D. is refunded only after this period. This is the period during which; there may be possibility that the work executed by contractor may show defect. It is generally six months or one mansoon which ever is more.

5. Time Period of Work

This time period allotted for any work is based on type of work, various items involved; time required for each item of work; total estimated cost of work. This varies from a few days in case of supply of materials to a few years in case of construction work of very large amount. In general it is about 15 days to 3 months for supply of materials and 6 months to 18 months for construction work of appropriate amount and size.

6. Cost of Tender Form

Each tender is made available to contractor by making payment to the department as the cost of tender form. This is basically to curtail fissiparous tendencies of contractor and secondly to recover the expenditure incurred in the form of advertisement and printing cost of tender documents. Department issuing tender form is not supposed to earn profit in the form of tender form; but department is supposed to recover its cost.

7. Duration of Project

It is also an important aspect of tender. The Escalation charges are generally not paid for first 12 months *i.e.* after the start of the works. Different type of civil works need different duration i.e. time limit for its completion. For example a road work needs less time in comparison to building. Further depending upon cost, a particular type of work also needs different duration for its construction. A building cost Rs. 50.00 lacks will need less time than that of a building having a cost of Rs. 5.00

crore. Therefore this duration of project also becomes an important item of tender. It depends on various factor such as type of work type of labour required and their availability; curing period in case of R.C. works, weather conditions, infrastructure required and available and so many other factors.

5.15 INVITATION OF TENDERS

There are three ways in which tenders can be invited.

1. Private Invitation
2. Public Notice
3. Negotiable Tender.

This depends upon type of work *i.e.* Government or Private. In case of private work generally known contractors are invited and asked to submit their tender in prescribed proforma and by discussion and negotiation one of the contractor is awarded that job. Of course; all necessary formalities are required to be completed to avoid any dispute or problem existing later on. However in Government job wide publicity is required to be made so that each registered and qualified contractor can bid his offer. This method opens chances for new contractors. Earlier contractors used to be non-technical persons having little or no knowledge of that job; tremendous problems have erupt due to this. But now-a-days qualified and technical persons having technical degree have entered in the field of contract. This has saved a lot of un-necessity disputable problem. The work in case of public notice is generally given to contractor bidding lowest amount. For small works generally petty contractors are insisted and by negotiations the work is given to that contractor. In case of construction of private individual work; labour contract is given generally by mutual negotiations only.

The minimum time gap between the last the date of issue of the tender form and the date for receiving and opening of tenders is generally as under and in no case; the tender form be issued during the gap period.

	Perticular of work	*Time gap*
1.	For all works; the acceptance of tenders of which is within the competence of Suptt. Engineer and below	4 days
2.	For all works; the acceptance of tenders of which is within the competence of Chief Engineer	7 days

5.16 TENDER DOCUMENTS

Before the work is given on contract; an agreement is prepared. This agreement should have following documents attached to, make it valid; further each paper of this document must have signature of both parties *i.e.* owner (department engineer or competent authority issuing tender) and contractor undertaking that work. These documents are issued to contractor in the form of tender form containing all the informations and conditions. It also includes drawings and specifications to have an idea regarding the work for which the tender is issued. These tender document include :

1. Title page : Title page should have title of work to be performed; reference member location of site and issuing authority of tender.
2. Index page : Giving list of contents in serial order with page number.
3. Printed form of agreement duly completed as necessary.
4. Brief tender notice as published earlier in the prescribed form.
5. Detailed notice to contractor in prescribed form (tender notice).
6. A complete set of drawings showing the general dimensions of the work proposed and so far as necessary, details of various parts as per specified rules.
7. Additional instruction to tender if any.
8. Statement of design details if required.
9. General specifications.
10. Detailed specifications for all items involved and other items.
11. Special specifications for concrete work.
12. General conditions of contract.
13. Special conditions if any.
14. Statement of machinery available with the contractor and machineries available with owner that can be given to contractor on hire for specified job with details of their charges; terms and conditions.
15. A certificate to the effect that the estimate made and the plans on which the tenders are based have been carefully examined and that change are called for the reason to be specified.
16. Declaration of contractor.
17. Schedule B.
18. Schodule BI.

5.17 PREPARATION AND SUBMISSION OF TENDER

The bidder who is interested to execute the job, submits his tender to the department. Once the tender notice is issued by the department, various contractors; interested for that job, procure it from department after making payment of cost of tender form. These contractors study the tender document visits and inspects the site; decide individually at wha\particular rate the tender can be executed earning reasonable profit. Generally two envelope system is adopted. The tender documents along with earnest money deposit is submitted to department in time as mentioned in tender notice and a receipt of acknowledgment is duly received.

5.18 SCRUTINY OF TENDER AND ACCEPTANCE

Once various tenders received by the department; next job is to scrutinize them. It is mainly to have minimum cost of project maintaining good quality work. Following factors are considered while doing scrutiny.

1. Name of contractor and his registration.
2. Earnest Money Deposit DD or challan whether attached, if this is not submitted the tender is rejected immediately.
3. Previous experience of contractor for similar jobs if any and total experience.
4. His previous record of works executed.
5. Financial position.
6. Machineries available with him in relation to the work to be performed.
7. Man power; skilled and unskilled available with him.
8. Rates quoted or percent rate quoted.

Earlier was practice to award tender to a contractor; has quoted minimum rates. But now-a-days along with his percentage rate; other factors given above are also considered. Now-a-days qualified engineers have entered into the field of contractor and it has been experienced that there are always less problems as engineering contractors can easily understand the drawings and specifications and knows the importance of quality control. This has also become point of consideration while doing scrutiny and award of tenders.

A comparative statement is prepared of contractors bid submitted and minimum bid with his previous experiences is taken into consideration; final decision is made. In case department feels, that contractor in consideration is not reliable; contractor; having second lowest bid is taken for consideration. If required by the department, or owner, the bidder is required to produce evidences of experience, ability and financial standing as well as statement of plant and machinery. The person whose tender is accepted is required to execute an agreement bond to complete the work as per terms and conditions mentioned in tender deed. He is required to start the work; immediately after issue of work order.

The acceptance of contract to a contractor is issued clearly indicating the details of work; its site; cost and rate quoted by bidder. The contractor then informs department; its acceptance and submits security deposit sign agreement deed; gets work order and starts the work.

5.19 REVOCATION OF TENDER

A tender becomes a contract if it is accepted. The right to reject one or all the tender without assigning any reason is the right of the owner or department. Similarly, before a tender is opened and accepted; bidder has right to withdraw his tender due to any reason known to bidder only. Section 4 and 5 of Indian Contract Act permits withdrawal of tender by the contractor. This is called as 'Revocation of Tender'. However it is permissible only before the opening and acceptance of tender and not later on. Once tender is opened and accepted, contractor has no right to get his tender revocated.

5.20 EXTRA ITEMS

Extra items are those which are not mentioned in the estimate and had to be provided as per the directives of the competent authority. Executive engineer in PWD/CPWD are authorised to sanction the execution of and the rates to be paid for extra item of works; not provided for in regular tender accepted by higher authorities by themselves or by their sub-ordinate officers subject to the following conditions :

1. The amount of an individual item does not exceed Rs. 10,000/-and total cost is within the amount upto which Executive Engineer is competent to accept.
2. The rate to be fixed; mutually agreed between engineer-in-charge and contractor if rate is not given in schedule of rate.
3. If no rate is available in current schedule of rate; it is to be analysed; before finally fixed.
4. If cost of any extra items exceeds Rs. 10,000/- approval from S.E. or other higher competent authority is required.
5. No extra item should be got executed from contractor on oral orders or in anticipation of sanction by the competent authority.
6. The rate of extra item be scrutinised by the divisional accountant before they are accepted and sanctioned by the competent authority.
7. No extra item rate list can be sanctioned against price work tender.

Many times contractor refuses to execute extra works if not included in tender and having lower rate in schedule of rates. Under such conditions either contractor is offered higher rates or as such new contract is required to be executed. These extra items are found due to :

1. Impractical planning and extra items are required to make construction possible.
2. Changes made by owner due to some reasons.
3. Due to unforeseen circumstances.
4. Certain items are not drawn on drawings by mistake and later on required to be executed.

5.21 ADDITION AND ALTERATIONS

During the process of preparation of drawings and estimates; it is always a good practice to take into account each item of that work accurately. There should be no ambiguity in specification and no omissions in drawing. As these, some time create problem in execution or become a point of dispute between contractor and engineer. However the engineer incharge reserves the right to make any changes in. Omission from, and addition to or substitution for; the original design drawings and specifications and instruction as are necessary in the opinion of the engineer in charge during progress of work and which may be given to the contractor in writing duly signed by engineer. Such changes are deemed to have formed as work included in the original tender and it is binding to the contractor to carry out such works. The time of completion if requied can be extended proportionately and rates for more additional items are given as per schedule of rates. However while making alteration in drawings in general and specifications in particular care should be taken so that there will not be considerable change in the cost of that work; further it may not be objectionable of contractor. In case of dispute the matter is generally referred to superintending engineer of that circle and is final binding on to the contractor. In case of any alteration; addition or substitution if employment of additional labour, material and equipment is required; the claim revision of rates must be specified by contractors within 7 days from the receipt of order. The Engineer in-charge may revise the rates if found justified. In no circumstance, however contractor has any right to suspend the work due to non-settlement of rates of items. Similarly in the case of curtailment of work, contractor has right to claim compensation as he is likely to loose his profit.

5.22 DEFECTS

It is responsibility of contractor to execute work as per drawings and specifications. Engineer incharge if finds any defective work not executed as per condition of contract; can order the contractor to dismantle and reconstruct the defective part. It is customary to contractor to dismantle that part and reconstruct it as per directive of the engineer in charge. No claim for dismantling and re-construction can be made by the contractor. In case contractor refuses to dismantle and reconstruct; engineer-in-charge has right to get the work dismantled and reconstructed on behalf of the department. The expenditure incurred will be deducted either from the running bill of contractor or from remaining bill of contractor or from security deposit as the case may be. This if required can be publicised to black list the contractor for further work. Further if the same is repeated; the department reserves the right to terminate the tender giving due notices to the contractor. Under such condition if the remaining work is less; department can take that work executed by employing labour or as the case may be and the amount can be debited from security deposit of the contractor. If remaining work is enough; department can get that work executed through some other agency (contractor) and S.D. of earlier contractor is forfeited.

5.23 LIQUIDATED DAMAGE

It is an amount of compensation payable by a contractor to the owner due to delayed construction. This amount of compensation is not related with real damage. In case, contractor fails to complete the work within prescribed time as mentioned in tender deed the contractors is required to pay to the owner a sum stated in the tender as liquidated damage for such default. It is not a penalty as imposed per day for delay. This liquidated damage does not relieve the contractor from his obligations and liabilities under the contract. In case if the work or premise is used by owner before its completion, the amount of liquidated damage is reduced in proportion in which the value of the part has been utilised. However this reduction is permissible only if owner or Government has issued the certificate of works occupancy.

5.24 UN-LIQUIDATED DAMAGE

This is known as ordinary damage having relation with the actual damage done. When a contract is broken, the party who suffers by such breach is entitled to receive from the party who has broken the contract.

5.25 ESCALATION OF RATES

Once the agreement deed is signed; work order is issued, contractor is required to complete job at specified rates in given project duration. He is not entitled to claim higher rates than those mentioned in tender deed, due to rise in market price of materials or labour. But with present inflation rate; it has become difficult to contractors to complete job, particularly when work is very large involving considerable time period. Hence to overcome this difficulty, contractor either quotes higher rates or asks for escalation. In absence of this many times contractor either abandons the work, stop the work or goes slow. Hence to overcome this; now-a-days contractor is allowed to take benefit of escalation and hence escalation clause is introduced in agreement in agreement deed. It is therefore now in practice to mention the rate of all important materials like cement; steel; wood; brick; pipes, fixtures in the agreement so as to decide escalation later on.

5.26 LETTING

As per contract deed and contract Act; a contractor assigned a work can not sub let or assign whole or part of that work to any other contractor without written permission from the engineer in-charge. This is permitted only under certain special condition when the original contractor has become ill or under gone some treatment which takes him unable to execute the job; however the responsibility of original contractor regarding quality work can not be avoided.

Sub-letting of work to petty contractor for some specialised job is a common feature. Generally main contractor assigns concreting work to some gangs specialised for the job; or electrical fittings, plumbing etc. are assigned to other contractor; such subletting is not in written form; but it is however permitted as the other party is not directly involved in contract deed.

5.27 INSPECTION OF WORK

The engineer in-charge of site is required to visit the site regularly and inspect the quality and progress of work and to instruct contractor time to time as and when required in case of defective construction. Engineer in-charge must ask the contractor to dismantle and reconstruct that defective part at his own cost and no claim should be made by contractor for this. In case of refusal of contractor, the engineer in-charge should get the work done and to recover the expenditure increased from contractor or can deduct this amount from his running bill. It is the responsibility of contractor or his representative to provide free access to department engineer at site as and when he visits. The engineer in-charge is required to maintain his daily diary; maintaining the work progress record.

5.28 LITIGATIONS

Arbitration means settlement of dispute between two parties involved; generally contractor and owner. This settlement is generally done not by court of law but by a few persons having complete knowledge of that work and deputed by parties involved. These person or persons appointed to settle the dispute are called 'Arbitrator' and process of setting dispute is called Arbitration'. Arbitrator gives decision on the basis of evidences available; inquiries made and his own skill and knowledge. His decision is generally accepted by both the parties and also incorporated into the decree of the court if required *i.e.* if earlier decision is not accepted by either of the parties and if matter is referred to court of law. Arbitrator has to take such decision which must be correct and acceptable by both the parties.

The essential requirement for a person to work as an arbitrator is that.

1. He is in a quasi judicial position, therefore, the first essential requisite in persons occupying that post is judicial impartiality and freedom from partiality and freedom from bias.
2. He should be honest; impartial and disinterested in any of the parties involved.
3. He must be qualified person having through knowledge of that subject matter under dispute.
4. He must be a person of high integrity and unshakable faith in justice.

Litigations in construction industries are increasing very large. Many of the problems can be solved by means of an Arbitration act; however, in most of cases they are not. Therefore litigations are increasing many fold. In most of the countries litigations and not arbitration is preferred. The main reason behind can be that arbitrators are also biassed; or the parties involved in may not be getting satisfactory results through a process of arbitration. In construction industry there are mainly two possible causes for litigation, bad construction or bad contractual arrangements.

The improper specification is one of the reason for arbitration which leads to litigations, A current practice to prepare specifications is cut and paste through computer. It makes so many mistakes that reaching to a amicable solution through arbitration becomes difficult if not impossible. Moreover specifications are a very important document of contract process which is given least attention. Many countries have prepared standard specifications, but these specification can not be followed for all type of construction works. An experienced contractor, especially, a wity one, will spot weaknesses in the specification that can be exploited.

There is another factor in the propensity to litigation–the relation between the engineer representing the designer on behalf of the building owner on the one hand, and on the other, the contractor. Under the contract, the engineer has to approve quite specifically much of that is done on the site, and has the power to direct the contractor to proceed in a particular manner, or even to remove work deemed to be unsatisfactory. This has to be well documented. All this is as it should be; but sometimes the correspondences between the engineer and contractor becomes progressively more acrimonious. At that stage, it becomes difficult to parties to have cool discussion to reach to a mutually acceptable solution. Threats are made, and eventually, the parties proceed to arbitration or litigation. In U.K. steps have been taken to obviate it. Many contracts now require the appointment of an adjudicator at the beginning of construction. As soon as there is disagreement; before matter develops into a formal dispute, the adjudicator steps in and takes a provisional decision. This allows the work to progress unhindered by soured relations. Later on, if either party is dissatisfied with the adjudicator's decision, it can proceed to arbitration.

5.29 SPECIFICATIONS

Specification embrace endless details, ranging over a scope which cannot possibly be converted in any text book. Certain funda-metial ideas have been dealt herewith in brief.

Legally the specifications form an integral part of the construction contract. It describes the work and materials to be used in construction in details. It indicates the nature, character and the form of the finished work and prescribe the special characteristics of the materials which can be used therein, more particularly about the data not shown on drawings.

A combination of drawings and specifications give the contractor the complete idea about the work he has to perform and materials he has to use in the construction.

Specifications in general should be definite and explicit. These should be lucidly written in the simplest possible language with clauses in a logical order. The description should be exact and complete A common fault in specifications is indefiniteness of description arising because of not having actual idea of work and materials is due to obscurity in the language used. Words like "Proper" or insufficient which do not give or "As far as possible" clear idea should not be used. Specifications should be written by an experienced engineer who has complete knowledge of construction further they should be concise as far as possible because fewer words will involve less risk of legal difficulties

Specifications are generally of two types:

(*i*) General or brief specifications or (*ii*) Detailed specifications.

General specifications states the clauses of work to be employed in various parts of the building or other work. These specifications only serve as the guideline to the person preparing estimate or defining the work. Hence though they are attached to the contract document but do not form the part of the contract document to be challenged in law.

Detailed specifications lay down the rules with regard the quality of material and workmanship. These form a part of the contract document.

QUESTIONS

1. Write a tender for construction of a building cost Rs.75.00 Lacs. Assume all other factors.
2. Name the factors that affect the economy in construction.
3. Write short notes on:
   (a) Security deposit.
   (b) EMD
   (c) Liquidated damages
   (d) Defect liability period
4. What do you mean by litigations in Contract? How these are resolved?
5. Explain the term arbitration in detail.
6. What is defect liability period and what is its importance in tender?
7. What is tender document, what precautions are necessary while preparing tender notice for any work?
8. Explain Indian Contract act briefly.
9. Write short notes on:
   (a) Escalation
   (b) Sub letting of contract
   (c) Termination of contract
   (d) Unliquidated damages
10. Explain the procedure of preparation of bill and its payment to the contractor in detail.
11. Write notes on:
    (a) Extra item in construction
    (b) Two envelop system
    (c) Advances to contractor
    (d) Breach of contract

Chapter

PROJECT MANAGEMENT TECHNIQUE (CPM/PERT)

6.1 INTRODUCTION

Project management, Project organisation, team concept, management by task force are all terms for one and the same form of organisation set up with the object of planning and implementing projects within a minimum of time and comprehensively in the general interest of a venture. The planning and control of complex projects is one of the major challenges to the managers of industry and business. The modern manager is baffled with the problem of completion of large and complex projects in the stipulated time and budget. Network techniques like CPM (Critical Path Method) and PERT (Program Evaluation and Review Techniques) aid the managers in planning, scheduling and control of large and complex projects and programmes keeping in view the constraints on the various resources required.

The Egyptian and Romans worked construction miracles in their days. Surviving ruins attest to the brilliance of their architecture, but of their construction planning and scheduling little. It can be thought that they might have solved their scheduling program by the "care a bigger whip" philosophy. History records much about the construction details, but littie records are available on the method of control. In the mid nineteenth century, at least one writer discussed a work-versus-time graphical representation very similar to today's bar charts. However, it remained for Henry L. Gantt and Frederick. W. Taylor in the early 1900s to popularize their graphical representations of work versus time. Their Gantt charts were the basis for today's bar graphs or bar charts. The work to Taylor and Gantt was the first scientific consideration of the problem of work scheduling. Although this work was originally aimed at production scheduling, it was readily accepted for planning construction and recording its progress. The bar graph is still and excellent graphic representation of activity. It is easily read and understood by all levels of management and supervision.

The interdependence of activities and the effect of resources limitations on constituent sections and on total programme could not be depicted on these charts. It was not possible to show the inter-relations and inter dependencies of different activities which control the progress of the project. Hence later techniques like curves, line of balance and mile stone developed and provided improvements. They took into account the inter-dependencies of activities. They made it possible to show inter-relations of different activities to a certain extent. But these methods did not give any contributory factors for practical utility and hence could not become popular.

With the growing complexity of works, rapidly changing technology and an intensified competitive environment, the problem of decision making assumed new dimensions. The time available for decision making shortened. Hence networks techniques like CPM and PERT developed to overcome these constraints. There are many other techniques such as GERT (Graphic Evaluation and Review Technique), RAMPS (Resource Allocation and Man Power Planning System) etc. are also becoming popular.

The use of the network techniques like CPM and PERT for construction, project planning, scheduling and control has grown fast in last few decades. It is an interactive time sharing programme oriented towards the educational application of the critical path method for solving project network problem.

Historically planning, scheduling and controlling operation in a systematic manner date from the early era of scientific management, especially with the introduction of Gantt charts. This graphical method designed, was particularly suitable for production programming and was basically time based plan on which actual progress could be recorded against targets along a linear scale. Probably, this was the beginning of scientific project management technique. The objectives of any project management techniques are:

1. To achieve a time cost balance relationship by proper planning and scheduling the project.
2. To equip the administrative with modern scientific techniques and methodologies available for implementation of projects and programs.
3. To develop capabilities for designing and operating project information and control systems for effective monitoring projects and programmes under implementation.
4. To impart organisational concepts and practices conductive to inter departmental co-ordination and administrative integration for effective implementation of projects and programmes.

CPM/PERT

The credit for introducing critical path method probably goes to Morgan B. Walker of E.I. Du Pont engineering services and James E. Kelly of Ramington Rand, who in 1957 developed a network and applied to a chemical refinery and found that a time-cost balance can be achieved by proper planning and scheduling the work through networks.

To get a quick field test, this method was applied to an overhead and maintenance shut down at one of the Du Pont Work. In this particular turn around, the product was a self-detonating neoprene intermediate. The turn-around had been done many times before and the average time for the shutdown was 125 hours. The CPM is credited with having reduced the shutdown period from 125 to 93 hours by better analysis through the management techniques. This was the starting, expediting and improving labour performance on critical activities, further cut in the shut down time. It was then reduced to 74 hours. This improved performance means an increase of a million dollars of production. Here onward many modifications have been made by many experts particularly in the cost and resource allocation areas. Inspite of so many benefits, it is difficult to realize that only a minority of the construction industry understands or utilizes this technique, it may be due to less publicity of the method.

In January, 1958 development of PERT, *i.e.* Program Evaluation and Review Technique was originated by Admiral W.F. Raborn of special project office of the Navy Bureau of Ordinance, who was charged with the overall management of the Polaris Missile Program. Earlier it was called as Program Evaluation Research Task then it was named as Program Evaluation and Review Technique. Many a times it is also called as Project Evaluation and Review Technique.

The basic difference between CPM and PERT is that the CPM uses only one fixed time for each activity hence is a deterministic network technique whilst PERT uses three times hence it is a probabilistic network technique.

Variation and modifications on CPM and PERT developed many other techniques like RPSM (Resource Planning and Scheduling Method) RAMPS (Resource Allocation and Manpower Planning System) MIS (Management Information System), RPC (Resource Planning and Control), GERT (Graphical Evaluation and Review Technique) etc. All these systems use the same basic concepts of CPM and PERT.

With continuous increase of activities of a project, it has become essential to use computer for network planning.

6.2 TRADITIONAL PLANNING

Historical planning scheduling and controlling operations in a systematic manner date from and early era of scientific manner date from and early era of scientific manager specifically with the introduction of Gantt Chart. This graphic was designed, particularly suitable for production programming and was basically a time based plan on which actual progress could be recorded against target along a linear scale.

Gantt Chart/Bar Chart

It was during 1910 that Mr. Henry L. Gantt developed a chart for U.S. Army for an improved and conducting the production of ordinance factories. It is a simple chart consists of bars of different activities which displays schedule as well as comparison between actual performance and the original schedule. Hence it is also known as bar charts.

In this type of chart, duration of an activity is represented by horizontal line on X-axis and activities are mentioned on Y-axis. Horizontal line of duration on X-axis are plotted to the scale. For each activity top portion of bar indicates total portion of that activity which has been completed whilst lower portion of bar indicates the duration of the activity. A Gantt chart is shown in Fig. 6.1.

Activity	*Time duration*
A	10 days
B	20 days
C	15 days
D	8 days

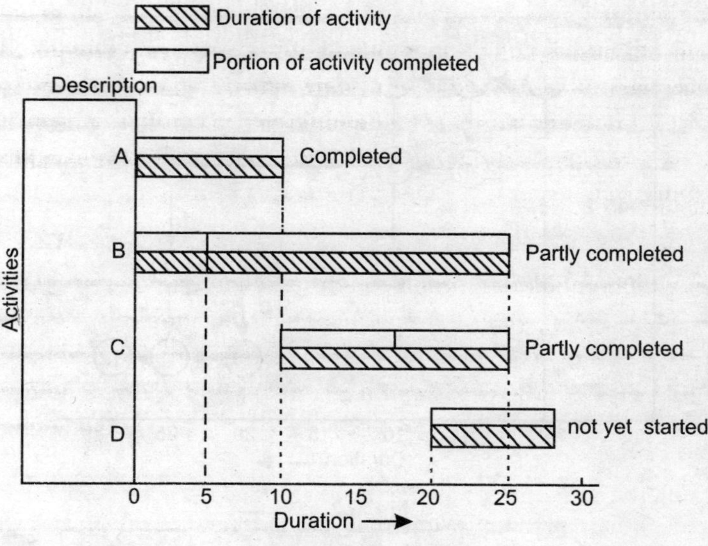

Fig. 6.1.

Salient features of Gantt/Bar chart

1. These charts are simple to prepare, easy to understand and can be drawn quickly.
2. Each activity is shown separately. Further comparison between actual performance and original schedule can be made very quickly hence modification can be carried out easily when required.
3. Inter-relationship of different activities can be done to some extent.

Along with some advantages as mentioned above, it has certain deficiencies also :

1. Inter-dependency of activities and their sequence is not defined. It is one of the major drawbacks with this chart.
2. It does not indicate the effect of any work behind or ahead of schedule on the project completion time.
3. It does not indicate the progress of the whole work at a glance. Further it does not reflect the tolerances in the duration times estimated for various activities.
4. It is suitable only for simple and small jobs.

Even with these limitations and certain deficiencies, Gantt/Bar charts are used to small jobs even today. These are also used at top level management to have on overall idea of project progress under chronological coincidences

Mile Stone Chart

This chart came as an improvement over Gantt/Bar chart. It was developed during 1940. In this chart a project is divided into activities which are further divided suitably. Events are arranged in chronological sequence and connected by means of arrows according to their logical sequence hence becomes a network. A mile stone chart for four activities *A*, *B*, *C* and *D* is shown in Fig. 6.2.

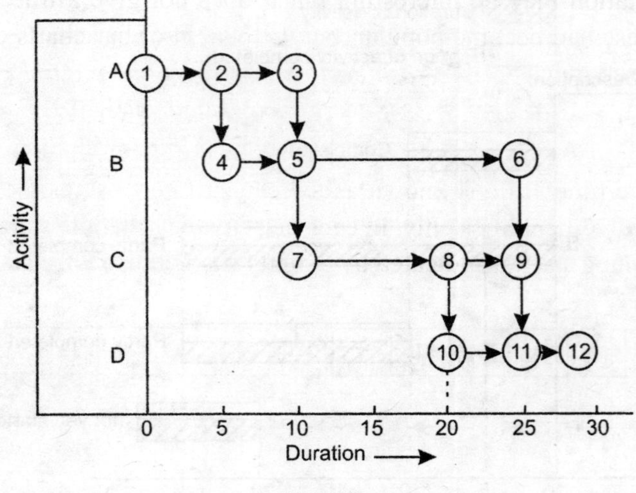

Fig. 6.2.

Activity *A* has been divided into 1-2 and 2-3 similarly activities *B*, *C* and *D* have been sub-divided. It indicates once 1-2 part of activity *A* is completed part 4.5 of activity *B* can be started and hence the project will be completed at (12). These circles are called as nodes.

These charts also have certain deficiencies like Gantt chart hence not used now-a-days.

S-Curve

The bar chart often suffers from morning-glory-complex. It blooms early in the project but found no where later on. When the project begins to take shape. These bar charts do not lend itself to planning revisions.

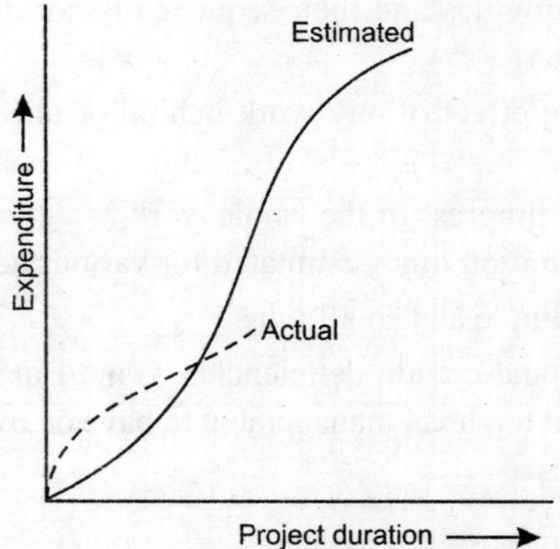

Fig. 6.3. S-curve expenditure vs durations.

Although progress can be plotted directly on the schedule bar chart, the S-curve has become more measuring progress. The usual curve consists of two plots : scheduled expenditure *vs* time and actual expenditure *vs* time. Similar curves can be prepared for man-hours, equipment and material acquisition, etc.

Though the presentation may be interesting but it does not give a true indication of project completion. Though these had become popular, but the benefit of bar charts could not be ignored. As such these are not used much.

Squared-network

A network drawn on the time scale is known as the squared network as the activities are shown mostly by horizontal and vertical lines only. In each activity the schedule start and finish are joined by solid lines, the remaining stack is indicated by dotted line. These are used at middle and top level management.

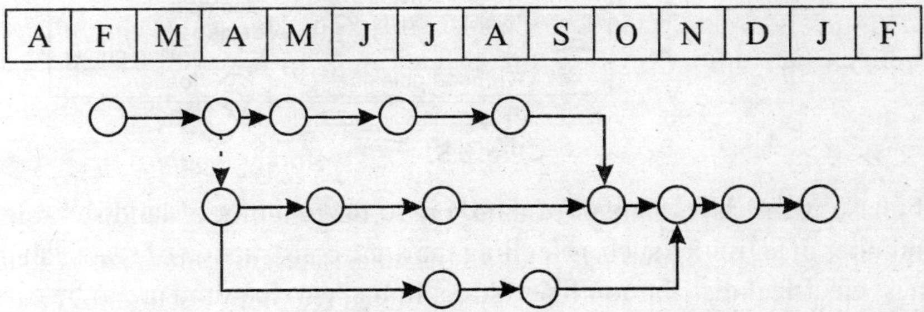

Fig. 6.4. Square network.

6.3 NETWORK PLANNING

The difference between success and failure, profit and loss, depends upon the outcome of well laid plans. The degree of details incorporated in the plans, varies considerably. As many persons are involved. Also the problem of communication of ideas from one individual to the other assumes importance. Therefore, it becomes necessary to place reliance upon formalized systems, written messages, codes and symbols to ensure that the meaning does not change as the information is passed on but represents what is to be done and the past each task play in the whole project.

The network planning concepts where plans are illustrated by arrow diagrams goes a long way towards meeting these requirements. It offers an integrated management control system, eliminates ambiguity and reduce misunderstanding between working groups. It assists all levels of management to define the work to be carried out, produce better work schedules based on knowledge of resources required and resources available, decide the best way of applying resources to achieve project cbjectives, monitor progress and detect in advance the points where delay would be caused. The Military Engineer Services have made it obligatory that all building projects are to be planned on the network technique. Though the network is simple to understand but without training and enough practice people find it difficult to construct network diagrams specially for projects where the inter-dependencies of activities is complex.

In general, any building project comprises three distinct phases. The first is the pretender planning which consist of selection of a suitable site and then its acquisition. After its acquisition, the planner prepares a layout plan showing the road pattern, the essential building, parts and open spaces etc. required. Water supply, sanitary requirements, electrification and all details drawings are prepared depending on project size of project. Hence a network for this can be standardized.

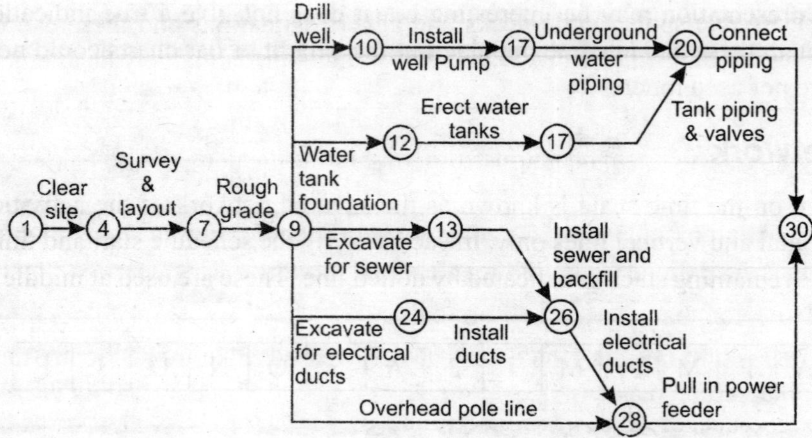

Fig. 6.5.

The second phase is the development of land, *i.e.* to make a plot of land suitable for building construction, a number of activities such as levlling the area, construction of roads, laying of sewers, water supply lines, electrical distribution lines, street lights, etc. required are to be carried out. The ingenuity in planning the sequence of these activities goes a very long way towards a satisfactory execution of the project both with respect to time and cost. In planning the development, the main consideration is to co-ordinate the working of various services in such a way that the construction of building can be started at the earliest and preferably go on side by side with the developmental work. A typical network for development of land for building construction project is shown in Fig. 6.5. This is a broad based activity breakdown network to illustrate the importance of pre-planning and that of network.

The idea of this network is to clearly show the inter-relationships and inter-dependencies between the various organizations. Hence as standard network for site development has been shown.

The third phase is the construction of building which comprises of several activities. The best method for planning such projects is to show detailed network diagram which will show the inter-relationships and inter-dependencies of the various activities and help in planning the resources of men and material. Here also a standard network can be developed as most of the activities in construction works are fixed. There may be change in quantity of work and duration, hence standard networks for this can also be developed.

6.4 ADVANTAGES OF NETWORK TECHNIQUES

The planning by network compels logical understanding of the project from start to completion. Since the planning of sequence and scheduling the time are done distinctly, full concentration is possible to be given on planning the sequence which is very important for achieving speed and economy for the project. The sequence of observations is established hence a time cost balance can be achieved by application of network techniques to complex project.

It informs an idea about the time status, cost status. Responsibility of delay for the project or increase in the cost of project is pin pointed. Managerial action is thus focused on exceptional problems, contributing to more reliable planning and more effective control. Network analysis helps in locating trouble spots, bottle-necks and adverse situations and these can be avoided by proper planning. Extra expenditure due to delay in completion can largely be avoided.

One of the attractive fracture of CPM/PERT is that they are useful tools for management at various levels within an organisation and at different times within the life cycle of a project.

Hence the benefits of network technique are :

1. The force the execution to plan the details.
2. They make it easier to analyse the effect of alternative decisions.
3. Helps in communication and co-ordination.
4. Helps in identifying the effects of delay and bottlenecks to enable the managements to take timely corrective action.

6.5 APPLICATION OF NETWORK PLANNING IN INDIA AND ITS LIMITATIONS

In the present stage network techniques are oriented only for training and education and is of little use for planning and scheduling real civil engineering projects because of size of the network. It may be due to large number activities which require many days for manual calculations which is a monotoneous job to certain extent. But by doing development and analysis by computers ; and developing standard computer programmes, its proper use can be made.

There are only a few projects which could be completed within the prescribed time period. The reason for not completing projects within the prescribed time may be like poor communication, incorrect use by management, shortage of trained practitioners, inadequate training facilities etc.

So far very little attention has been given for reducing the duration of the project and the cost of the project. The delay of the project means wastage of time, money and loss of production. The proper application of network technique and its systematic progress can bring down the project duration by about 15 to 20 per cent of substantial benefits.

This could not be tried in factories also so far. A few production units have used these techniques and have found it very successful and useful. But still sufficient attention is yet required to be given.

Network techniques may be used for diverse kind of projects and programmes such as

1. Construction of irrigation and power projects.
2. In the construction and erection of new factories and for expansion of existing factories.
3. For annual shutdown and maintenance works.
4. For production planning.

6.6 HIERARCHY OF NETWORKS AND MANAGEMENT

The networks are evolved showing different degree of details to suit different levels of management. Normally there will be 3 or 4 levels of networks/bar charts to aid different levels in management.

1. Milestone charts derived from networks for top management.
2. Master control network for chief executives and senior managers.

3. Plant wise, section wise and discipline wise networks for middle and junior levels of management.
4. Bar charts derived from networks for supervisory staff.

6.7 NETWORK TECHNIQUE

The main object of any project management is to minimize the capital expenditure consistent with good performance and reliability of the plant and machinery installed and to commission the project in the short possible time.

The use of the network techniques like CPM and PERT for construction, project planning scheduling and control has grown fast in the last few years. It is an interactive time sharing program oriented towards the educational applications of the CPM and PERT for solving project network problems. Achievement of a favourite time cost balance is possible only by proper planning, careful scheduling and co-ordination of labour, equipment and contractors and by maintaining the supply of materials properly. Network technique is the only method in the present circumstances which can control all the aspects and it cannot be foreseen that many other new methods have developed a few more new methods will emerge and replace these methods in the near future but CPM/PERT Techniques are still working effectively and efficiently.

Whereas PERT is normally applied to projects under research, the CEM deals Projects of known magnitude.

As for a critical path method is CPM is there are two methods to draw network.

1. Activity on Node (AON) method
2. Activity on Arrow (AOA) method

In cite engineering construction, AOA method is most popular whenves AON is also equality powerful tool. The reason may be availability of computers to resolve any complicated problem. AOA can be solved by use of computers easily and therefor dealt in detail AON is also explained briefly to have an idea, its advantages, deficiencies etc. later on

6.8 FUNDAMENTAL OF NETWORKS TECHNIQUE (CPM)

A "network technique" is a flow diagram consisting of the activities and events that must be accomplished to attain the program objectives' showing their planned sequence of accomplishments, inter-dependencies and inter-relationship.

"CPM" originally "Critical Path Method" "may be named "Complete Project Management" technique as the primary object is to complete the project by proper management of all the activities and events. It is a cyclic, directed and non-stochastic process deals with fixed time duration for each activity.

6.8.1. Activity

The backbone of the CPM is a graphic model of project. The basic component of this model is an arrow which represents one activity in the project.

It is defined as a time consuming task to be executed in the project for execution of a truss, fixing shutters, oiling an engine etc. It is denoted by an arrow which connects two nodes in the form of circles as shown in Fig. 3.6. It is not being a vector quantity not drawn to the scale.

Fig. 3.6 indicates that A is an activity which will consume 6 units of time duration for its completion. Tail of this activity represents the start of the activity and head the finish of the activity. Circle indicates node numbers. This activity A has node number as 4-5. These nodes of any activity must be in increasing order. Preceding node is called *I* node and succeeding node as *J* node.

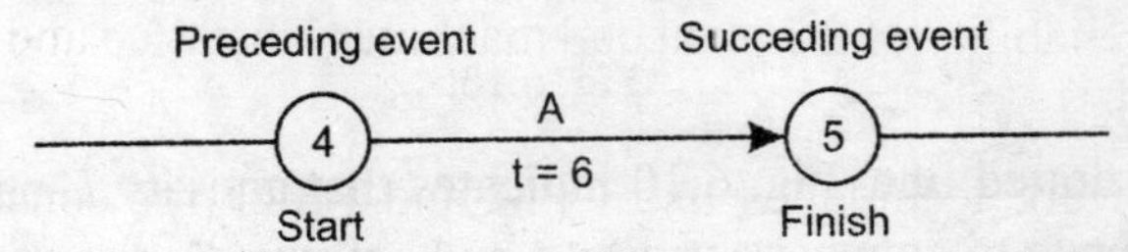

Fig. 6.6.

The arrow is always from left to right as shown above and the length of the arrow has no significance. These arrows are arranged in logical sequence in which the activities of the project are to be accomplished. Activities are classified as

(*a*) **Sequential activities.** Where the starting of an activity is dependent upon the completion of some other activity (Fig. 6.7)

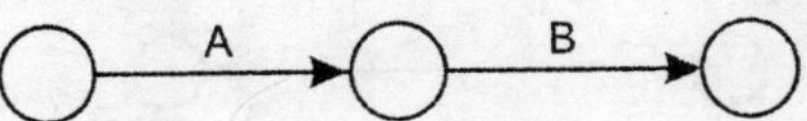

Fig. 6.7. Sequential activities.

(*b*) **Concurrent activities.** Activities which are independent and can occur simultaneously are called concurrent activities (Fig. 6.8)

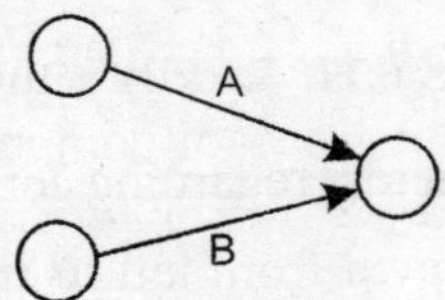

Fig. 6.8. Concurrent activities.

(*c*) **Pre-requisite activity.** Activity which immediately precedes the one being considered.

(*d*) **Post-requisite activity.** One which immediately follow the one being considered (Fig. 6.9)

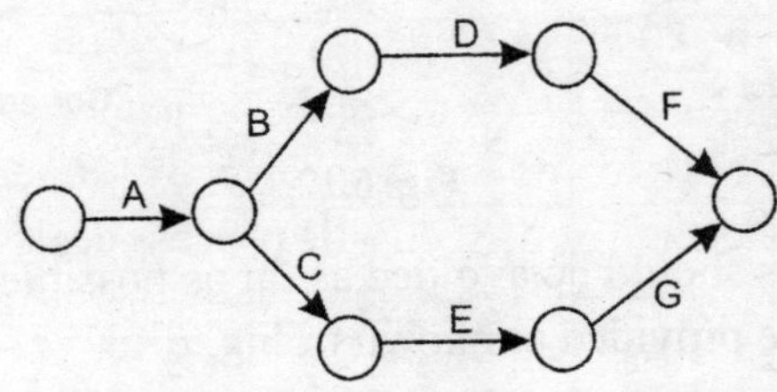

Fig. 6.9.

In Fig. 6.9, *A* is pre-requisite activity to activity *B* and activity *C* similarly activity *B* and activity *C* are post requisite activities to activity. *A*.

(*e*) **Dummy activity.** An activity with zero duration is called dummy activity. It simply represents the interdependencies of activity. It helps keeping the logic of the network intact.

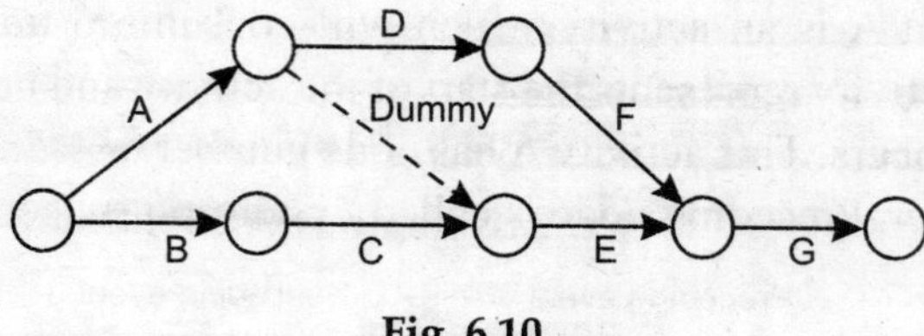

Fig. 6.10.

It is represented by dotted line. Fig. 6.10 indicates that activity *E* not only depends on *C*, but it also depends on *A*. Hence to connect activity *A* and activity *E*, dummy activity has been drawn whose duration is zero.

Dummy activities should be avoided as far as possible because they create confusion.

(*f*) **Ladder activities.** Certain types of activities in a project such as laying of underground pipelines, underground cables, manufacturing of parts/products can be conveniently represented by ladder type activities are started such that on activity is started after partial completion of the other.

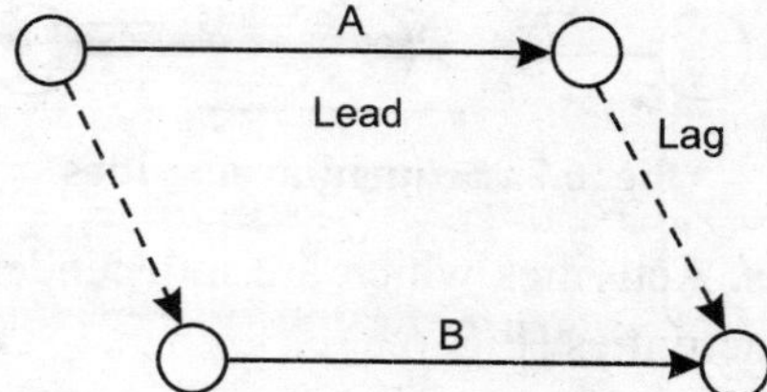

Fig. 6.11. Ladder activities.

While drawing a network certain factors regarding activities representation must be considered.

1. Activity arrows should be drawn from left to right always. Backward arrows should be avoided as they are confusing.

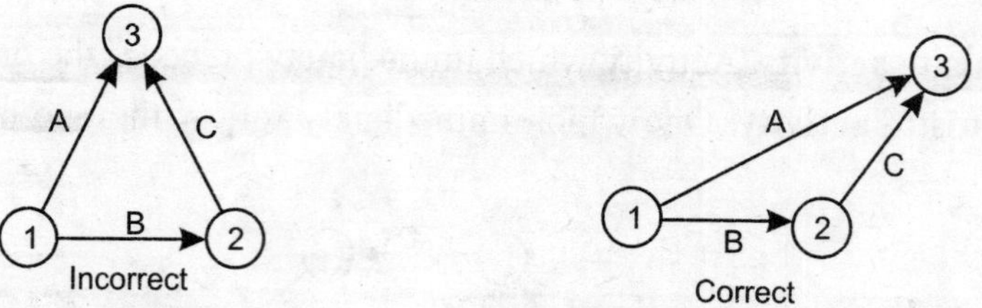

Fig. 6.12.

2. Crossing of activities should be avoided as far as possible (Fig. 6.13). If it is not possible then hooks should be provided as shown in Fig. 6.13

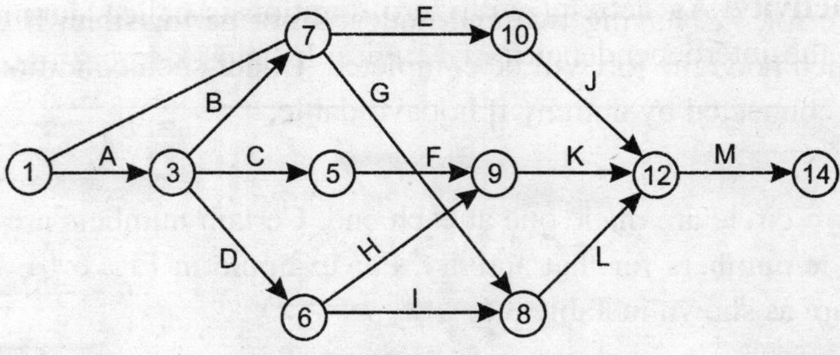

Wring representation.

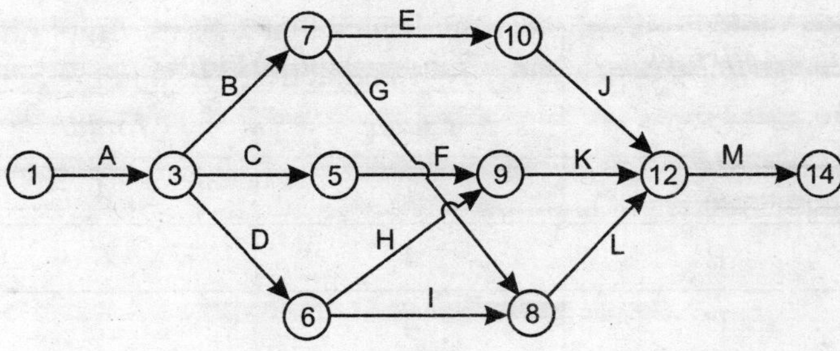

Fig. 6.13. Correct.

3. The node numbers must be in increasing order. Further a few numbers should be omitted as shown in Fig. 6.13. It is simply for the convenience that if any activity forgotten earlier if required to be added, it will be possible then to allocate node number for this additional activity in increasing order.

Many a times these arrows, indicating activities, are drawn in double line. If done so, the double line arrows help in making the progress of work during control monitor phase.

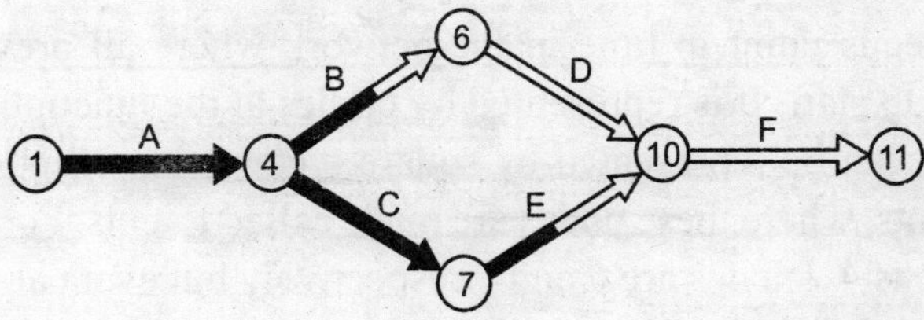

Fig. 6.14.

In Fig. 6.14, each activity has been shown in double line. This figure further indicates that activity *A* and *C* have already been completed and activity *B* and E are in progress. But it is certainly tedius to some extent.

4. There should be no formation of loop as shown in Fig. 6.15. Such mistake occurs mainly in PERT network or in very large network.

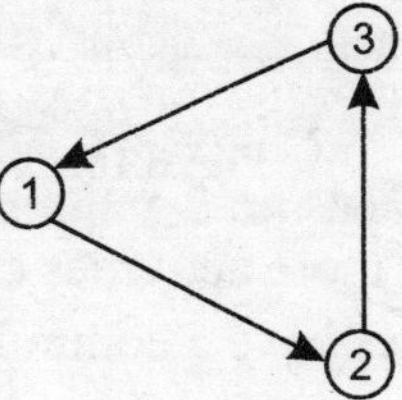

Fig. 6.15.

5. Denoting error, *i.e.* having two end nodes is not permissible. It does not give an idea that to which node the job will be completed. Under such conditions these two activities should be connected by dummy if not avoidable.

Node Number

For each activity, two circle are made one at each end. Certain numbers are given in these circles which represent node numbers for that activity. For example in Fig. 6.14. The node number for different activities are as shown in Table 3.6.

Table 6.1.

Activities	*Node Number*	
	I node	*J node*
A	1	4
B	4	6
C	4	7
D	6	10
E	7	10
F	10	11

It has been further illustrated in next article *i.e.* Event.

Event

'Event' is an instantaneous point in time in the network where all previous jobs are completed and following ones are to start. It is represented by circles at the junction of activities as indicated by arrows. However. it can be represented by square rectangular, elliptical etc. These circles are called as nodes and points, where these nodes are made called Events. For example in Fig. 6.14. for activity *B* and *C*, *I* node is 4 *J* nodes are 6 and 7 respectively but event at 4th node is a point which indicate that activity *B* and *C* cannot be started unless preceding activity *A* is complete. It indicates what are the activities preceding to and what are, the activities following to this event.

The starting event is also known as the tail end event or proceeding event and the ending^event is known as head event or succeeding event (Fig. 6.16).

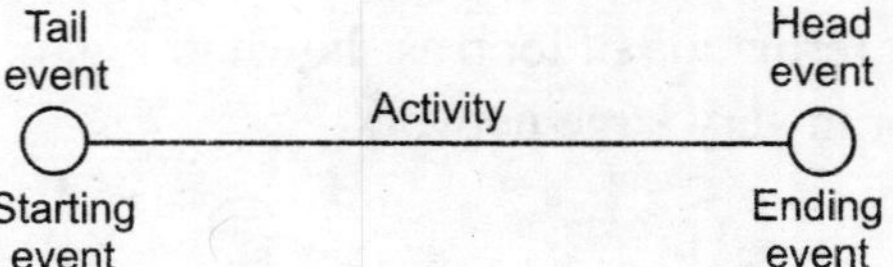

Fig. 6.16.

A 'merge event' exists when two or more activities converge at single event. Similarly a "burst event" exists when two or more activities have a common beginning event Fig. 6.17.

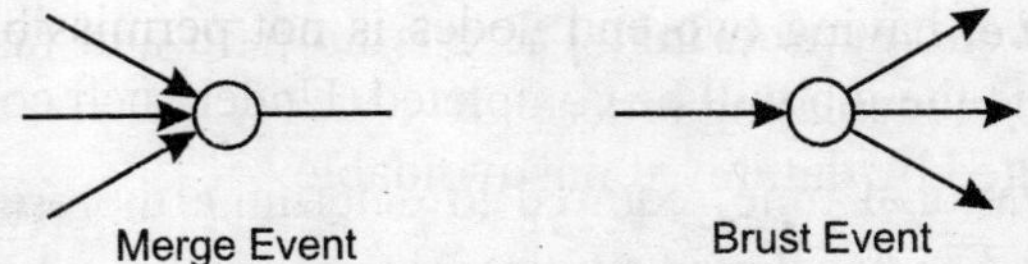

Fig. 6.16.

An event may be a 'merge and burst' event simultaneously as with respect to some activities it can be a merge event and with respect to some other, it may be burst event (Fig. 6.18).

A "Milestone Event" represents an important stage of completion of the project and is represented as follows (Fig. 6.19).

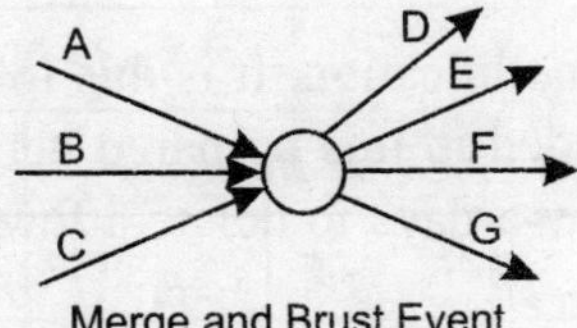

Fig. 6.18.

Rest Code	Event No.	Cost Code
Description		
TE	Activities	TL

Fig. 6.19.

"Interface event" provides the link for the sub-networks and the main network and represented on event node as shown in (Fig. 6.20).

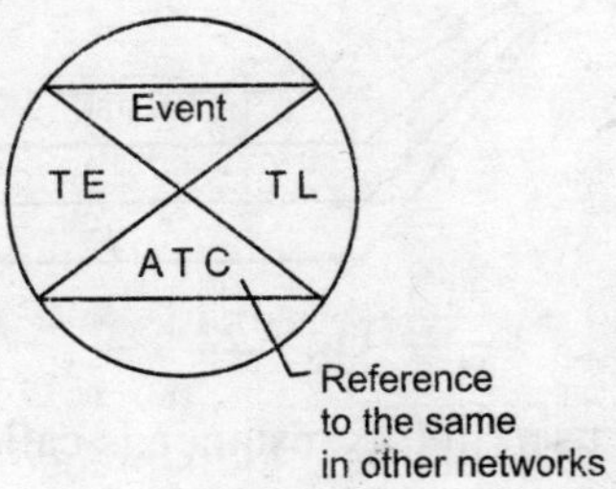

Fig. 6.20.

The normal symbol used to represent an event is a circle, but for specific events other symbols can be adopted (Fig. 6.21)

Fig. 6.21.

Fulkersons rule of event numbering. There is no hard and fast rule for event numbering except that these should be in increasing order.

To reduce the computational time required to determine the critical path by use sequential algorithms, Fulkerson D.R. has developed a numbering algorithm which proceeds as follows :

1. Define an initial event as one which has emerging activity only. Lebel it 1.
2. Delete all activities from numbered events. This creates at least one initial event. Lebel it 2.
3. Repeat this step until all events have been numbered in increasing order.

Then stop.

But this method needs certain modification. It is due to that, if some activity omitted earlier by mistake and if required to be added to this prepared network, all event numbers will have to be changed. Hence it is practice now-a-days to delete a few numbers as shown in many network earlier.

Bus Bar

As the project starts a number of activities usually origin. The result often looking like a traffic jam as shown in Fig. 6.22.

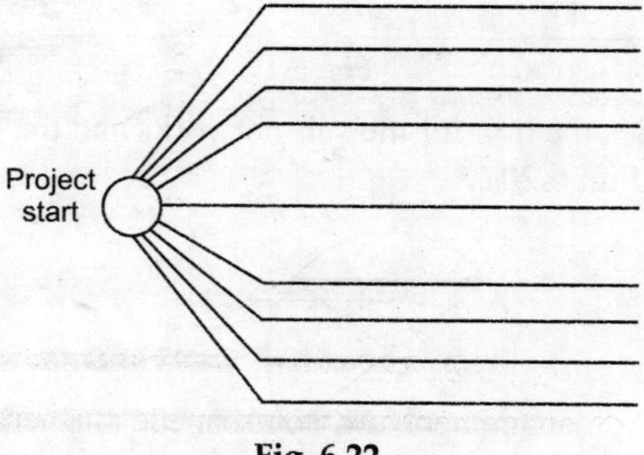

Fig. 6.22.

A useful technique sometimes used in this instance is called the bus bar as shown in Fig. 6.23.

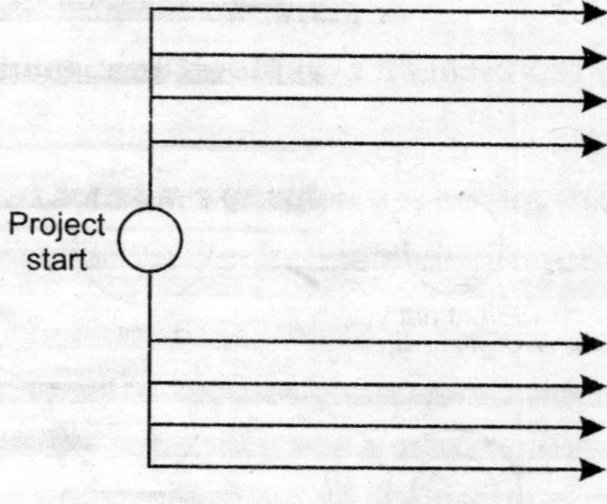

Fig. 6.23. Bus Bar.

The advantage of this technique is to reduce unproductive congestion of the network at project start or at project finish. Though it creates intersection of arrows at points which are not events and may create a confusion but still it is becoming popular because easy to understand.

Network

Network is a schematic graphical representation of all the activities denoted by arrows and events, comprising project in order of precedence and shows their inter-relationship. It is also called as flow diagram. Fig. 6.24 shows a network and events. Table shows the detail.

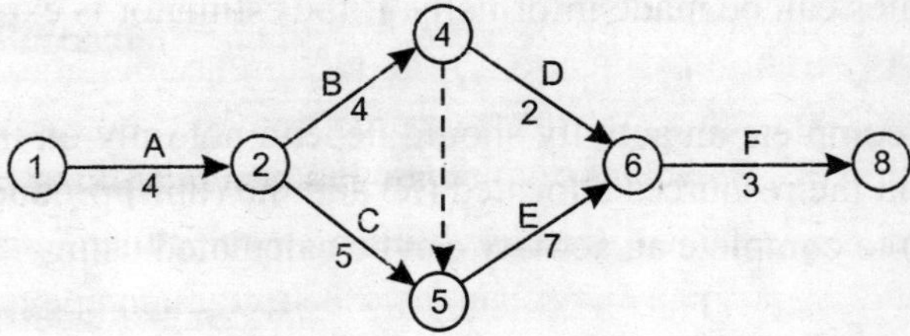

Fig. 6.24. Flow diagram (Network)

Table 6.2.

Activity	*Nodes*		*Duration in days*	*Activity which*		
	I node	*J node*		*Preceds*	*Follows*	*Occur simultaneously*
A	1	2	4	none	B, C	–
B	2	4	4	A	D, E	C
C	2	5	5	A	E	B
D	4	6	2	B	F	–
E	5	7	7	B,C	F	–
F	6	8	3	D, E	none	–

In this figure encircled 2 indicate an event. At this event, activities *B* and *C* cannot be started unless activity *A* is completed. Hence six number of events are their in this network right from\ starting to finish work. Each event gives an idea about the work already completed and next work to be started.

"Duration" of an activity is the estimated time required to complete an activity. It is generally denoted below the activity in the network.

Time Estimates

The preparation of the arrow diagram furnishes a number of advantages, including

1. A disciplined method of preparing a plan
2. A method that considers the project in detail.

The time dimension used for CPM analysis is project time. Any convenient unit can be used but it must be consistent throughout the network. The unit usually used is either minutes, days or weeks. In civil engineering works, the common unit is days or weeks. In CPM network we generally use full time units such as 3 days, 8 days or 6 minutes or 3 weeks etc. not like 4 days and 5 hours or 6.4 minutes. To estimate the time duration for an activity, the estimator assumes a normal work crew, depending on type of work and experience.

One method of estimating activity duration is to estimate quantity of the activity and divide that figure by the assumed work crew. It is of course, not possible to make an accurate time estimate for the entire project, but when the project is properly broken down into its discrete activities. Very accurate time estimates can be made informally, if the estimator is experienced in the type of project being planned.

The time required to complete an activity should depend not only on the quantum of work to be xecuted (Q), but also on the resources allocated (R) and the (unit) productivity of the resources (P). The time required (T) to complete an activity can be calculated using

$$T = \frac{Q}{R \times P}$$

Statistical data stored over a period of time for a number of similar construction projects plays an important role in the estimation of the three parameters Q, R and P, and consequently helps in estimating the duration of an activity. The duration for an activity can be estimated using several approaches. For example, we can use "time-study" approach. In this approach, the time,

$$\mathrm{T} = \frac{Q}{p \times n}$$

where Q = total quantity of work, p = productivity factor, and n = normal size of crew. Note that Q, p and n are all dependent on the availability of the information or data. At the time of estimate, information on all the three variables would be difficult to get.

In practice, instead of trying to accurately estimate the duration, a reasonable duration is allocated and commitment obtained from the agency that will be held responsible for its implementation. If the duration is not acceptable to any agency, then it may be changed.

The project time estimate for the activity is usually referred to is the activity duration and is shown below the arrow.

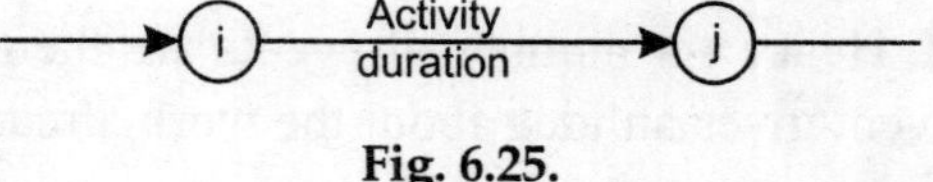

Fig. 6.25.

Many a time, it is not possible to forecast the time required for an activity, particularly in earthwork in cutting or foundation excavation etc. In such cases it is quite proper to add some contingency time. More the uncertainties of job, more the contingency time included. A proper break down of the project into various activities reduces the contingency time.

An example to time estimate is given below in tabular form. Crew size assumed is also shown below :

S.No.	Activity	Quantity	Particulars	Duration
1.	Concreting 1 : 2 : 4	10 m^3	@ 2.5 m^3/day 2 masons and 6 helpers	4 days
2.	Brick masonry in 1 : 6 cm	10 m^3	@ 2 m^3/day 2 mesons and 6 helpers	5 days
3.	13 mm thick cement plaster in 1 : 4 cement sand mortar	100 m^2	@ 10 sq m/day by 1 mason and helpers	10 days
4.	Overhead line	450 metre	@ 50 m/day on 16 polesby 3 wiremen	9 days

Thus it is possible to find the duration required for all the activities in a project. After computation of activity duration is time estimates for different events and activities are computed. They have been dealt in subsequent paragraphs

Event Times

For each event two types of times are expressed depending on the occurrence of that event. These times are Earliest occurrence times or Early occurrence time (*EOT*) and Latest occurrence time or Late occurrence time (LOT). *EOT* of any event is defined as the first day on which no more works need to be done on any of the activity entering into that node or event (events are denoted by node number). Similarly *LOT* is that first day on which no more work should be required on any of the activities entering into that node if no delay is to be caused in completion of the project.

In case of more than one activities leaving an event *EOT* is taken maximum of all activities whilst *LOT* is taken minimum of all activities leaving that node. For critical-event *EOT* = *LOT*

EOT for an event = *EOT* for preceding event + Duration

LOT for an event = *LOT* of following activity – Duration.

It has been illustrated by an example later on.

Following are the methods employed to calculate *EOT* and *LOT* of the events :

Forward-pass-calculations. To find the duration of the project we have to find the earliest occurrence time of the last event in the network. This is calculated by forward pass calculation. In the forward pass calculation the EOT for each event in the networks is to be calculated.

The first event in the network occurs at zero time and the subsequent event occurrence time can be calculated with this. The earliest event occurrence time for any event in the network is equal to the EOT of the preceding event plus the duration of the intervening activity.

$$\text{EOT}_{(J)} = \text{EOT}_{(i)} + t_{(ij)}$$

In the case of merge events, EOT is calculated by all the possible paths and whichever value is maximum that is considered as the EOT for that event. By this formula, EOT for all the events can be calculated and the project duration can also be calculated.

Backward-pass-calculations. This helps in finding to what extent the occurrence of different events can be delayed without affecting the overall duration of the project.

Unless there is an imposed duration on the project, for the last event in the network, the latest allowable event occurrence time is taken as the same as the earliest event occurrence time, where there is an imposed date or duration that is considered as *LOT* for the last event in the network.

The latest allowable event occurrence time for any event in the network is the latest allowable event occurrence time for the succeeding event minus the intervening activity duration. Denoting the same symbolically.

$$\text{LOT}_{(i)} = \text{LOT}_{(J)} - t_{(ij)}$$

In the case of burst events where more than one activity is leaving, the value is calculated by all the possible paths and the value whichever is minimum is taken as LOT for that event.

Using above formula LOT for all the events can be calculated.

This further has been illustrated by an example.

Example. Draw an arrow diagram and find critical path, *EOT* and *LOT* of all the events.

Table 6.3.

Activity	***Duration***	***Activity which***	
		Follows	***Precedes***
A	4	None	B, C, D
B	6	A	E
C	8	A	E
D	7	A	F, G
E	6	B, C	I
F	3	D	I
G	5	D	H, J
H	8	G	I
I	6	E,	F, H
J	5	G, I	None

Solution. A network drawn is shown in Fig. 6.26 and *LOT* are shown in tabular form.

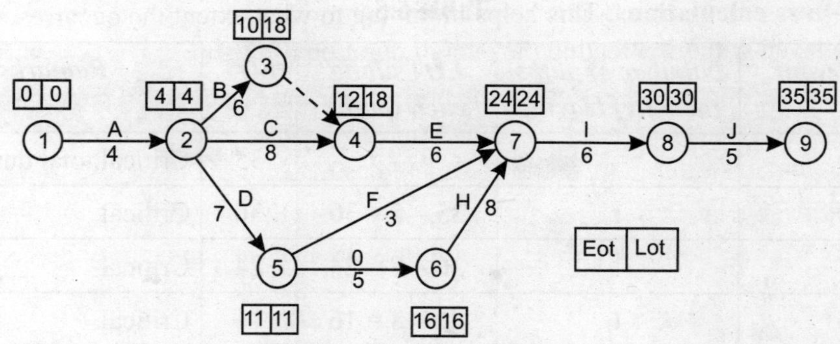

Fig. 6.26.

Critical path —1 — 2 — 5 — 6 — 7—8 — 9

Critical Activities — A, D, G, H, I, J

Critical period — 35 days

EOT for different Events are as shown below in table 6.3

EOT for different events are

Table 6.4.

Node/Event	*Number of paths*	*EOT of all paths*	*EOT of event*	*Remarks*
1		0	0	Critical
2	1	0 + 4 = 4 (1 – 4)	4	Critical
3	1	4 + 6 = 10 (2 – 3)	10	
4	2	10 + 0 = 10 (3 – 4) 4 + 8 = 12 (2 – 4)	12	Maximum
5	1	4 + 7 = 11 (5 – 5)	11	Critical
6	1	11 + 5 = 16 (4 – 7)	16	Critical
7	3	12 + 6 = 18 (4 – 7) 11 + 3 = 14 (5 – 7) 16 + 8 = 24 (6 – 7)	24	Maximum Critical
8		24 +. 6 = 30 (7 – 8)	30	Critical
9	1	30 + 5 = 35 (8 – 9)	35	Critical duration of project

LOT for difference Events are

Table 6.5.

Node/event	*Number of paths (activity) leaving*	*LOT along each (path)*	*LOT*	*Remarks*
9	–	–	35	Critical,total duration
8	1	35 – 5 = 30	30	Critical
7	1	30 – 6 = 24	24	Critical
6	1	24 – 8 = 16	16	Critical
5	2	24 – 3 = 21		Minimum
		6 – 5 = 11	11	Critical
4	1	24 – 6 = 18	18	
3	1	18 – 0 = 18	18	
2	3	18 – 6 = 12	12	Minimum
		18 – 8 = 10	10	–
		11 – 7 = 4	4	Critical
1	1	4 – 4 = 0	0	Critical

1. Slack. It is a measure to find out to what extent the event occurrence time can be delayed without affecting the project duration or end date denoted by symbol *S*.

$\therefore \quad S = LOT - EOT$

The slack for any event may be positive, zero or negative.

2. Activity times. For the activities in the network, the following times are calculated :

EST – Early start time or Earliest start time

EFT – Early finish time or Earliest finish time

LST – Late start time or Latest start time

LFT – Late finish time or Latest finish time.

EST is the earliest possible time, an activity can start without changing the sequence of the activities in the network. *EFT* is the earliest time by which an activity can be completed. Similarly *LST* is the latest time by which an activity can be started without delaying the project duration and *LFT* is the latest time by which an activity can be completed without delaying the project.

An activity can commence only when its preceding event is achieved and that an activity must finish before its succeeding event can be achieved. Thus the earliest possible start of an activity is dictated by the earliest date of its preceding event and the latest possible finish is obtained from the latest date of the event succeeding the activity.

Following equations show the relationship, between various activity times :

EST of an activity = *EOT* of tail event

EFT of an activity = *EST* of that Activity + Duration

LFT of an activity = *LOT* of head event

LST of an activity = *LFT* of that activity – Duration

Floats. It is a measure to find out to what extent an activity can be delayed without affecting the logic of the project or its total duration.

Four types of float have been identified. They are total float (TF), free float (FF), Interfering float (IF) and independent float (FIND).

Total float (TF) is the difference between the maximum time allowed for an activity and its essential duration. Hence it is the time by which a particular activity can be delayed without causing any effect on the completion period of the project. The term free float (FF) is the amount of time by which the start of the activity may be delayed without interfering with the start of any succeeding activity. Interfering float (IF) is the difference between TF and FF. If full use of IF is made, subsequent activities become critical and if it is exceeded, the total duration of the project will increase. Independent float is the earliest starting time for the following activity minus the latest completion time for the preceding activity minus the duration of the activity.

Hence TF = LFT – EFT

or = LST – EST

FF = EOT of following event – EFT of activity or EST of following activity – FET of activity (If there is burst, max. EST of all the activities should be taken)

IF = TF – FF

F_{Ind} = EST of following activity – LFT of preceding activity (maximum of all) — Duration of this activity.

Floats are used in scheduling to show where flexibility exists. They have more meaning while discussing a large one-time project such as building a highway, developing a new aircraft or missile etc.

Of these four floats only two are found practical. Total float and free float. other float do not prove useful in actual usage, its use is no longer recommended.

3. **Calender time.** Project time is just sufficient for manual computation. However, when the information is put to use in an actual project, we are interested in dates. A project calender which converts project days into calender dates is more useful. As such these days/dates are converted to calender dates/time.

Errors in Network Planning

1. **Looping Error.** In construction of network, some times, such a situation arises when all activities are inter-related to form a closed loop or cycle as illustrated in figure. It is an impossible situation to represent the activity. It is called looping in networks and is always avoided. Thus one property of a correctly constructed networks diagram is that it is non-cyclic. (Fig. 6.27)

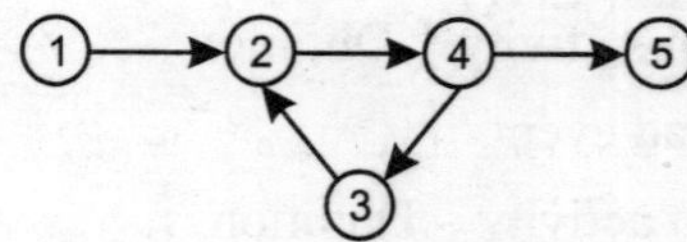

Fig. 6.27.

2. **Dangling Error.** In networks planning their is always a unique start and unique end event. Some times such situation arises that two or more activities terminate at different events. This type of situation is called Dangling in networking. Therefore it is essential to take care while drawing networks. Also called as Denting (Fig. 6.28)

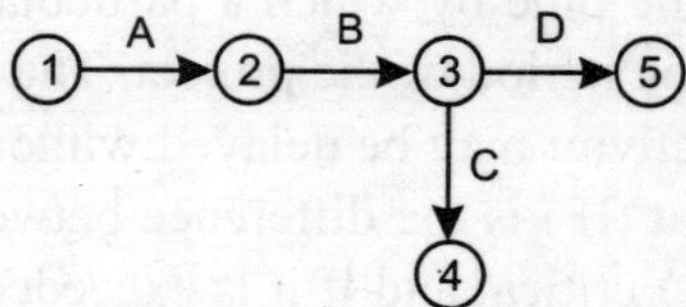

Fig. 6.28.

3. **Waqon Wheel Error.** As fingure represents in case of merge and burst that activities D, E and F (all the three) depend on A, B and C (i.e. all the three). It may be possible that activity D and E may be depending on A, B and C but F may be depending only on A and B and not on C. This situation is called wagon wheel and error as wagon wheel error. (Fig. 6.29)

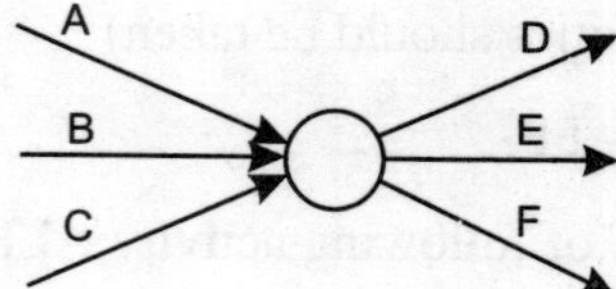

Fig. 6.29.

6.9 CRITICAL PATH AND CRITICAL ACTIVITIES

The critical path has been defined as the longest path through the network. All events on the critical path have early times equal to their late times. Events to lying on critical path must be reached on schedule, if the project is to be completed in the minimum time. Activities lying on this path are calculated critical activities. Delay of these activities will cause the delay of the whole project by equal duration. Hence there should be no delay on these activities. These activities have zero float.

There can be any number of critical path through the network. One path can spread out into a number of paths. A number of critical paths can converge into one. However, the critical paths must be a continuous chain of activities. It cannot be intermittent. Of course there is no hard and fast rule but if the critical path is drawn as a straight line, it makes it convenient to understand it easily.

A numerical shown here illustrates all the basic fundamentals of the network theory explained above.

Example. Draw the arrow diagram, mark critical path and find EST, EFT, LST, LET, TF, FF, IF for activities listed below :

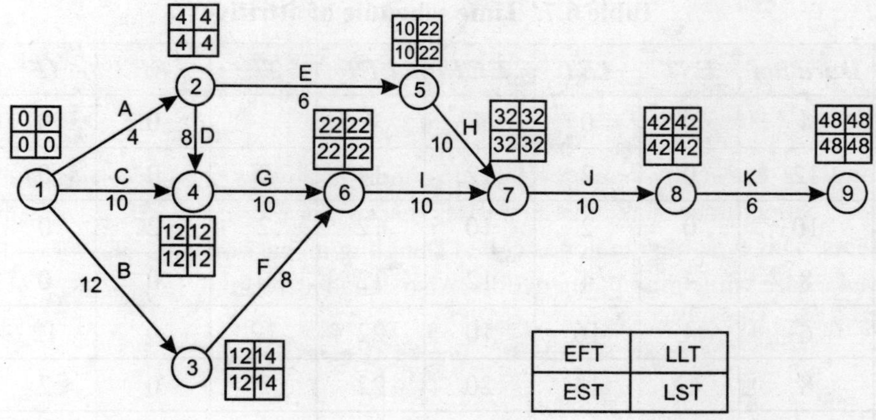

Critical Path : 1 – 2 – 4 – 6 – 8 – 9
Critical Activities : A, D, G, I, J, K
Critical Time : 4 + 8 + 10 + 10 + 10 + 6 = 48 days

Fig. 6.30.

Table 6.6.

Activity	***Duration in days***	***Activity which***	
		Precedes	***Follows***
A	4	None	D, E
B	12	None	F
C	10	None	G
D	8	A	G
E	6	A	H
F	8	B	I
G	10	C, D	I
H	10	E	J
I	10	F, G	J
J	10	H, J	K
K	6	J	None

Solution. Network is as shown in Fig. 6.30

Table 6.7. Time schedule of attritive

Activity	*Duration*	*EST*	*LST*	*EFT*	*LFT*	*TF*	*FF*	*IF*	*Remerks*
A	4	0	0	4	4		0	0	Critical
B	12	0	2	12	14	2	0	0	
C	10	0	2	10	12	2	2	0	
D	8	4	4	12	12	0	0	0	Critical
E	6	4	16	10	22	12	0	12	
F	8	12	14	20	22	2	0	2	
G	10	12	12	22	22	0	0	0	Critical
H	10	10	22	20	32	12	12	0	
I	10	22	22	32	32	0	0	0	Critical
J	10	32	32	42	42	0	0	0	Critical
K	6	42	42	48	48	0	0	0	Critical

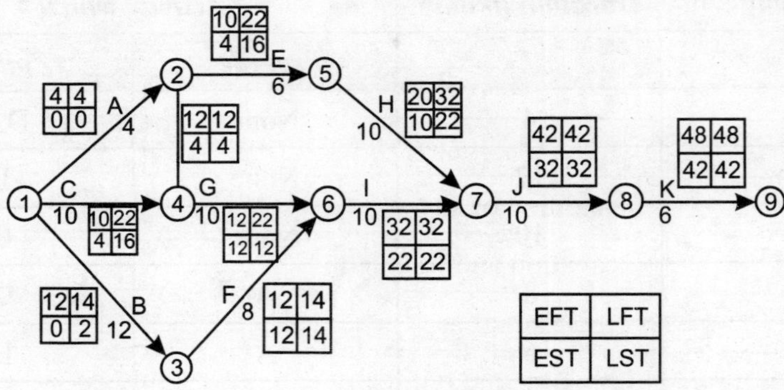

Fig. 6.31.

Event times and activity times are also as shown in Fig. 6.30 and Fig. 6.31.

6.10 NETWORK REPRESENTATION

There are two systems of representation of networks. as explained ealer

1. Activity on Arrow system i.e. AO A technique.
2. Activity on Node System, *i.e.* AON technique.

In activity on arrow system, as explained earlier, each activity is represented by an arrow, the tail end of the arrow represent the start and head of arrow represents the start and head of arrow represents the finish or end of the activity. Description of activity is written above the arrow and its duration below the arrow.

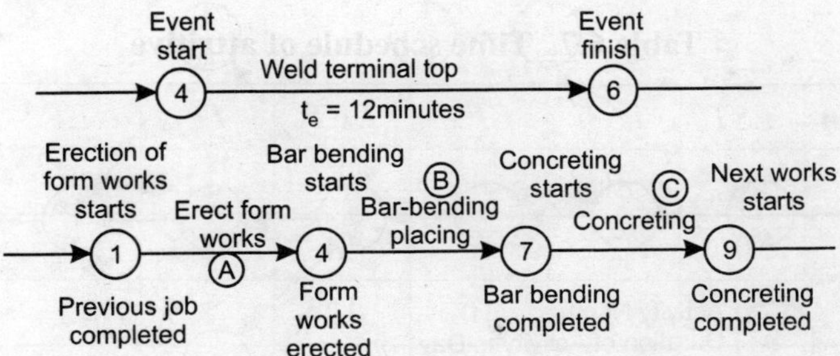

Fig. 6.32. Activity time

In AON system, these activities are represented in the form of circles or nodes and arrows are used simply to show their inter-dependency. This above network of Fig. has been shown by this system also in Fig. 6.33

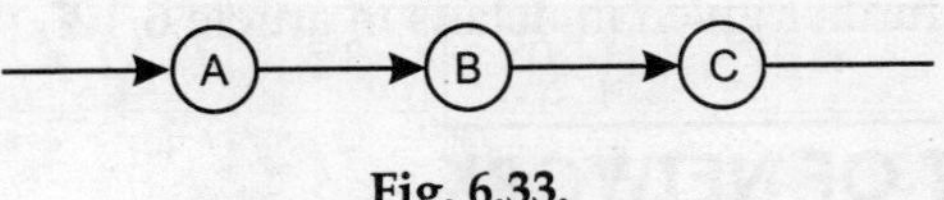

Fig. 6.33.

This gives an idea that unless event A *i.e.* erection of formwork is complete, we cannot go to event *B*, i.e. we cannot start the work of bar bending and placing.

Following example illustrates an AON system.

Table 6.8.

Activity	*Description*	*Duration*
A	Preparation of site and layout	1
B	Excavation and foundation	1
C	Superstructure	7
D	Formwork and reinforcement placement	4
E	Casting of roof slab	7
F	Plastering of walls	12
G	Flooring	7
H	Wood work/Steel work	10
I	Roofing	7
J	Plumbing/Sanitary fittings	7
K	Electrical fittings	7
L	Finishing work	7
M	Clean up of site.	3

A network has been developed and drawn as shown in Fig. 6.34

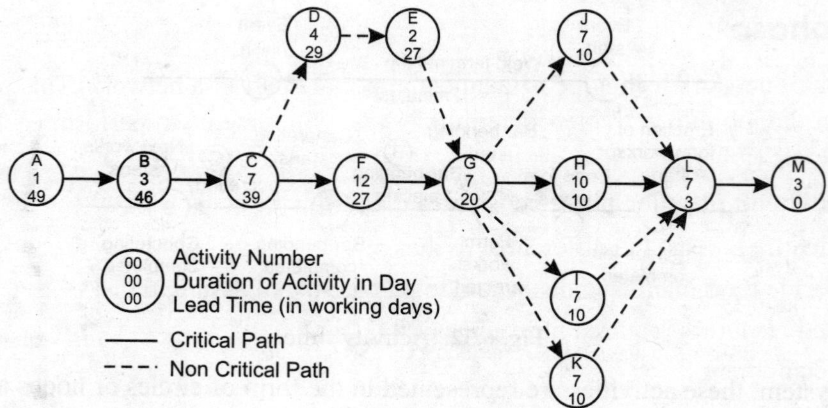

Fig. 6.34. A network diagram for a house

This Activity on node is further given in details in article 6.12.

6.11 DEVELOPMENT OF NETWORK

The main object of any project management is to minimise the capital expenditure consistent with good performance and reliability of the plant and machinery installed and to commission the project in the short possible time.

The traditional methods for construction, projects place heavy reliance upon experience and very little is generally done for planning. In the present day pattern of increasing project complexity and carrying out large number of projects, relevant experience is not always available. The changing construction methods and working conditions often invalidate past know how. This calls for planning methods which can be successfully applied by groups of organisations, each organisation being responsible for separate parts of the projects. This can be carried out by the network techniques also called as CPM, PERT etc.

Before drawing up of a network of before development of a network, all the activities in the project are carefully listed out. The order of the activities is arranged keeping the following in view:

(*a*) What activities must be completed before this activity can start ?

(*b*) What activities can be started after this activity is completed ?

(*c*) What activities can run simultaneously with this activity ?

On the basis of the answers of the above questions, an arrow diagram can be developed.

After the development of network, realistic activity duration is incorporated. Then time schedule is prepared, critical path is marked and total duration of project is determined.

Three Phases of CPM. Three phases of network planning are

1. Planning phase.
2. Scheduling phase.
3. Control monitor phase.

Fig. 3.35. shows flow diagram for these three phases of a project.

Planning-phase

Planning phase of network technique is the most important part of a network. This phase includes breaking down the job into activities, determining the requirements to perform each activity and preparing the logic (job-sequence).

The procedure for planning phase consists of the following steps :

(*a*) To identify and list the different activities.

(b) To decide their interdependency and to prepare list of assumptions.

(*c*) To find resources required for each activity, *i.e.*

(*i*) equipment.

(*ii*) workmen–skilled and unskilled and resources available

(*d*) To decide duration of each activity,

(*e*) To develop arrow diagram.

(*f*) To find cost for each activity.

To discusc this is detail, one example is given herewith.

Let us assume that concreting of roof for a building is to be made. Hence first step is to prepare the list of activities and to decide their interdependency which has been shown in tables 6.8 and 6.9. During preparation of list certain assumptions are to be made. In this particular case certain assumptions are made. They are

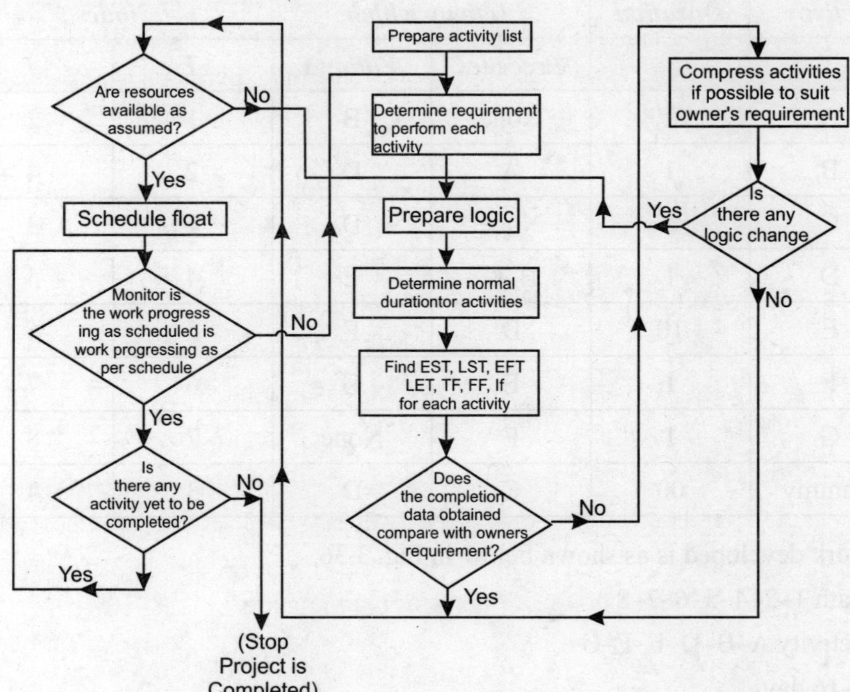

Fig. 6.35. Flow diagram for planning, scheduling and control of a project

(*i*) Concreting is to be done in one phase.

(*ii*) Masonry work upto roof level and beams if any at roof level are completed.

(*iii*) shuttering is available at site.

Activities involved and their duration are —

Table 6.9.

	Activity	***Duration***	***Symbol***
(*i*)	Erection of Shuttering	2 days	A
(*ii*)	Bar bending and placing	1 days	B
(*iii*)	Order and supply of concreting materials and equipment.	2 days	C
(*iv*)	Concreting	1 day	D
(*v*)	Curing of concrete	10 days	E
(*vi*)	Removal of shuttering	10 days	F
(*vii*)	Cleaning of site	1 day	G

These activities are listed below as per their interdependency.

Table 6.10.

Activity	***Duration***	***Activity which***		***Nodes***	
		Precedes	***Follows***	***I***	***J***
A	2	None	B	1	2
B	1	A	D	2	4
C	2	None	D	1	3
D	1	B, C	E	4	5
E	10	D	F	5	6
F	1	E	G	6	7
G	1	F	None	7	8
Dummy	00	C	D	3	4

The network developed is as shown below in Fig. 3.36.

Critical path 1–2–4–5–6–7–8

Critical activity A–B–D–E–F–G

Duration–16 days

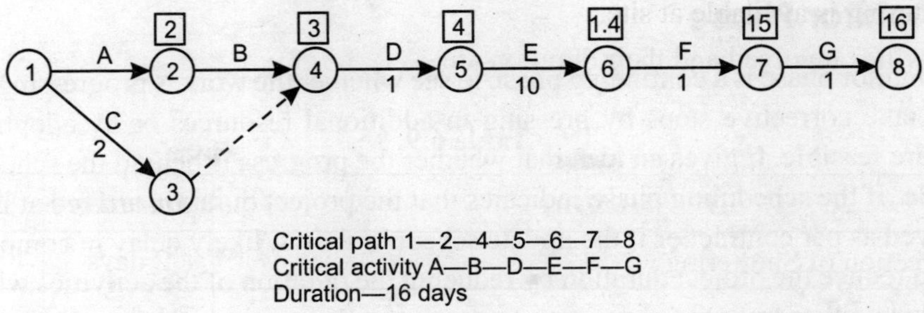

Fig. 6.36.

Scheduling

The scheduling phase includes determining normal duration for each activity, preparing time schedule, *i.e.* to find *EST*, *EFT*, *LST*, *LFT*, *TF*, *FF*, *IF*, for each activity and *EOT* and *LOT* for each event, comparing project completion period obtained as per network with that specified by the owner, compressing the schedule to suit completion date required by the owner by activity compression if possible or by logic change, if necessary. At this stage the assumptions of resources can be scheduled otherwise logic may have to be changed to suit available resources.

The main objective of the sequencing problem is to make an efficient assignment of operations and it is to be efficient in the sense of a minimum ratio of idle time to work time. In the network, we want to be efficient in the sence of minimum elapsed time or minimum delay of the project. In the first case (sequencing) a level work load is the prime concern ; in the second case, a minimum or one schedule completion time is the prime concern.

Three criteria that can be used in establishing a schedule are :

1. Earliest start for all activities.
2. Latest start for all activities.
3. Minimum-maximum manpower.

This has been more illustrated in chapter Resource Planning and Scheduling.

The reason for deviation in the schedule of any project are illustrated below.

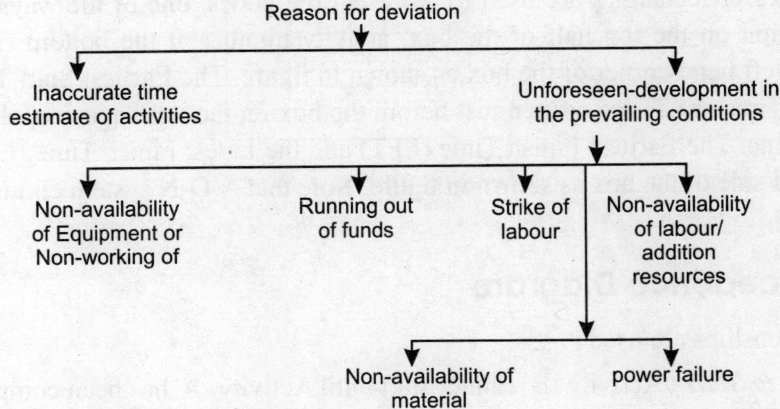

Under such cases networks needs revision or updating.

Control-Monitor Phase

The control monitor phase is a continuing phase to see whether the work is progressing as scheduled and if not to take corrective steps by pressing in additional resources or by adopting any logic changes that are feasible. It gives an idea that whether the progress is behind the schedule or ahead of the schedule. If the scheduling phase indicates that the project duration arrived at is greater than the time allowed as per contract or if the update schedule shows, likely delay in completion, then it is necessary to reduce the project duration by reducing the duration of the activities where possible. This can be done either by logic change or compression. It is necessary to crash the activities on the critical path and also to take care of any other activity that might become critical as a result of this crashing. While compressing the activities, their cost time relationship should be kept in view. After considering several alternatives that may be available, the one that results in reduction of the project duration to a specific time at least cost has to be arrived at.

Hence it is regarded as "the underline managerial function" begins with reviewing the difference between schedule -ana actual performance. Once the project has, begun, the analysis and correction of this difference, forms the basic aspect of control. It requires use of computer in large projects. A flow diagram of planning, scheduling and control of a project is shown in Fig. 6.35.

Representation of Node in A-O-N Network

Nodes are usually represented by squares or rectangles but circles and other convenient geometrical shape can also be used. A common type of node representation is shown in Fig. 6.37.

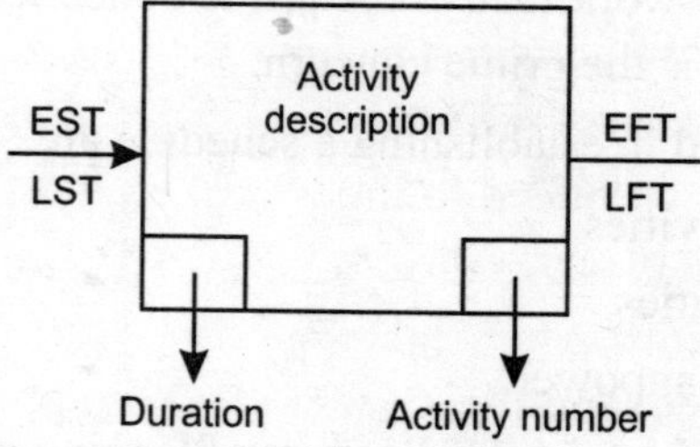

Fig. 6.37.

When squares or rectangles are used to represent the nodes, one of the ways is to write the activity description on the top half of the box, activity number at the bottom right corner and duration on the left hand corner of the box as shown in figure. The Earliest Start Time (EST) and the Latest Start Time (LST) are written just before the box on the left handside, above and below the connecting line. The Earliest Finish Time (EFT) and the Latest Finish Time (LFT) are written on the right hand side of the box as shown in figure. Note that A-O-N system eliminates the use of dummies.

Logic in Precedence Diagram

Four basic relationships are used :

(*i*) ***Finish to Start*** **:** Activity 'B' cannot start until Activity 'A' has been completed as shown in Fig. 6.38. (*a* and *b*)

Fig. 6.38. (*a*) Fig. 6.38. (*b*)

(*iii*) ***Finish to Finish*** : Activity B cannot be finished until Activity A has been finished as shown in Fig. 6.39

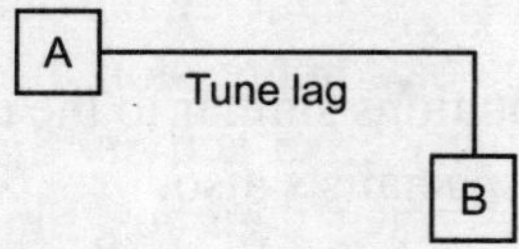

Fig. 6.39.

(*iv*) ***Start to Start combined with finish to finish*** : Activity B cannot be started and finished earlier than the stated time lag after the start and finish of 'A' as shown in Fig. 6.40.

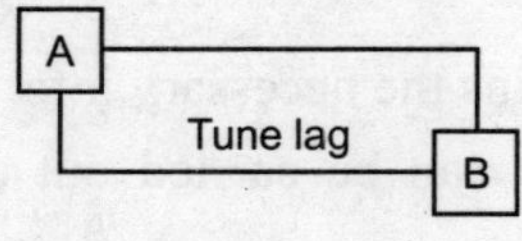

Fig. 6.40.

Drawing A-O-N Networks

- Events have no places while activities have their places with durations. Numbering of the activities will be the same as in A-O-A network.
- Start of an activity must be linked to the end of all activities which must be completed before that start may take place.
- The connecting lines run in the left-right direction. A few vertical lines, drawn to connect the appropriate nodes, should preferably join to the left hand side of the box and as close to it as possible. This helps in the diagram to be clear, easily understandable and manageable in size.
- When a number of activities start simultaneously, an activity called DEBUT (D) is provided as the first activity, though such an activity may not exist in the list. Debut activity has zero-duration (D_0). Likewise, a finish activity (F) is provided as the last activity and it also has zero duration (F_0) as shown in Fig. 6.41.

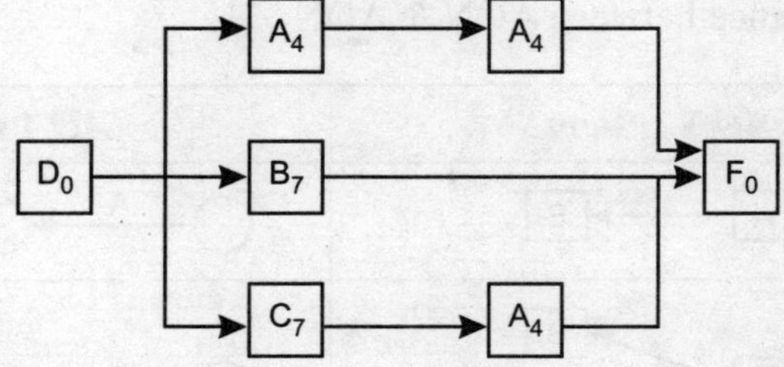

Fig. 6.41. AON diagram with critical path A-P

each other or must be separated by a time delay.

- □ This system is self sufficient as it contains all activity times (EST, LST, EFT, LFT) on the diagram itself. This facilitates efficient scheduling and control.
- □ Revisions and modifications can be carried out easily without affecting most of the activities, *i.e.*, few activities only have to be altered to incorporate changes in duration, logic and activity sequence.
- □ Pre-operations and post-operations of the activity under consideration are distinctly visible.
- □ This system adopts simple notations similar to the engineering flow charts and hence can be easily understood by non-specialists also.
- □ A-O-N system of the network completely eliminates the use of dummy activities.

Advantages of Precedence Diagram (A-O-N)

1. They can show activities which should be allowed to overlap each other or must be separated by a time delay.
2. It is self sufficient as it contains the necessary information regarding the project.
3. Revisions and modifications can be carried out easily without affecting most of the activities.
4. No arrow is required to show the logical sequence of activities in a precedence network.
5. It adopts simple notations similar to engineering flow charts. Hence, it can be easily understood.

Comparison of AOA and AON Systems

- □ The use of AOA seems to be more convenient when creating a simple network manually, while AON is more convenient if the network is large, complex and with many relationships. Construction projects like highways, pipelines, multistoreyed buildings etc., are such that several parallel strings of repetitive and sometimes overlapping activities are involved. Hence, for the projects involving highly repetitive nature of series of operations, precedence network (AON) is preferred over A-O-A network system.
- □ AON system is much easier for setting up computation.
- □ AON is simple to draw and revise when compared to AO (though computer programmes have made it simple).
- □ AON is simpler to explain and can be understood even by a non-technical persons

Some- Schematic difference between AON & AON

S.No	*AON system*	*AOA system*
1.	A → B	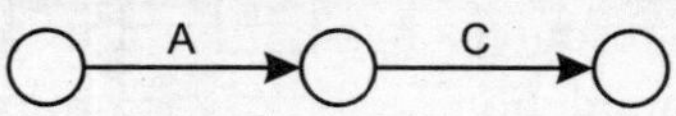 ○ —A→ ○ —C→ ○
2.	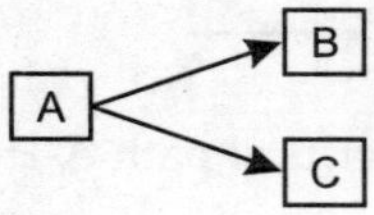 A → B; A → C	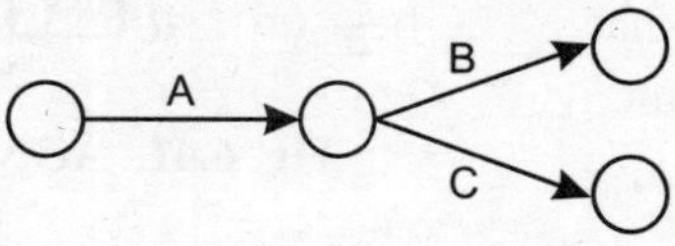 ○ —A→ ○; —B→ ○; —C→ ○

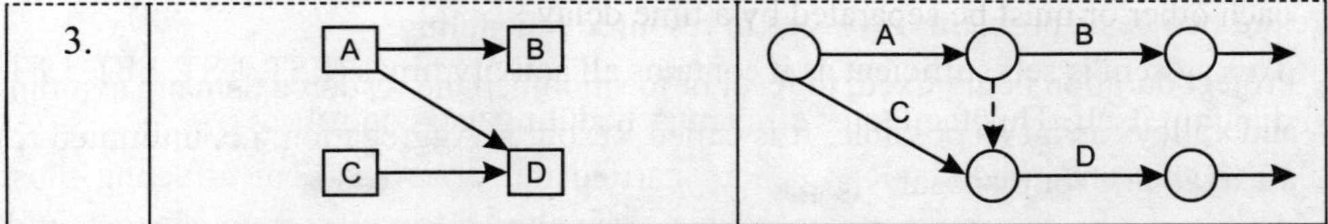

Representation of Relationships

This are shown in Fig. 6.43 and Fig. 6.44.

- Activity A and B go on concurrently
- Activity C succeeds activity A
- Activity D succeeds activity B, and
- Activity D is dependent on activity A also.

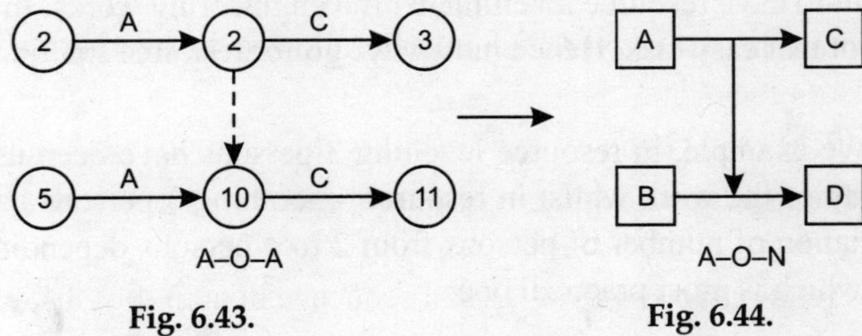

Fig. 6.43. Fig. 6.44.

6.12 RESOURCE SCHEDULING

The main objective of any management, *i.e.* to achieve a time-cost balance relationship by proper planning and scheduling the project has already been explained in previous clauses. Man power and equipment have been assumed to be available as and when they are needed, which is of course not usual. The basic planning problem is, therefore, how to decide between alternative ways of using resources (Equipment, material and manpower) such that the best profits are obtained. These decisions are perhaps the most difficult that the management has to face. It is for often necessary to make decisions which will affect a significant proportion of the organisation's money flow for many years ahead and to make those decisions at a time when there is paucity of sound information upto which to bear them. Previous experiences is often used as a basis for evaluation with modifications, if required.

The "resource analysis report" allows a manager to best use all means at his disposal. It is a technique to level out the peaks and the valleys that prevail in the managers industry. In planning resource requirements, therefore, the following key questions have to be answered.

(*a*) What are the total requirements of resources for a project over its duration ?

(*b*) If resources available are insufficient than what is the minimum duration by which the completion period of the project can be delayed ?

(*c*) What is the best utilization of the available resources to carry out the project in a fixed time.

There are two basic problems involved in resource scheduling.

1. Project duration being fixed, to level or to smoothen the resource demand avoiding peaks and valleys as far as possible. It is called Resource Aggregation, i.e. unlimited resources are available for disposal.
2. With fixed availability of resources, to minimize the project duration, i.e. Resource allocation. It occurs when there is scarcity of resources.

Resource-Levelling/Resource Smoothing

There are two basic possibilities of utilising the available resources. In case of resource levelling, available resources are fixed crew right from beginning to the end, whilst in resource smoothing, there is of course variation in number of crews to be used but the variation is not large. Fig. 6.45 shows these two cases.

It is rather difficult to have resource levelling in civil engineering works. In other works also to achieve levelling is not that easy task. Hence mostly we go for resource smoothing where variation is very less.

As shown in above example, in resource levelling 3 persons have been used right from very first day to the last day of the work whilst in resource smoothing 3 persons have been used for 6 days and there is variation of number of persons from 2 to 4 persons depending upon the type of work to be executed which is most practical one.

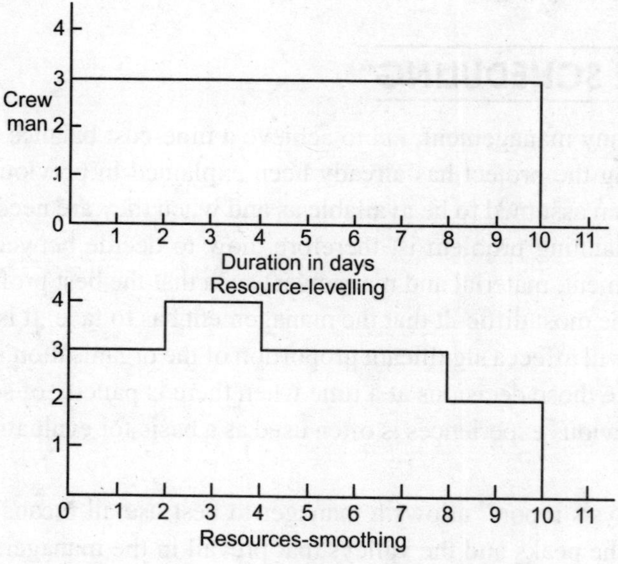

Fig. 6.45.

Resource-Aggregation

The resources for activities on the critical path are accumulated in the time period allocated to each critical activity. Further activities having float time available, however, there is a choice between aggregating resources somewhere in between the earliest time to latest time. This has been illustrated by an example.

Fig. 6.46. shows a network involving 5 activities. Crew size required for each activity is also shown in the same figure. and 6.46.

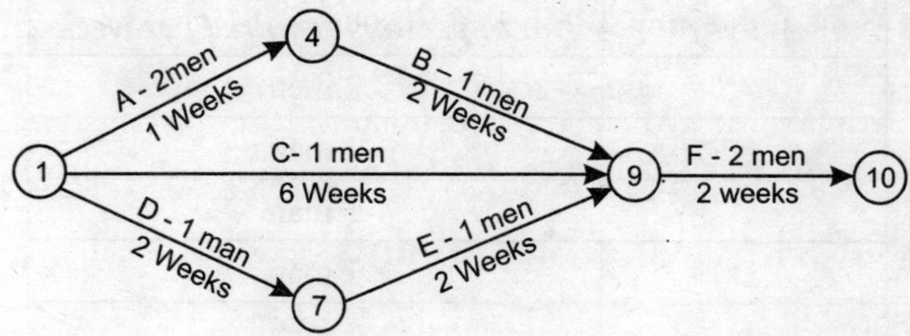

Fig. 6.46. Activity C critical critical path 1-9 Period 6 weeks.

shows the time-schedule for each activity. Figs. 6.45 and 6.46 indicates the resources required of each week for above network if they are performed at the earliest time which can be allowed to each activity. Similarly Figs. 6.47 and 6.48 indicate the resource required per week for the same network if the activities are scheduled at their latest time.

These figures indicate that extra resources are required either in the beginning or in the later period. Though total working hours are same, but extra resource requirements could be met but probably it may cost extra. It may be due to overtime working of regular employees or due to engaging temporary staff. Hence both these schedules, *i.e.* scheduling activities either at the earliest period or latest period ; may cost extra.

Instead of scheduling all activities at their earliest period or latest period, if these are staggered, *i.e.* activities are scheduled in between their earliest period to latest period, better distribution can be achieved. Hence though there may resource aggregation but it will be less in number. Resources required for complete project duration may be smooth. It will have less extra cost. This has been illustrated for the same example in Figs. 6.50 and 6.51.

Thus, instead of merely aggregating resources against previously determined time schedule, it is necessary ; if minimum costs are sought, to allocate the work to be done against the resources available.

Table 6.11.

Activity	*Duration*	*EST*	*EFT*	*LST*	*LFT*	*TF*
A	1	0	1	3	4	3
B	2	1	3	4	6	3
C	6	0	6	0	6	0 critical
D	2	0	2	2	4	2
E	2	2	4	4	6	2
F	2	6	8	6	8	0 critical

Table 6.12.

Crew size required

Activity	*No. of persons required per week*
A	2 men
B	1 man
C	1 man
D	1 man
E	1 man
F	2 men

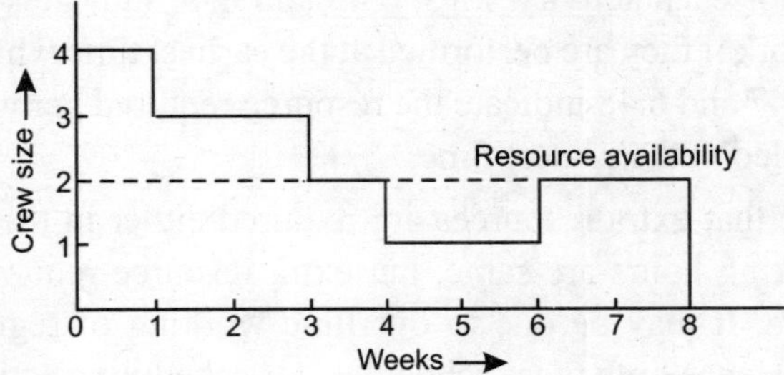

Fig. 6.47. Schedule of resources if scheduled all activities at earliest dates

Table 6.13. Scheduled at Earliest Dates

Activity	*Persons required per week*							
	1	*2*	*3*	*4*	*5*	*6*	*7*	*8*
A	2	–	–	–	–	–	–	–
B	–	1	1	–	–	–	–	–
C	1	1	1	1	1	1	–	–
D	1	1	–	–	–	–	–	–
E	–	–	1	1	–	–	–	–
F	–	–	–	–	–	–	2	2
Total	4	3	3	2	1	1	2	2

Table 6.14. Scheduled at Latest Dates

Activity	*Persons required per week*							
	1	*2*	*3*	*4*	*5*	*6*	*7*	*8*
A	–	–	–	2	–	–	–	–
B	–	–	–	–	1	1	–	–

C	1	1	1	1	1	1	–	–
D	–	–	1	1	–	–	–	–
E	–	–	–	–	1	1	–	–
F	–	–	–	–	–	–	2	2
Total	1	1	2	4	3	3	2	2

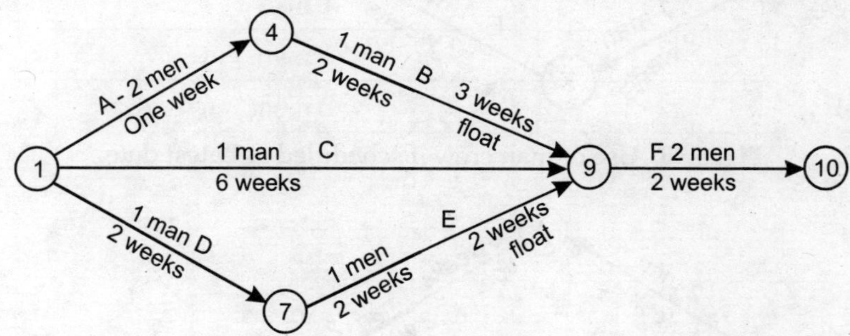

Fig. 6.48. Use of man crew if scheduled at Earliest dates.

Table 6.15. Resource Smoothening Based on Earliest and Latest Schedule

Activity	*Weeks*							
	1	***2***	***3***	***4***	***5***	***6***	***7***	***8***
A	–	–	2	–	–	–	–	–
B	–	–	–	1	1	–	–	–
C	1	1	1	1	1	1	–	–
D	1	1	–	–	–	–	–	–
E	–	–	–	–	1	1	–	–
F	–	–	–	–	–	–	2	2
Total	2	2	3	2	3	2	2	2

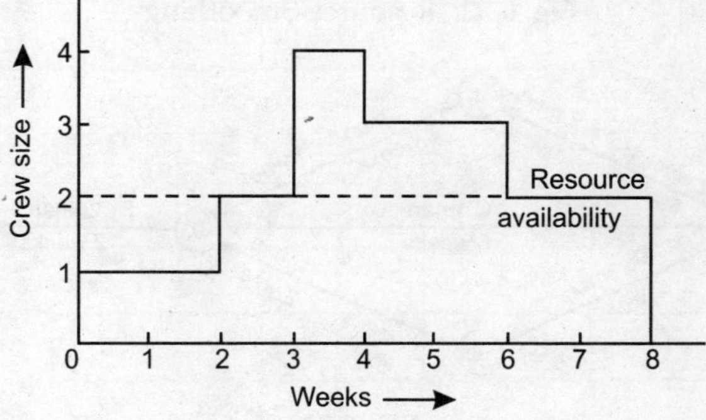

Fig. 6.49. Resources if scheduled at latest date.

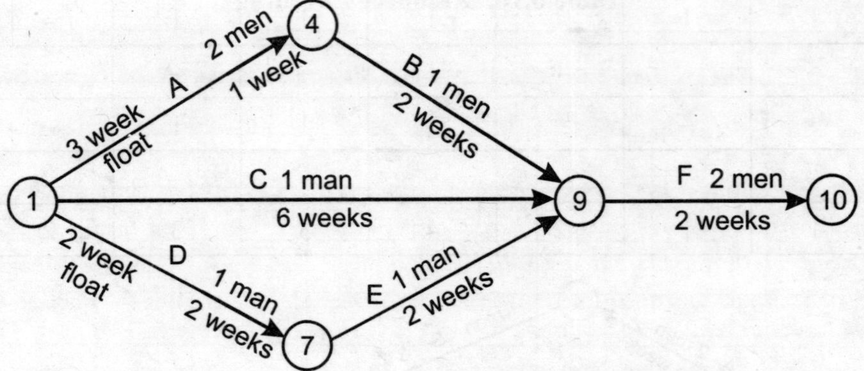

Fig. 6.50. Use of man crew if scheduled at Latest date.

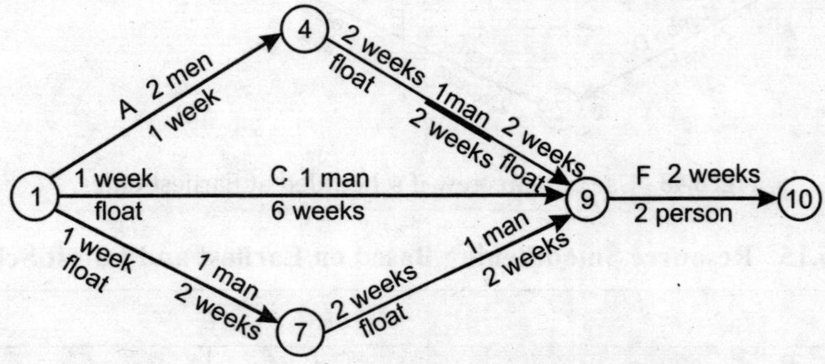

Fig. 6.51. Resource allocation to have resource smoothing.

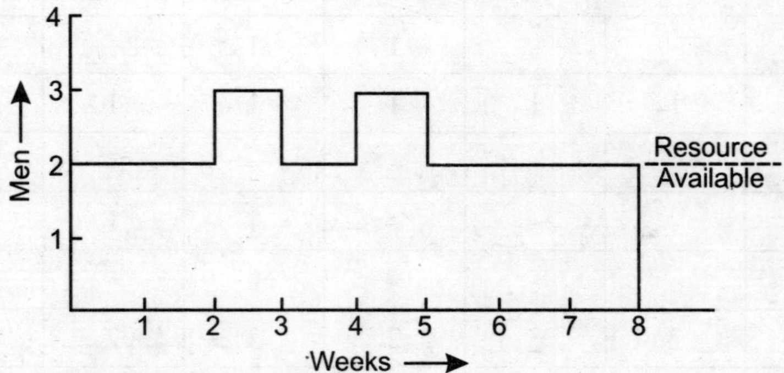

Fig. 6.52. Resource-Smoothing.

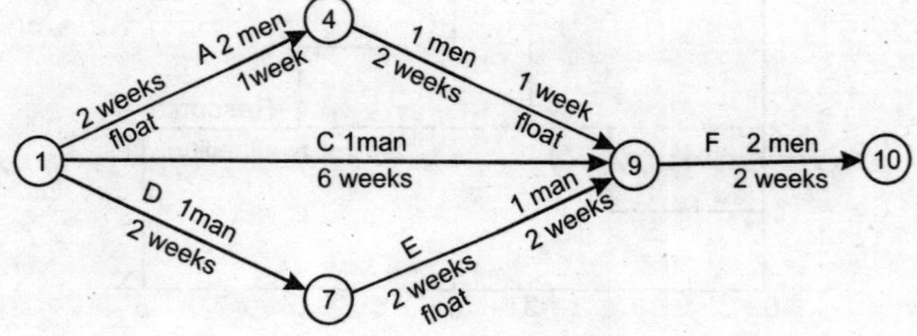

Fig. 6.53.

Table 6.16. Resource Levelling

Activity	*Weeks*								
	1	*2*	*3*	*4*	*5*	*6*	*7*	*8*	*9*
A	2	–	–	–	–	–	–	–	–
B	–	–	–	1	1	–	–	–	–
C	–	1	1	1	1	1	1	–	–
D	–	1	1	–	–	–	–	–	–
E	–	–	–	–	–	1	1	–	–
F	–	–	–	–	–	–	–	2	2
Total	2	2	2	2	2	2	2	2	2

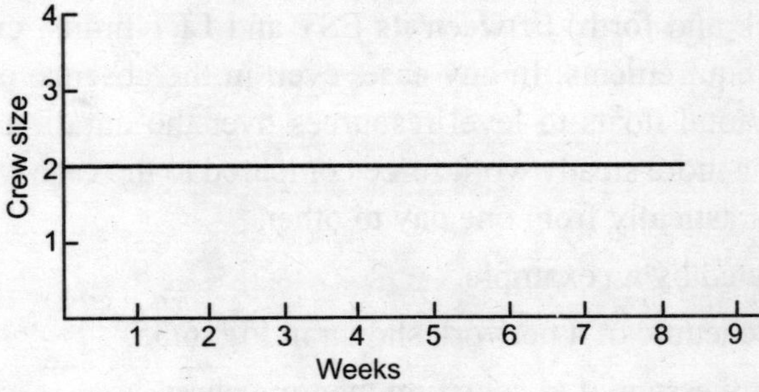

Fig. 6.54. Resource-levelling.

Resource Allocation

The planning network provides an ideal basis for the allocation of resources as the network itself defines the sequence in which the work must be carried out and the time analysis show, by the amount of total floats the relative priority attached to each activity.

Allocation means deciding what resources are required for each activity. The usual scheduling method is then to allocate the available resources to activities on a time unit basis. As there will be several activities in list, and they can only be dealt with one at a time, the next step is to arrange them in order of priority. The sequence in which activities are scheduling is important as it is a question of "first come first served" the earlier activities having the peak of available resources and later ones may be delayed if required.

It might at first sight appear that critical activities or activities with small total float should be scheduled first. But it is found that there is always advantage if shortest activities, i.e. the activities which consume less duration are scheduled first. There are many other factors which influence the choice of scheduling priority.

In resource allocation period where limited and fixed number of resources are available, many a time it become essential to increase the project duration as insufficient resources are available in a particular period of time. This has been illustrated in above example in Fig. 6.54. The duration of project extends by 1 week. This levels the use of resources hence a Resource levelling case.

If the allocation is carried out with a fixed time limit, and no delaying of activities is permitted beyond the amount allowed by the available float time, it is inevitable that the level of available resources is exceeded but it is so arranged that this will take place at the point where it makes the minimum increase above the present level. It is also illustrated in above example. There indicate that the level of resource availability has been exceeded in two places only.

It has also been illustrated by another example shown in subsequent paragraphs.

Construction of the Time Chart and Resource Levelling

The end product of a network calculations is the construction of the time chart (or schedule). This time chart can be converted into a calender schedule convenient for use by the personnel who will execute the project.

The construction of the time chart must be made within the limitations of the available resources since it may not be possible to execute concurrent activities because of manpower and equipment limitations. This is where the total floats for the non critical activity become useful. By shifting a non-critical activity (back and forth) between its EST and LFT limits, one may be able to lower the maximum resource requirements. In any case, even in the absence of limited resources, it is common practice to use total floats to level resources over the duration of the entire project. In essence this would mean a more steady work force compared to the case where the work force (and equipment) would vary drastically from one day to other.

This has been illustrated by an example.

Table 6.17 shows a schedule of a network shown in Fig. 6.55.

These informations are essential to construct the time chart.

Table 6.17. Schedule

Activity	*Duration*	*Nodes*		*EST*	*EFT*	*LST*	*LET*	*TF*	*FF*
		I	J						
A	3	1	2	0	3	0	3	0	0 Critical
B	3	2	3	3	6	3	6	0	0
C	2	2	4	3	5	4	6	1	1
Dummy	0	3	4	6	6	6	6	0	0 Critical
D	3	3	5	6	9	10	13	4	4
E	7	4	5	6	13	6	13	0	0 Critical
F	5	4	6	6	11	14	19	8	8
G	6	5	6	13	19	13	19	0	0 Critical
H	2	3	6	6	8	17	19	11	11

Critical path 1–2–3–4–5–6

Critical activities A–B–Dummy–E–G

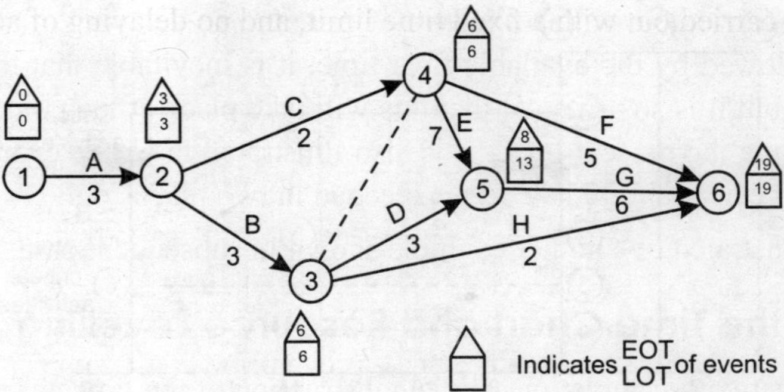

Fig. 6.55. Network.

Fig 6.56 shows the time chart corresponding to above example. The first step is to consider the scheduling of the critical activities and their after non-critical activities are considered. The critical activities are shown with solid lines. The time ranges for non-critical activities are shown by dotted lines, indicating that such activities may be scheduled anywhere within these ranges provided the precedence relationships are not disturbed. For non-critical activities, duration is shown as including float available to each of them.

The dummy activity (3, 4) consumer no time hence as shown by a vertical line. The numbers shown with the non-critical activities represent their durations. If resources are not an effective factor, each non-critical activity should be scheduled as early as possible. This allows taking the utmost advantages of the float in case the execution of any of these activities is delayed unexpectedly.

Now consider that the following manpower requirements are specified for different activities and it is required to develop a time schedule which will level the manpower requirements during the project duration.

Table 6.18. Man Power Requirements

	Activity	*Number of men*	*Duration*
1–2	A	5	3
2–3	B	7	3
2–4	C	3	2
3–5	D	2	3
4–5	E	2	7
4–6	F	5	5
5–6	G	6	6
3–6	H	1	2

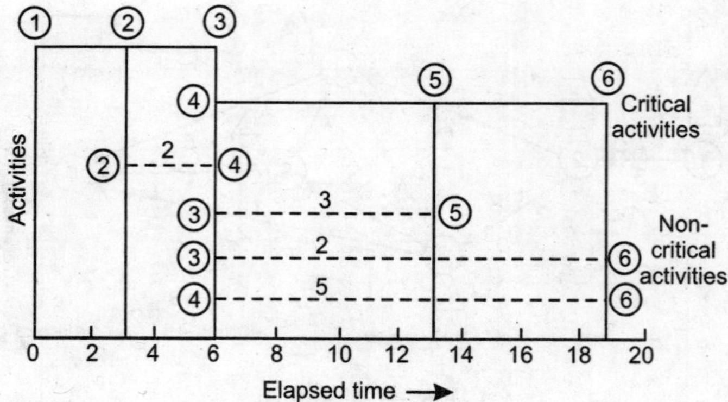

Fig. 6.56. Time-Chart corresponding to Fig. 6.55.

Fig. 6.57 shows the manpower requirements over time if the non-critical activities are scheduled as early as possible while Fig. 6.57 shows the requirements if these activities are scheduled as late as possible. The dotted line shows the requirements for the critical activities which must be satisfied if the project is to be completed on time.

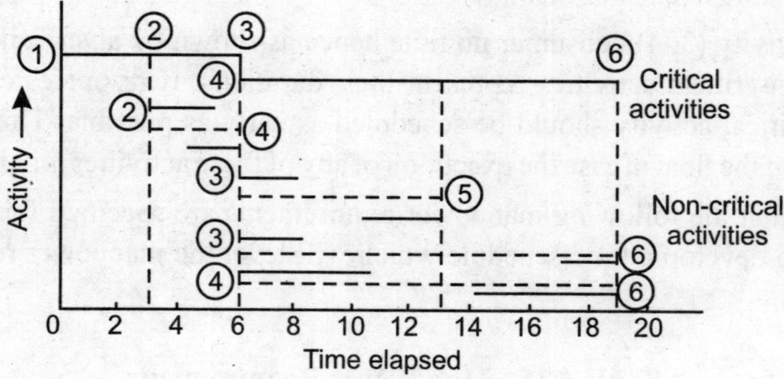

Fig. 6.57.

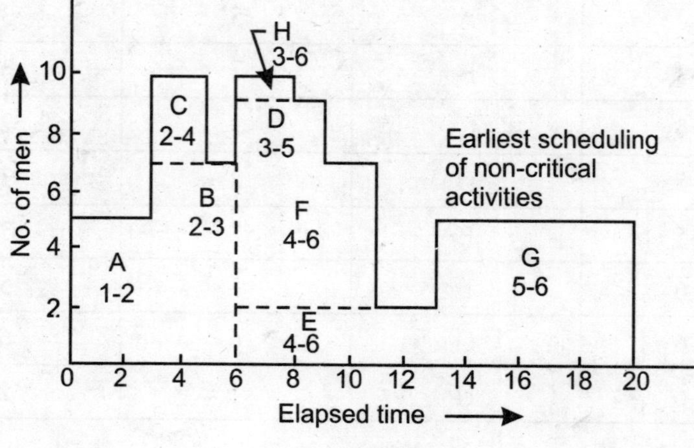

Fig. 6.58.

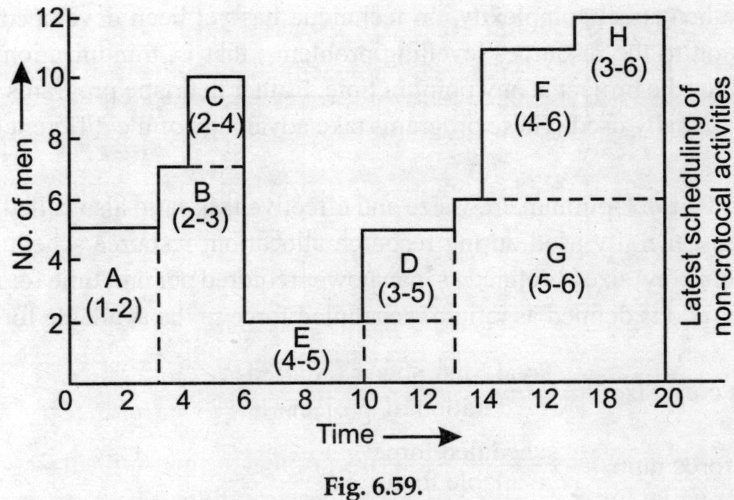

Fig. 6.59.

The project requires at least 7 men as indicated by the requirements of the critical activity (2-3). The earliest scheduling of the non-critical activities results in a maximum requirement of 10 men while the latest scheduling of the same activities sets the maximum requirements at 12 men. This illustrates that the maximum requirements depend on how the total floats of the non-critical activities are used. However regardless how of the non-critical activities are used, however regardless how the floats are allocated the maximum requirement cannot be less than 10 men. This follows since the range for activity C (2–4) coincides with the time for the critical activity B (2–3). The man power requirement using the earliest scheduling can be improved by rescheduling activity D (3–5) at its latest possible time and activity H (3–6) immediately after activity F (4–6) is completed. This new requirement is shown in Fig. 6.60. The new schedule has now resulted in a smoother allocation of resources.

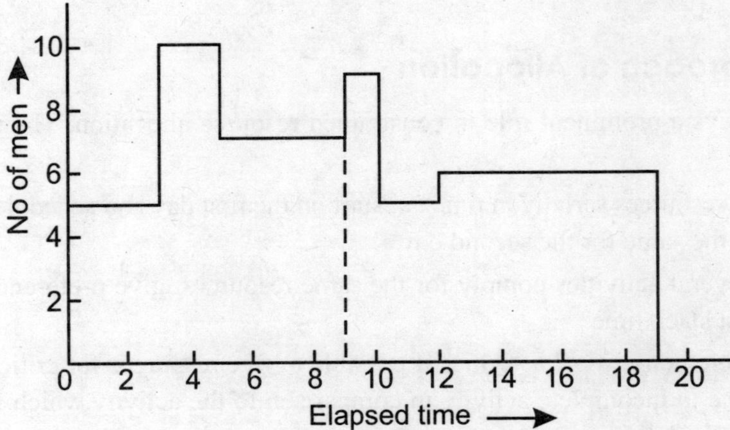

Fig. 6.60. Improved man power requirements.

In some projects the objective may be to keep the maximum resource utilization below a certain limit rather than merely levelling the resources. If this cannot be accomplished by rescheduling the non-critical activities, it will be necessary to expand the time for some of the critical activities. This however may necessitate that some jobs be executed in segments rather than as an entity.

Because of mathematical complexity, no technique has yet been developed which will yield the optimum solution to the resources levelling problem ; that is, minimization of the maximum required resources for the project at any point in time. Rather heuristic programs similar to the one outlined above are actually used. These programs take advantage of the different floats for the non-critical activities.

Effective Task Ratio. Optimum crew size and effective task ratio also called as effective force ratio, are two terms generally used during resource allocation, resource scheduling and resource smoothing. Optimum crew size is defined as man power required per unit time for resource levelling and effective force ratio is defined as ratio of scheduled force to the available force.

$$\therefore \quad \text{Optimum crew size} = \frac{\text{Total Men power required}}{\text{duration of project}}$$

$$\text{Effective force ratio} = \frac{\text{scheduled force}}{\text{available force}}$$

In above example optimum crew size

$$= \frac{5\times3+7\times3+3\times2+2\times3+2\times7+5\times5+6\times6+1\times2}{20}$$

$$= \frac{125}{20} = 6.25 \text{ persons/day}$$

Now 6.25 persons can not employed hence to have resource levelling 7 persons are employed for 20 days.

$$\therefore \quad \text{EFR} = \frac{125}{7\times20} = \frac{125}{140} = 0.89 \approx 0.90$$

This shows that effeciency of labour used is only 90 per cent. The EFR should be as far as possible unity. If EFR = 1. This indicates all the-labours employed are used most efficiently and effictively.

Heuristic Approach of Allocation

This procedure plays a prominent role in constrained resource allocation. The method consist of three rules :

(*a*) Allocate resources serially in time *i.e.* start on the first day and schedule all jobs possible. Then do the same for the second day.

(*b*) When several activities comply for the same resources, give preference to the activities with least slack time.

(*c*) Reschedule non-critical activities if possible to free resources for critical activities. Give preference to incomplete activity in comparison to the activity which is yet to be given, while rescheduling.

A flow diagram illustrating this is shown in Fig. 6.61.

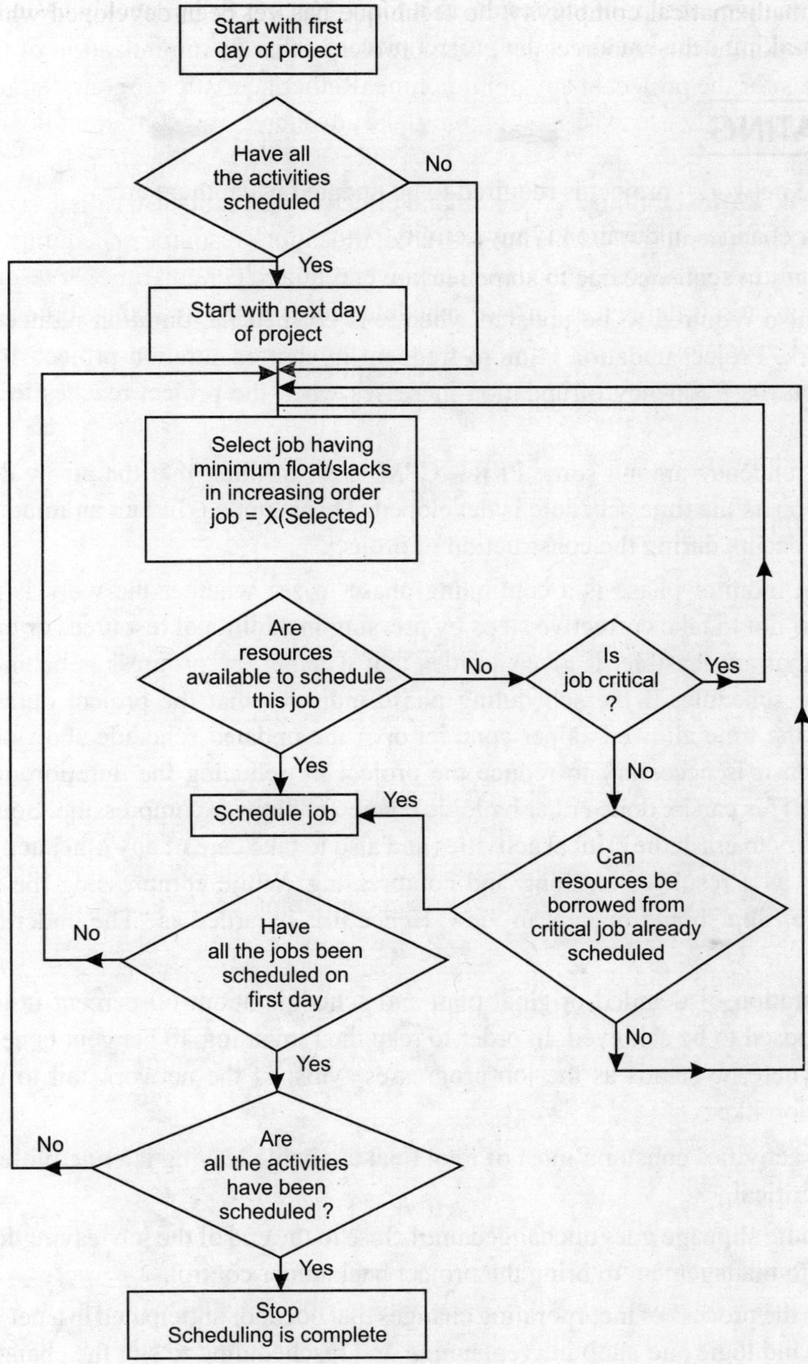

Fig. 6.61.

Computer in resources allocation. Use of computer in resource allocation can be made when there are somewhat over-whelming complexity of resource allocation task. A computer can very quickly calculate the best manner of allocating available resources so that particular project, objectives can be met within the priorities set by management. It can produce a variety of simulations of possible alternatives and print the results in a form which- can be quickly assimilated by those responsible for taking decision about the project plan.

6.13 UPDATING

When to update network– project is required to be updated when there are

(*i*) Major changes in duration if any activity.

(*ii*) Changes in sequence due to some reason.

Project is also required to be updated when it is crashed i.e. duration reduced due to some urgency of work. Project updation is more frequent in shorter duration project. In case of large duration project, the frequency of updation increases when the project reaches to its completion stage.

There is a tendency among some PERT–CPM users to think that the arrow diagram can be discarded as soon as the time schedule is developed. This is not so. In fact an important use of the arrow diagram occurs during the construction of project.

The control monitor phase is a continuing phase to see whether the work is progressing as scheduled and if not to take corrective steps by pressing in additional resources or by adopting and logic changes that are feasible. It gives an idea that whether the progress is behind the schedule or ahead of the schedule. If the scheduling phase indicates that the project duration arrived at is greater than the time allowed as per contract or if the updated schedule shows likely delay in completion, then it is necessary to reduce the project by reducing the duration of the activities where possible. This can be done either by logic change or activity compression. Sometimes it may become necessary to crash the critical activities and also to take care of any other activity that might become critical as a result of crashing and compressing. While compressing the activities their cost time relationship should be kept in view. Hence it is regarded as "The underline managerial function."

With preparation of detailed original plan and schedule about 60 percent utility from CPM network is supposed to be achieved. In order to reap the remaining 40 per cent benefit from it, one should know where he stands as the job progresses. Most of the network fail to monitor which results in situation like :

(*i*) Early activities consume'most of total float available leaving later activities all critical or near critical.

(*ii*) Schedule slippage goes unchanged until close to the end of the job leaving few alternatives open to management to bring the project back under control.

Updating is the process of incorporating changes that occur or anticipated in a net work plan with respect to time and logic and adopting replanning and rescheduling to suit the changed conditions. There are no hard and fast rules regarding the frequency of updating. On a refinery working three shifts a day even daily updating may be required otherwise even monthly updates may also not be necessary. It is upto the person in charge to update the network when he feels it necessary.

One should consider updating :

(*a*) each time where there is a change in the scope of work.

(*b*) when float activity has consumed the available float completely,

(*c*) If there is a major set back on critical path.

The informations required for an update are

(*i*) Activities completed.

(*ii*) Time required for all activities in progress.

(*iii*) Any new information on activities yet not started.

(*iv*) Change in logic if any.

(*v*) Any new activity if any and its logic.

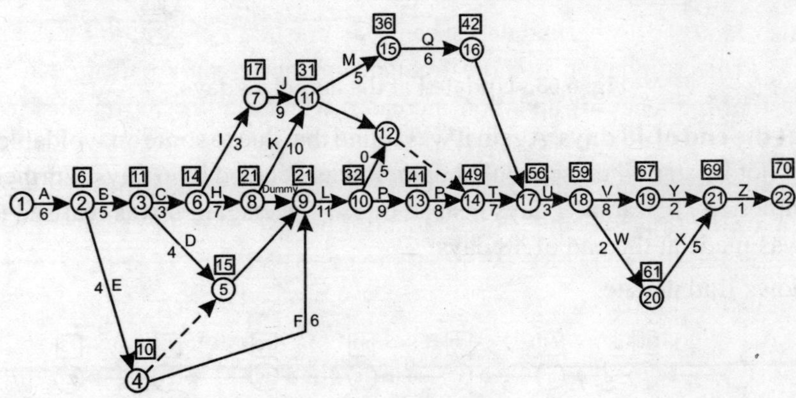

Critical path–1–2–3–6–8–9–10–13–17–18–19–21–22.

Critical period–70 days.

Critical activities–A, B, C, H, L, P, R, T, U, V, Y, Z.

Fig. 6.62.

The objective of updating is to arrive at a rescheduled completion date so that corrective measures can be taken in time to bring the project back on schedule. Thus a control monitor phase provide a tool to keep him accurately advised on the rapidly changing situation called his project and it it is properly utilised, he is able to stay "on top" of his project, both for costs and time.

An example of updating is illustrated herewith. A network has been shown in Fig. 6.62 which has a total duration for its completion as 70 days. This network was updated twice during its progress. First update was made at the end of 18 job days.

Update-1 at the end of 18 job days. Here it was found that due to some unavoidable reasons, activity *C* could not be completed by the end of 14 days. It took 4 days extra than that of the period scheduled for its completion. Hence it was realised to revise and update it at the end of 18 days.

Fig. 6.63 gives the updated CPM drawn taking into account the deviation required to complete the activities in progress at that time and any changes expected in the activities yet to be started. In this first update the critical path was not changed. The overall change was to increase the project duration by 3 days *i.e.* from 70 to 73 days. Unless corrective steps were taken, certain activities crashed, project was likely to be delayed by 3 job days.

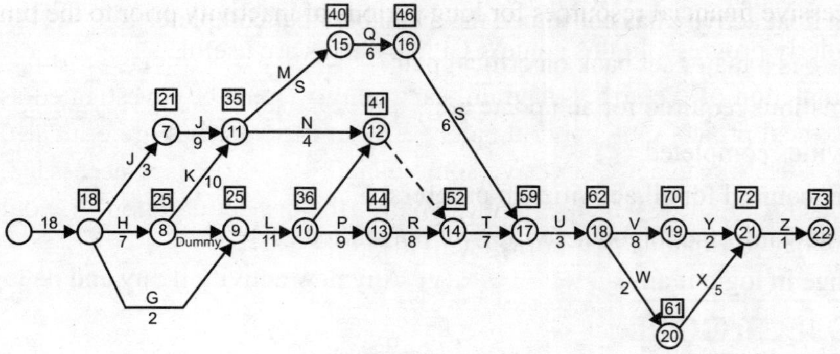

Fig. 6.63. Update I at the end of 18 days.

Update-II at the end of 48 days. Again it was found that due to some unavoidable circumstances activity M could not be started on scheduled date. It was delayed by 6 days. Further it took 2 days extra for its completion *i.e.* it took 7 days instead of 5. Hence again it was realised to update it and second update was made at the end of 48 days.

Fig. 6.64 shows IInd update.

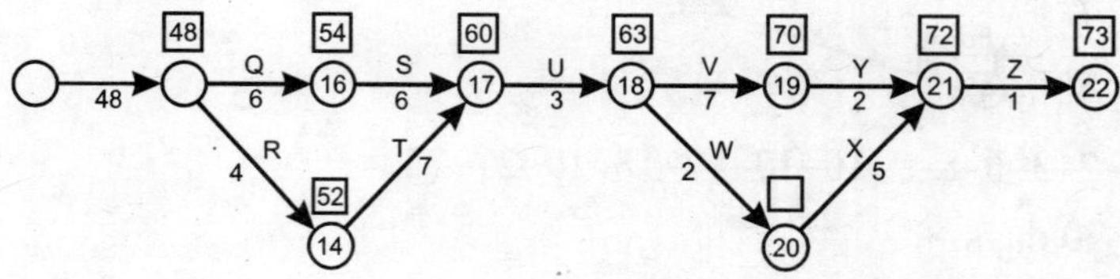

Fig. 6.64. Updaie-II at the end of 48 days.

It was realised to avoid the increase in the project duration activity-V (18–19) was crashed by one day. This changed the critical path as shown. Thus the project was behind by 3 days only. It was possible to crash other activities by 3 days to bring down the project duration to its original schedule of 70 days but neither it was found to be economical nor there was any such urgency.

Hence control monitor phase controls the overall project.

6.14 COST CONTROL

All owners, whether in Public sector or Private sector, have certain monetary limitation, the investment value is of primary concern for all programs. The initial investment value is usually measured in currency which is largly controlled by design, but the initial cost also affected by time available for accomplishing the work, timely completion and delays. Some investment values are measured by media other than money. The value of many projects is measured in national defence capability or protection of human life.

Few investments are of value to an owner until the facility is built and ready for its intended use. Many facilities are needed before planning is started, and the earliest practicable completion date is great importance to the owner. During the planning phase, the owner is always concerned with financial resources to be needed during the various time periods throughout the development of a program. He must make available sufficient resources to avoid costly delay without unnecessarily reserving excessive financial resources for long periods of inactivity prior to the time, the funds are needed for orderly progress. In this context CPM systems are useful.

Upon completion of the arrow diagram, each activity must be investigated as to its duration and cost. In typical project only approximately 30% of the activities are critical. This means that approximately 70% of the activities have some quantity of float. It is necessary to expedite the project, it is useful to know which activities control the project duration. It would be useless to spend extra money to expedite non-critical activities.

6.15 PROJECT COST

In a project there are two type of costs are involved

1. Direct cost
2. Indirect cost

The direct cost is the amount charged to individual activity. It consist of direct investment made for labour, material and equipment. The direct cost for various activities increases as we compress the time. The costs which are related to control and direction of the project and consist of overheads such as supervision charges, office establishments, indirect labour and material etc are called indirect cost. The indirect costs of project increases with increase in the duration of an activity and that of whole project.

The manager would be interested to complete the project at the optimum cost where the sum of direct and indirect cost is minimum as shown in Fig 6.65. The time corresponding to optimum cost (minimum cost) is called optimum duration of the project.

It can be seen from the curve that total cost of a project goes on increasing beyond optimum one Normal duration may not give optimum cost.

If the network analysis indicates that the project duration and the project completion date do not meet the requirement, the manager has to consider the following alternative courses of action.

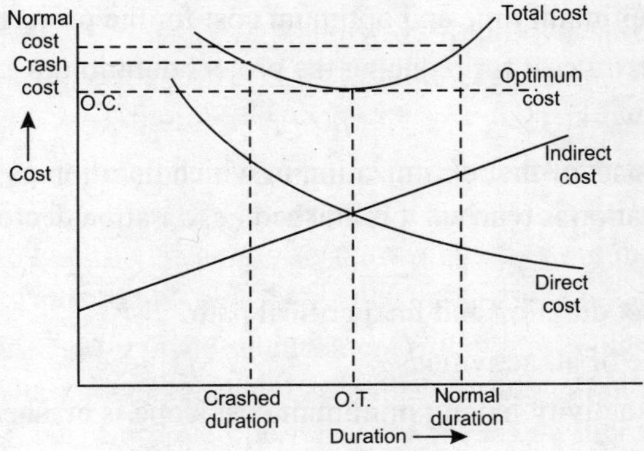

Fig. 6.65.

(*i*) The estimated time for the activities to be examined to check whether the duration for certain activities can be reduced without any extra resources and cost.

(*ii*) The logic of the network to be examined and to check whether some of sequential activities can be converted into concurrent activities.

(*iii*) Some of the activities can be crashed by deploying more resources which involve extra cost.

6.16 COST SLOPE

The term cost slope indicates the rate of increase or decrease in the cost of an activity per unit decrease in time. Assuming direct cost to be linear, time cost curve slope gives an idea about how much extra cost, one may have to spend for saving time.

$$\text{Time cost curve slope} = \frac{C_c \, \text{Ð} C_n}{d_n \, \text{Ð} d_c}$$

where C_c = Crash cost of the activity

C_n = Normal cost of the activity

d_n = Normal time of the activity

d_c = Crash time of the activity

This crashing helps the management in deciding the completion time of the project. Many a times extra cost involved in early completion of work may cause realisation of extra profit due to early completion of the work.

6.17 CRASHING OF NETWORK

Many a times it becomes necessary to decrease the scheduled duration of project. This decrease in time causes in increase in the cost of project beyond optimum limit. This increase in the cost due to decrease in the time beyond optimum limit is called crashing of the network.

Crashing of the project network can be carried out—

(*i*) For finding the minimum duration of the project and the corresponding cost.

(*ii*) For finding the optimum time and optimum cost for the project.

(*iii*) For finding the extra cost for reducing the project duration to a particular level.

Steps in Crashing Network

Steps in crashing consists of first optimization in which duration as well as cost both may get reduced. Once the optimization is reached it is crashed i.e. duration decreases with increase in cost. These steps consists of

1. Estimating project duration and final critical path.
2. To find cost stope of all activities
3. Then the critical activity having minimum cost slope is crashed in first stage and path is revised.

   In second/next stage, crashing of the activity on critical path having second lower cost shope is crashed.

   while doing crashing the critical path is likely to change Therefore after each crashing new network is drawn and path is checked.

In case of more then one critical path: activities of each critical path having minimum cost slope will have to be crashed simultaneously. Total cost of project is also calculated after each crashing.

4. The crashing *i.e.* step 3 is repeated till all possible activities can be crashed are crashed and when there is no possibility of further crashing.

   It should be noted that only critical activities of critical path are crashed as crashing non critical activity will not reduce time however will increase the project cost.

While crashing the activities of the networks we should examine whether the non-critical activities in the network are becoming critical. In crashing the project duration the focus should be on the critical activities as critical path only affects the project duration.

Example. A CPM network consists of the following activities :

Table 6.19.

Activity	*Node*		*Duration*		*Time cost curve slope Rs/day*
	I	*J*	*Normal*	*Crash*	
A	1	3	7	3	50.00
B	1	2	9	7	30.00
C	3	5	4	1	75.00
D	2	5	5	3	125.00
E	2	4	3	1	20.00
G	4	6	2	1	500.00

Schedule 1

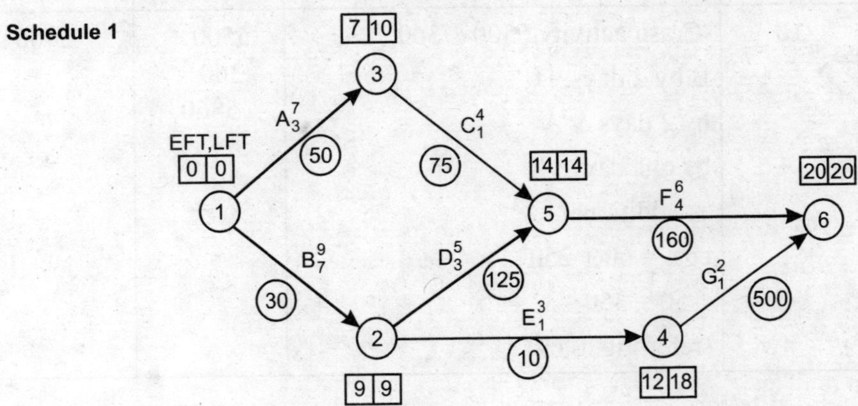

Critical path–1–2–5–6

Critical activities–B, D, F

Critical period = 20 days.

Fig. 6.66.

Find –

(*a*) Critical path.

(*b*) If all normal conditions exist, total cost of all the project activities is Rs 5500.00. Indirect cost is Rs. 150/- per day. Calculate all possible schedule and their total cost.

(*c*) Which is most economical project duration.

Solution. Network is as shown below in Fig. 3.67.

Suffix at activities indicate normal and crash time figures in the circle below each activity indicate time– cost curve slope for that activity.

Table 6.20 shows different crash schedules of network.

Table 6.20.

Schedule	*Duration in days*	*Remarks*	*Direct cost Rs.*	*Indirect cost @ Rs. 150/-*	*Total cost Rs.*
1	20	All Normal (refer Fig. 6.66)	5500	3000	8500
2	18	Crash activity B by 2 days ∴ additional cost = 2 × 30 = 60 (refer Fig. 6.67)	5500 + 60	2700	8260
3	13	Crash activity B by 2 days and D by one day ∴ additional cost = 2 × 30 + 125 = 185 (Fig. 6.68)	5500 + 185 = 56858	2550	8235
4	16	Crash activity 5500 + 360 B by 2 days, D by 2 days & A by one day ∴ additional cost = 60 + 250 + 50 = 360 (refer Fig. 6.69)	5500 + 360 = 5860	2400	8260

Since activity *B* and *D* can not be further crushed hence event (5) onward can not be started before 10 days hence there is no significance of crashing activity *A* and *C*. Further it is clear that by crashing activity *C* or *G* there would not be any reduction in time and by crashing activity *F* the total cost will be still more.

Schedule 2

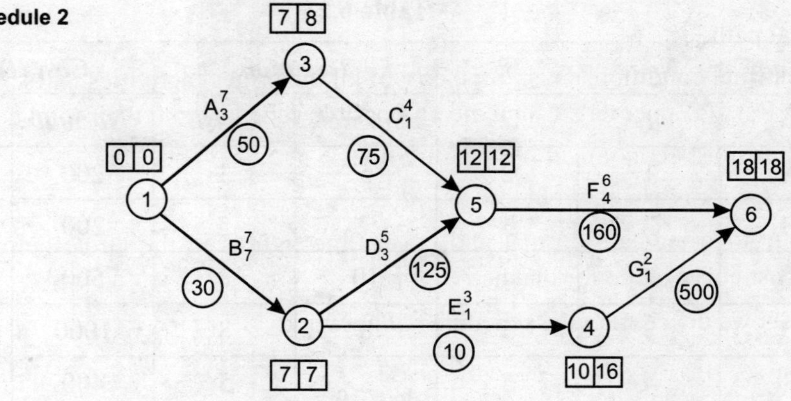

Fig. 6.67. 18 days.

Schedule 3

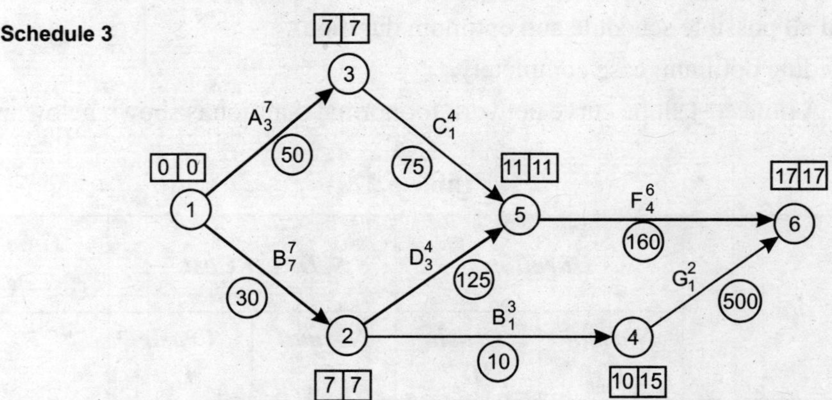

Fig. 6.68. Duration 17 days.

Schedule 4

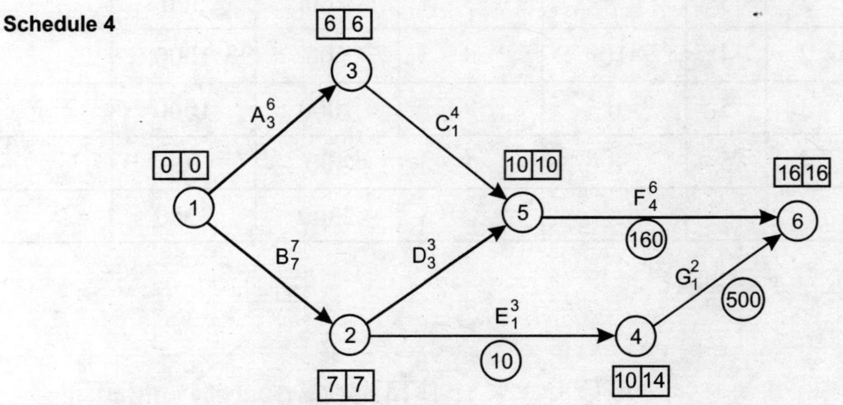

Fig. 6.69. Duration 16 days.

Hence schedule-3 with a duration of 17 days is most economical.

Example. A network CPM consists of the following activities :

Table 6.21

Activity	*Node*		*Duration*		*Cost (Direct)*	
	I	*J*	*Normal*	*Crash*	*Normal*	*Crash*
A	1	2	5	3	200	250
B	2	3	7	5	200	300
C	2	4	10	8	500	1000
D	3	5	8	8	1000	1000
E	4	5	8	5	400	700

If indirect cost is Rs. 150/- per day of project duration—

(*i*) Find all possible schedule and optimum duration,

(*ii*) Schedule optimum case completely.

Solution. A time cost slope curve network for normal duration is shown below in Fig: 6.70 and Table 6.22.

Table 6.22.

Activity	*I*	*J*	*Duration*		*Direct Cost*		*Time cost curve* $(C_c - C_n)/(D_n - D_c)$
			Normal D_n	*Crash* D_c	*Normal* C_n	*Crash* C_c	
A	1	2	5	3	200	250	25
B	2	3	7	5	200	300	50
C	2	4	10	8	500	1000	250
D	3	5	0	8	1000	1000	00
E	4	5	8	5	400	700	100
Total					2300	3250	

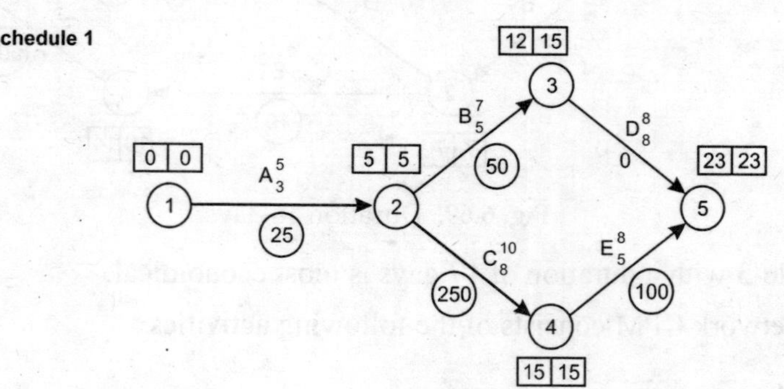

Fig. 6.70.

Schedule 2

Activity A is crashed by 2 days

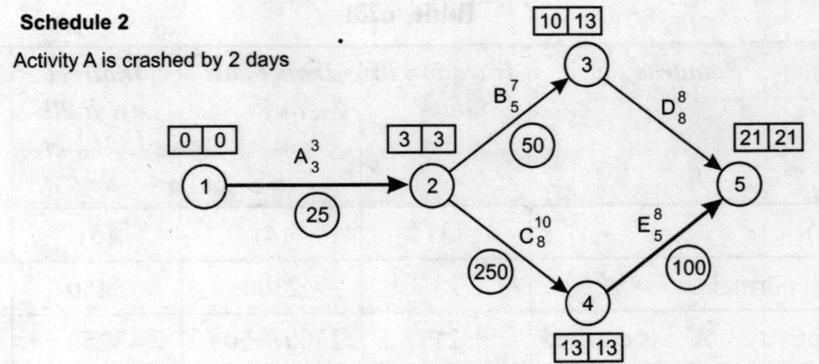

Fig. 6.71.

Schedule 3

(Activity A is crashed by 2 days and E by 3 days)

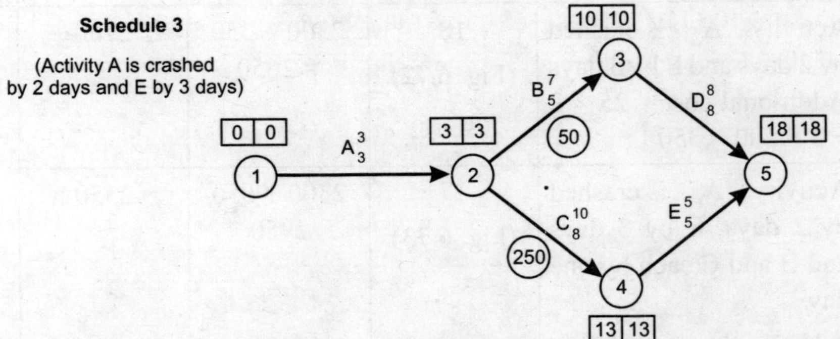

Fig. 6.72.

Schedule 4

(Activity, A is crashed by 2 days, E by 3 days and B&C by one days)

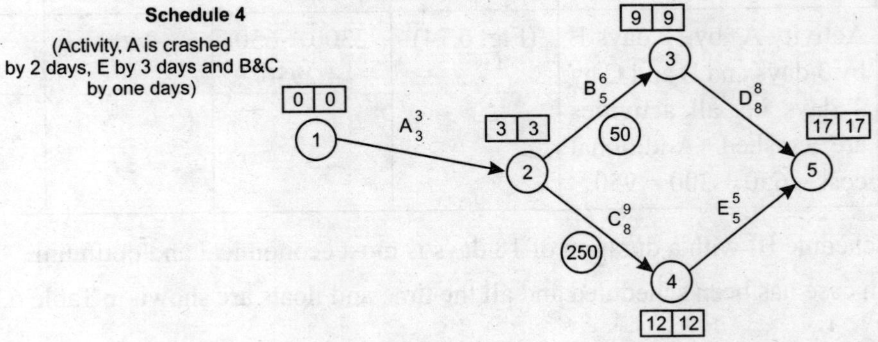

Fig. 6.73.

Schedule 5

(Activity, A is crashed by 2 days, E by 3 days and C by 2 days)

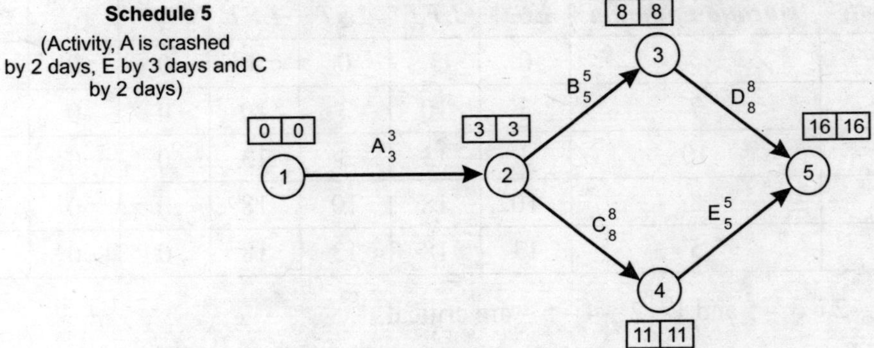

Fig. 6.74.

Table 6.23.

Schedule	*Remarks*	*Duration in days*	*Direct cost in Rs.*	*Indirect cost @ Rs. 1501-per day*	*Total cost Rs.*
(1)	(2)	(3)	(4)	(5)	(4 + 5)
1.	All normal	23	2300	3450	5750
2.	Activity A is crashed by two days. Additional Cost = 25 × 2 = 50.	21 (Fig. 6.71)	2300 + 50 = 2350	3250	5600
3	Activity A is crashed by 2 days and E by 3 days Additional cost = 25 × 2 + 3 × 100 = 350	18 (Fig. 6.72)	2300 + 350 = 2650	2700	5350 (obtention)
4	Activity A is crashed by 2 days, E by 3 days and B and C each by one day. Additional cost = 350 + 50 + 250 = 650	17 (Fig. 6.73)	2300 + 650 2950	2550	5500
5.	Activity A by 2 days E by 3 days and B and C by 2 days i.e. all activities are crashed. Additional cost = 650 + 300 = 950	(Fig. 6.74)	2300 + 650 = 2950	2400	5650

Hence Schedule III with a duration of 18 days is most economical and optimum.

Optimum case has been scheduled and all the time and floats are shown in Table 6.25.

Table 6.24.

Activity	*Duration ptimum*	*EST*	*EFT*	*LST*	*LFT*	*TF*	*FF*	*JF*
A	3	0	3	0	3	0	0	0
B	7	3	10	3	10	0	0	0
C	10	3	13	3	13	0	0	0
D	8	10	18	10	18	0	0	0
E	5	13	18	13	18	0	0	0

Path – 1 – 2 – 3 – 5 and 1 – 2 – 4 – 5 – are critical.

All the activities are critical.

6.18 NETWORK FOR PROJECT COST CONTROL

Network can be efficiently used for planning and control of project cost. Network provide the basis for budgeting. In large projects it may not be possible to find out the costs relevant for each activity. The concept of work packages or cost packages can be efficiently used to overcome this problem. The work packages or cost packages help in clubbing the costs for a group of allied activities basing on certain criteria. There can be activity oriented and event oriented packages depending upon the conditions. The work packages and cost packages will help both for budgeting and control of the project costs.

6.19 LINE OF BALANCE

Line of balance is a managerial tool applying the concept of 'management by exception to ascertaining and maintaining a balance in the components of the production process. It provides the project manager with a tool for efficient plaining and monitoring the construction of a big project which essentially involves repetitive construction/production units like construction of many houses which are identical.

This technique was developed by the U.S. Navy during World War II. It is an extension of critical path method. It provides the project manager with a chart that shows at a glance the status of any activity for every unit of the project at any designed time.

In applying line of balance method, a study is made as of a given point of time. The purpose of study is to determine.

1. Where all the phases of production should be at that point if the project delivery schedule is to be maintained.
2. Where all phases of production actually are; and compare the two in order to find those phases requiring attention.

Thus the management by exception comes into play, since those phases of the production phases which are in step with the required delivery schedules require no corrective action. Managerial energy, therefore, is not unnecessarily spent investigating such operations but may be concentrated on those areas which are current or potential trouble spots.

When a project consists of repetitive construction units, say 50 housing units, there comes a special problem. For each housing unit, the necessary sequence of activities is defined. The sequence in which individual housing unit is constructed within the context of the entire project is not a matter of necessary sequence, but a matter of management decision. For example, the owner of the housing project may wish to complete housing units in a sequence that will permit prospective buyers a choice of styles and location.

If the project manager attempted to implement this decision by preparing a network diagram covering the entire project, he would, in effect, be repeating the diagram for the single housing unit. This would result in an unnecessarily complicated and un-widely network diagram. A more practical method of planning and controlling project involving repetitive construction units is provided by the "line of balance" (LOB) technique. The LOB technique provides the project manager with a chart that shows at a glance the status of any activity for every unit of the project at any desired time. By highlighting behind schedule activities, this chart pinpoints to the manager the problem areas that could delay the complition of the project, and make him to take suitable corrective measures so as to complete the project in time.

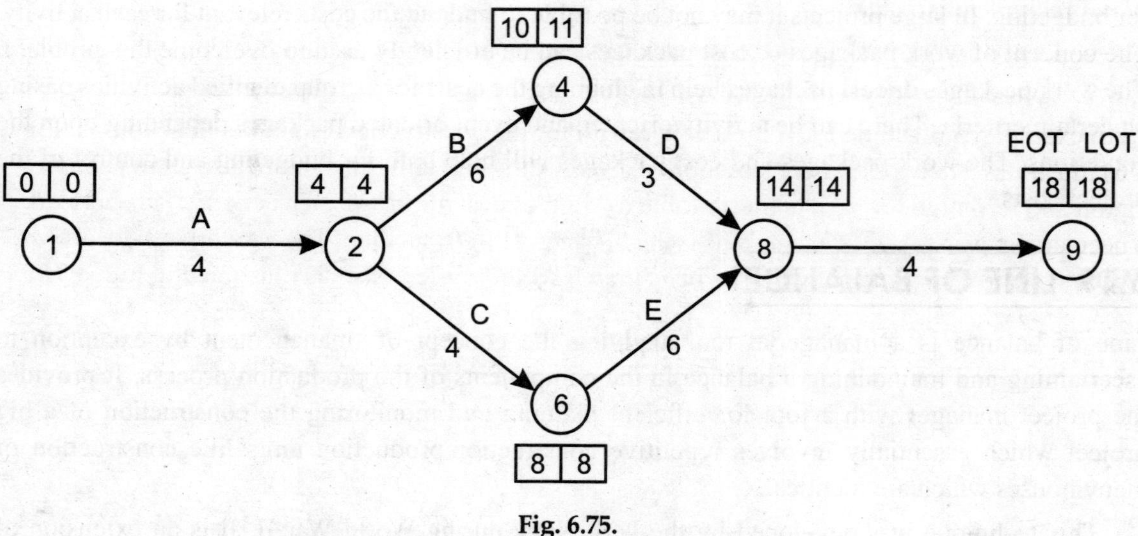

Fig. 6.75.

Generally important activities of a housing unit are listed and an arrow diagram is developed for a single unit. For manually prepared charts, the level of detail is usually limited to about 30 activities. Where more detail is desirable, supplementary lists and charts can be made up for sub-system. If required use of computer also can be made.

Lead time. Lead time is the time between activity completion and completion time of the housing unit, or project. Lead time for different activities of a network as shown in Fig. (6.75) is shown in table below :

Lead time for activity $A = 18 - 4 = 14$

$B = 18 - 10 = 8$

$C = 18 - 8 = 10$

$D = 18 - 13 = 5$

$E = 18 - 14 = 4$

$F = 18 - 18 = 0.$

Lob Technique

The LOB technique involves the following steps :

1. To list the activities of a single repeating unit.
2. To develop a network diagram for a single unit showing activity relationship and lead time for each activity.
3. To convert the network diagram to a time scaled network.

4. To construct an objective chart, i.e. a chart showing a cumulative calender schedule for the construction of housing units for the entire project.
5. To construct and maintain a progress chart showing (for each unit of the project) activities completed to date.
6. Lastly to construct LOB for the date of the progress report.
7. To use the analysis of the progress report as the basis for management action.

An example to illustrate LOB is dealt herewith for the construction of 40 identical houses. Fig (6.76) illustrates a network for construction of a single housing unit. This figure also shows activity duration and lead time for each activity. Table 6.25 gives description of each activity. This network has been drawn using AON system. This is the first step of LOB method. The network can be drawn by AOA system. The same network has been drawn by this system also as illustrated in Fig. 6.75

Table 6.25.

Activity	*Description*	*Events*
A	Perparation of site and layout	1
B	Excavation and foundation	2
C	Superstructure	3
D	Form work and reinforcement placement	4
E	Casting of roof slab	5
F	Plastering of walls	6
G	Flooring	7
H	Wood work/Steel work	8
I	Roofing	9
J	Plumbing/Sanitary fittings	10
K	Electrical fittings	11
L	Finishing work	12
M	Clean up of site	13

The next step is to draw a time scaled diagram for a single repetitive unit as shown in Fig. 6.76 for above example. This is plotted to conform with the lead time scale at the base of chart. For convenience, the critical path is plotted as a horizontal line. The resulting time scaled diagram shows activity dependence and lead times for a single housing unit.

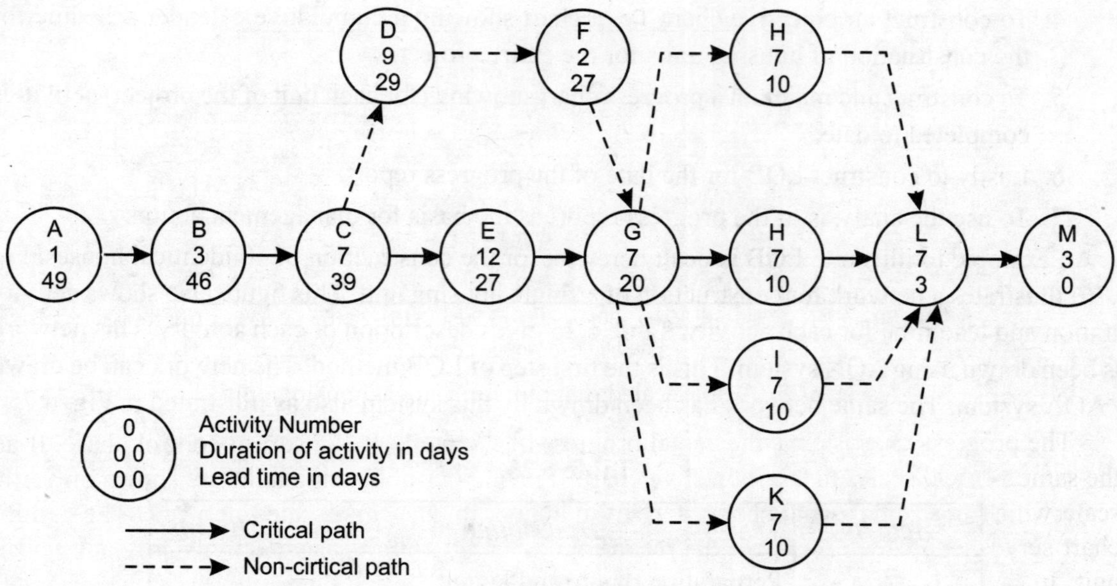

Fig. 6.76. A network diagram for single housing unit.

Then the objective chart is drawn to show the cumulative contract schedule for completion of the entire repetitive type project. For this example an objective chart is as shown in the figure. The object of this chart is to know well in advance the availability of different units at different time and total duration of project. This curve representing the contract schedule can take any shape depending on the wishes of the owner and on contractual requirements. As indicated on the chart the curve beings at the point of completion of the first housing unit and ends where the all units are completed. Hence person developing an objective chart should have experience of construction.

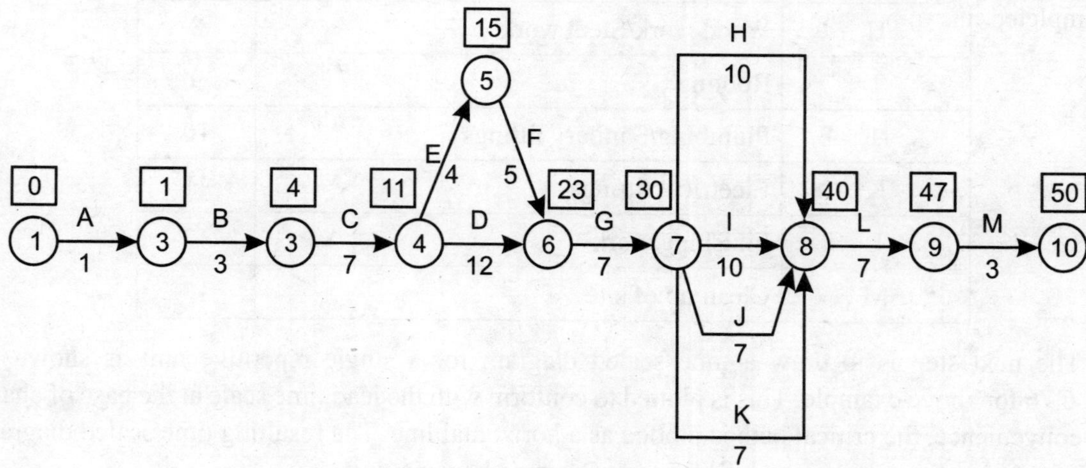

Fig. 6.77.

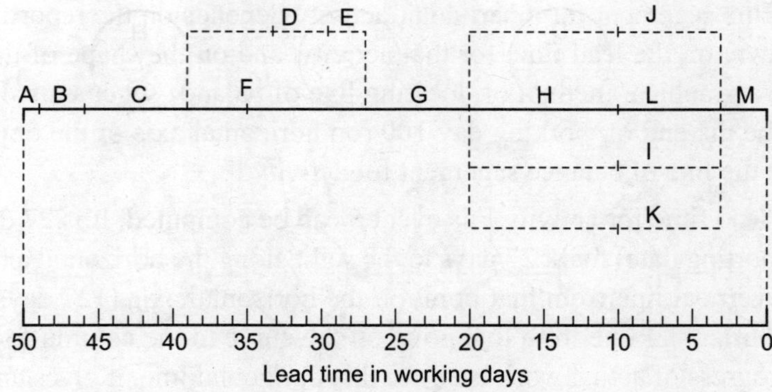

Fig. 6.78. A time scaled diagram for a single housing unit.

The progress chart shows the actual progress of the work. It is a sort of control phase. It uses the same vertical scales that of objective charts, *i.e.* number of housing units are shown on vertical scale whilst horizontal scale shows activity numbers for the time scaled network. The resulting chart serves as a check off sheet for recording the completion of each activity for each housing unit. In fact it is a kind of vertical bar chart showing, for a particular reporting date, the status of each activity for each housing unit. If desirable manager can use symbols or colour codes for representing the progress on progress chart for each activity.

Then the line of balance is drawn which gives and idea about actual progress of each activity on specified period. It enables manager to know whether he is going ahead or lagging behind for different for completion of the entire housing project. The progress chart serves as a check of sheet for recording the completion of each activity for each housing unit. The line of balance permits comparisons between scheduled and actual progress activities for different units. For convenience the line of balance is superimposed on progress chart. The manager develops the line by drawing a segment for each activity, then connecting these segments with verticals to form the steps of the completed line of balance.

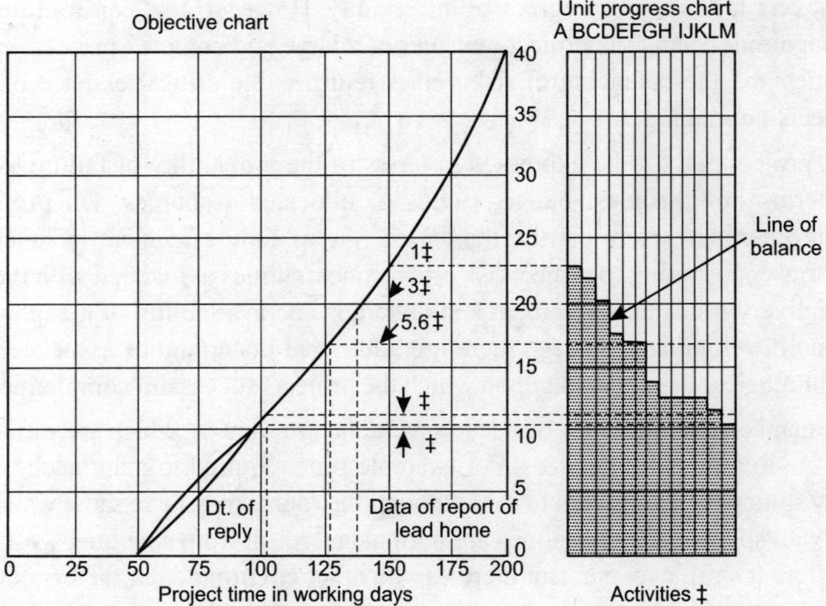

Fig. 6.79. The objective chart shows the cumulative contract schedule

The location of the segement for a particular activity depends on the reporting date for which the segement is drawn, on the lead time for that activity, and on the shape of the objective chart. Fig. 6.79 illustrates a graphical method of, locating line of balance segements. Here the reporting date is assumed to be the end of working day 100 (on horizontal axis of the objective chart). The problem is to locate the line of balance segement for activity F.

From Fig. 6.79 lead time for activity F is event 6 can be computed. It is 27 days. Starting with the date 100 (the reporting date) mark 27 days to the right along the horizontal axis of the objective chart. Then draw a vertical line from that point on the horizontal axis (127 days) to the objective chart curve, then a horizontal line from that point on the curve to the column for activity *F* on the progress chart. If progress of actual work is above this horizontal line, it gives an idea that activity F is going ahead of schedule for these many number of units. This horizontal line for activity *F* on progress chart is called line of balance for activity *F* only. If actual progress is below this line it gives an idea that the progress of activity *F* is lagging behind for these number of units.

Hence such segments are drawn for all the activities and line joining all these segments on progress chart is called Line of Balance of project on that prescribed date. A LOB for 100 days (reporting date) is shown in Fig. 6.79.

These comparisons provide the manager with needed information in making decisions about working overtime, transferring men, increasing the size of crew or employing additional equipment. Since the LOB for a particular date of reporting can be drawn at any time, the LOB technique can and in project planning and scheduling. Problem areas can be identified while there is still time to take corrective steps and the effect of any change can be evaluated. Finally the technique provides the project manager with a means of co-ordinating the works of sub-contractors to ensure that repetitive construction projects are completed on schedule.

6.20 PERT

All decision makers face varying degrees of uncertainty. However, few confront uncertainties of the type and magnitude facing the project manager of a large and complex project. In fact, dealing with uncertainties and the balancing of risk verses return is the critical essence of most project management decision-making.

In general, project risk can be expressed in terms of the probability of failure to meet project objectives in terms of schedules, budget, funds or allocated resources. On project involving research and development, there can be the added risk of failure to meet technical objectives. Among the factors contributing to project risk are the uncertainties associated with the project task, performance, delivery times for equipment and materials and availability of manpower resources. For research and development projects, there are additional uncertainties associated with rate of progress of technological development upon which the projects successful completion may depend.

Task performance variability may be due to several factors, among which are variation in human behaviour and environmental influences. For example time required to paint each room of a new hostel may vary somewhat from room to room though the rooms may have same wall configuration and amount of wall space. These variations are normal and result from changing worker motivation and fatique, different skill level etc. But there can be other environmental factors, such as relative humidity of air from day to day which affects the rate of drying, temperature variation, rains etc., which may contribute to delay in the start of some tasks.

Uncertainties of nature of work also contribute to project task variability. An example to show the variation in the time duration due to nature of work is an excavation task. It highly depends on composition and condition of the subgrade material to be excavated which may not become completely known until the excavation is already under way.

This uncertainty is of measure importance in projects involving research and development. This then depends on probability of occurrence of each event/activity in prescribed duration. It may be difficult for project manager to assess the probabilities related to completion of each task. Under such cases PERT which takes care of certain probabilities being a three time estimate network can be applied more effectively. There are many other methods also which take probability into account such as PROMT (Project Management Technique) can be used but they have beyond the scope of this book hence not covered.

The credit of introducing PERT originally known as Program Evaluation Research Task goes to Special Project Office (SPO) of the Navy Bureau of Ordinance who was charged with the overall management of the Polaris Missile Program. At present this method is called as Program Evaluation and Review Technique.

Basically there is not much difference between CPM and PERT. It consists the number of events arranged in a sequential and systematic order, i.e. basically PERT is an event oriented network. The first element in the PERT system is "section of specific identifiable events that are planned to occur in a Project." These events are also known as milestones in PERT. Secondly is to arrange these events in a systematic order depending upon their interdependencies. Next job of a PERT planner is to estimate the durations required to move from event to event together with a measure of uncertainties involved. The fourth element is the analysis of the network which highlights the critical events. Hence the basic methodology of CPM and PERT is the same.

It is a probabilistic, cyclic, and directed network. It is mainly used in Research Oriented Jobs, where duration for each activity is not fixed and mainly depends on theory of probability.

PERT Time

PERT uses three time-estimate for an activity viz. optimistic time, pessimistic time and normal time, *i.e.* most likely time.

Optimistic time 'a' is the estimated minimum time required by an activity if all the essential conditions required for completion of this activity are exceptionally favourable. The most likely time 'm' is one which is likely to occur more often than any other value. It is based on experience and judgment of similar activities being executed under similar conditions. This is the same as the CPM estimate, as one would make for an activity.

Passimistic time 'b' is the estimated maximum time for an activity if all the essential conditions are unfavourable.

This range of uncertainty associated with the actual time required to perform the activity. On the basis of these estimates, it is possible to derive the probability that a certain project would be completed before a specified date. The PERT group assumed that these three estimates would fall on a bell shaped (beta) curve. There was no proof available for this assumption, but lacking on empirical basis for choosing a specific distribution. The beta distribution was historically accepted as a mathematics model for activity duration times as shown in Fig. 6.80.

Based on Beta distribution, following formula is used to express the t_e, *i.e.* equivalent time or expected time of activity

$$t_e = \frac{a + 4m + b}{6}$$

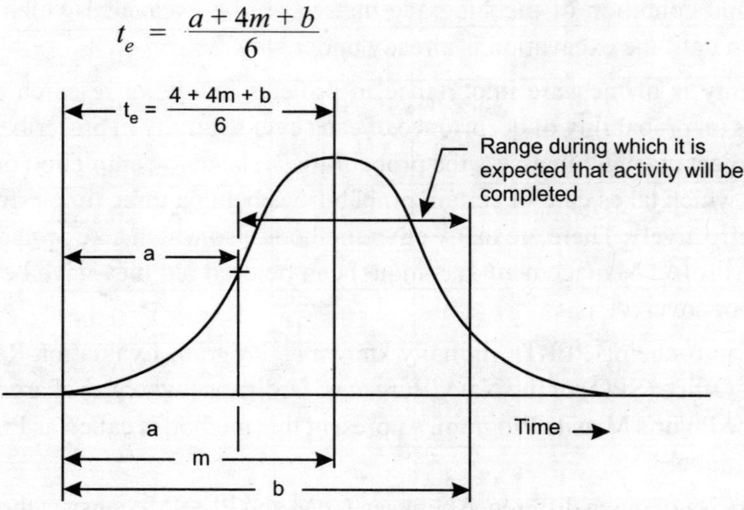

Fig. 6.80. Beta distribution curve.

Hence we reduce three times to one equivalent time. Then a variance is computed for each activity which is a function of range $(b - a)$ during which an activity is expected to be completed. Quantitatively if $(b - a)$ is small there is more certainty of meeting the time estimate and if $(b - a)$ is large, there is less chance. With this difference, a standard deviation, and variance are calculated for the entire project.

Hence standard deviation $= \sigma = \frac{b - a}{6}$

and variance $v = \sigma^2 = \left(\frac{b - a}{6}\right)^2$

By using these values, the probability of meeting a schedule event time is calculated on the following basis :

$$\frac{\text{Schedule date} - \text{Expected date}}{\sqrt{\text{Variance of event involved}}}$$

This value is then used for a statistical probability table. If an answer of 0.5 is achieved, there is a reasonable probability that the date will be met. A high probability of meeting the schedule is not necessarily good, for it indicates that the schedule is unrealistic. Of course, a low probability of meeting the schedule is not good either. Probability varies from 5 to 95%.

The new formula used for variance is based on research work held by Moder and Roggters. Accordingly we have

$$v = \sigma^2 = \left(\frac{b - a}{3.2}\right)^2$$

Table 6.26. (Probability Table)

Z	*Probability of Meeting Scheduled Date*	*Z*	*Probability of Meeting Scheduled Date*
2.8	0.997	– 0.2	0.421
2.6	0.995	– 0.4	0.345
2.4	0.992	– 0.6	0.275
2.2	0.986	– 0.8	0.212
2.0	0.977	– 1.0	0.159
1.8	0.964	– 1.2	0.115
1.6	0.9.45	– 1.4	0.085
1.4	0.917	– 1.6	0.055
1.2	0.885	– 1.8	0.036
1.0	0.841	– 2.0	0.023
0.8	0.788	– 2.2	0.014
0.6	0.726	– 2.4	0.008
0.4	0.656	– 2.6	0.005
0.2	0.579	– 2.8	0.003
0.0	0.500		

Multiply by 100 to find percentage probability.

Example. Following are the data for small project having 13 activities. What will be the probability schedule date (*a*) 40 days (*b*) 35 days.

Solution.

Table 6.27.

Activity	*Duration*			t_e
	a	*m*	*b*	
A	2.5	4	5.5	4
B	2	3	5	3
C	17	21	25	21
D	2	4	6	4
E	1	2	3	2

F	3	5	7	8
G	2	4	6	4
H	3	5	7	5
I	4	6	8	6
J	4	5	6	5
K	6	8	10	8
L	4	6	8	6
M	4	6	8	6

$$t_e = (a + 4m + b)/6$$

The network is plotted as shown below and standard deviation σ and variance v are calculated in table.

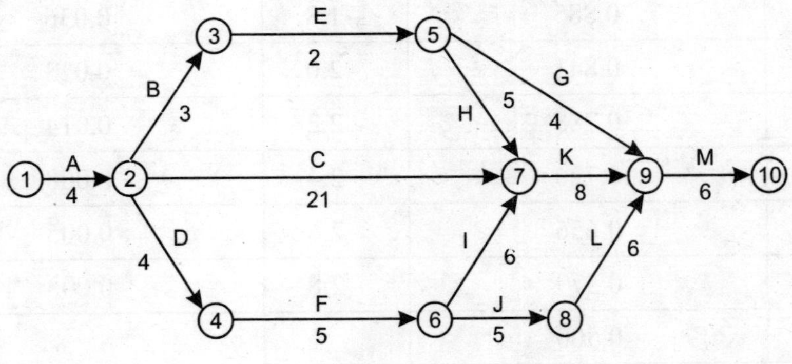

Fig. 3.81.

Duration – 39 days

Critical Activities – A, C, K, M

Critical Path -1 – 2 – 7 – 8 – 10

Sum of variance along critical path = 0.25 + 1.77 + 0.25 + 0.45 = 2.72

(*a*) If scheduled in 40 days

$$Z = \frac{40 - 39}{\sqrt{2.72}} = \frac{1}{2.72} = 0.60$$

∴ Probability (from table 6.26) = 72.6 %

Table 6.28.

Activity	*t*	$S.D.\sigma = \frac{b-a}{6}$	*Variance*, $v = \sigma^2$
A	4	0.5	0.25
B	3	0.33	0.1089

C	21	1.33	1.77
D	4	0.5	0.25
E	2	0.33	0.089
F	5	0.67	0.45
G	4	0.67	0.45
H	5	0.833	0.69
I	6	0.5	0.25
J	5	0.33	0.1089
K	8	0.5	0.25
L	6	0.5	0.25
M	6	0.66	0.45

is scheduled *t* be competed in 35 days,

then $$Z = \frac{35-39}{\sqrt{2.72}}$$

$$= 2.42$$

Therefore, probability of completing the job in 35 days is 7.7%. (Interpolated from probability table)

6.21 COMPARISON BETWEEN CPM AND PERT

Special mention should be made of a somewhat different types of application known as CPM and PERT. Both represents out growths of a production control chart to permit more effective management control of massive engineering projects. CPM is job or task oriented and has been associated primarily with construction and process industries. It is essentially a technique. PERT has evolved in military programs and is more event oriented.

The basic difference of CPM from PERT is in the method of estimating time for carrying out a particular activity. CPM times are based on personal experience, PERT times are based on statistical frequency distribution. Although project monitoring was PERT's original purpose, but now-a-days it is also used to plan works. CPM is a directed acylic-non-stochostic network whilst PERT is a probabilistic acyclic directed network. CPM is activity oriented technique. PERT originally was event oriented but now has given a way to activity orientation.

Table 6.29.

	CPM	*PERT*
1.	Activity oriented	Event oriented
2.	Single time estimate	Three time estimate

3.	Deterministic approach	Probabilistic approach
4.	Used for repetitive type of work in general.	Used for research type of work
5.	Project cost is important criteria	Cost is secondary orate
6.	Probability of completion of activity is 100	It varies from 0.01% to 0.999
7.	follows Normal distribution	Follow probabilities tribution

Example: for practice is given below.

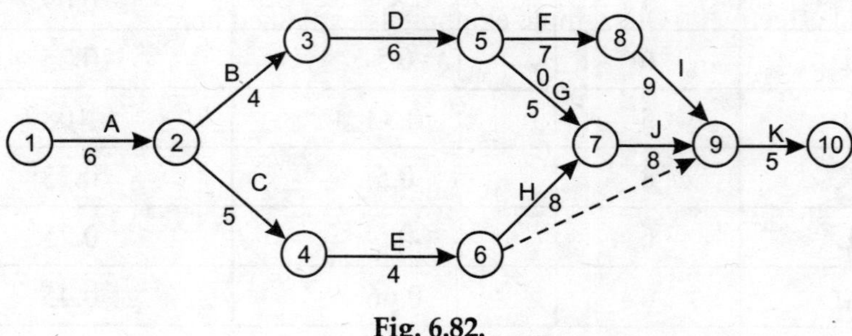

Fig. 6.82.

6.22 GRAPHICAL EVALUATION AND REVIEW TECHNIQUE (GERT)

It is a network analysis technique used in project management that allows probabilities treatment both network logic and estimation of activity duration. This technique was first described in 1966 by Dr. Alan B Pritsker of Perdue University and WWHapp. Compared to other techniques GERT is only rarely used in complex system, through the GERT approach addresses the majority of the limitations associated with CPM and PERT techniques GERT allows loops between tasks. The fundamental drawback associated with GERT is the complete programme (Monto Carlo simulation) required to model the GERT system. Development of GERT includes Q-GERTS allowing user to consider queuing within the system.

It is extensively developed and widely studies system employing probabilitic networks.

In GERT, nodes are considered to have an input and an output side, each characterised by certain logical relations with respect to connecting jobs. Two types of output sides are used which determine the type of branching that occurs from the node. Thus, deterministic branching is used when all activities preceeding from a node will be undertaken and probabilistic branching is used only when one of several emanating activities are to he performed.

6.23 ADVANTAGES OF GERT NETWORKS

1. The GERT network is quite useful in showing inconsistencies, inefficiencies, reduncancies in the procedure or policy in question.
2. In executing policy, the GERT network is quite useful in designing for the information flow procedures.
3. It is also quite useful in speeding up action at each branching node in the network.

4. Another potential use of GERT is in the planning and scheduling of the activities. They are associated with the extremely expensive programs carried out in the hostile environment of space or on the ocean floor.

6.24 MATRIX CPM CALCULATIONS

Many a times the matrix method is used to calculate the event times and critical path. The first manual solutions were by matrix. Even for computation of project times by computer, matrix method can be used effectively. One simple example is explained here.

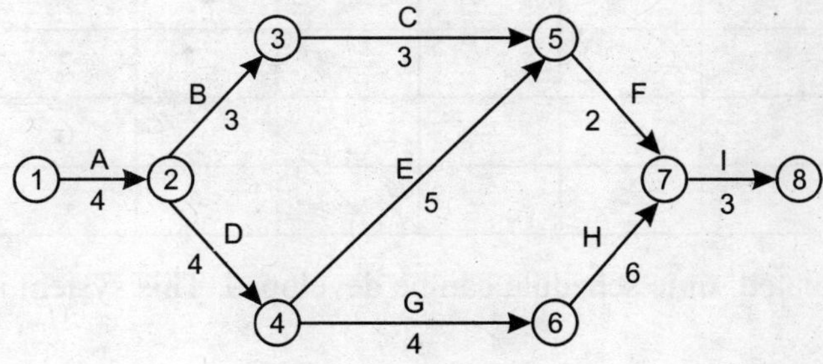

Fig. 6.83.

One network consisting of 9 activities is shown in Fig. (6.83). Duration of each activity is also shown below them.

The first step in this solution of CPM network is the preparation of matrix. The matrix consists of a size $(m - 1) \times (m - 1)$. Where m is equal to the number of events. Hence in this network $m = 8$ hence matrix grid will be of size $(8 - 1)(8 - 1) = 7 \times 7$. The matrix is prepared by listing J events on the horizontal axis and the i events on vertical axis as shown.

Table 6.30.

I/J	2	3	4	5	6	7	8
1	A						
2		B	D				
3				C			
4				E	G		
5						F	
6						H	
							I

The next step is to add the activity duration to the matrix. The activity duration is placed at the intersection of its i and j event lines. EOT at event 1 is 0. To compute the early event times for the next event, go to event 2 in the j. column.

Table 6.31. (Duration of activity)

I/J	2	3	4	5	6	7	8
1	4	–	–	–	–	–	–
2	–	3	4	–	–	–	–
3	–	–	–	3	–	–	–
4	–	–	–	5	4	–	–
5	–	–	–	–	–	2	–
6	–	–	–	–	–	6	–
7	–	–	–	–	–	–	3

Once EOT is calculated, time schedule can be developed. This system is useful for computer users.

6.25 ADVANTAGES OF CPM NETWORKS

There are many advantages of using networks planning as explained earlier. Here the advantages during various stages of planning and execution of any works are explained briefly.

- **Planning stage:-** planning is process comparing of methods or order of works to be used for any project in a sequential manner it can be executed. It is a schematically execution of works. Networks planning helps in estmating time of start and finish of various activities involved and then project as a whole. It further helps in identifying activities which are critical and at the same time the activities to become critical if proper attention is not given.
- **Scheduling stage:-** During preparing scheduling; networks helps in focussing attention on specific objectives like purchase of materials. assenting labour in advance to avoid works to get delayed. In case of more then, one project with contractor it helps contractor to make project and efficient use of available resources like men and machineries at different project sites. Monitoring available funds is also possible to contacting firm.
- **Organizing stage:-** It helps owner in awarding contract to the efficient and competitive contractor. If contract is awarded to a good competitive contractor various problems of litigations and disputes are minimized and get avoided. Now a days, qualified engineers have entered into contracting system, such they are aware of networks planning and therefore execute works sincerely and efficiently. They can understand network and give attention to proper execution of works

  Controlling and Monitoring stage: Once work order is issued to contractor their work starts. It is responsibility of owner and contractor to execute works properly with understanding and good consideration. The contractor is required to maintain quality the owner (engineering firm) is responsible for timely payment of bill and monitoring quality contraction to finish work as per required time schedule. Arrangement of resources as and when required should be made to avoid delay of work, increasing un-necessary cost of project Attention on critical activities by owner and contractor are essential requirement of this stage.

Once the project is complete, duration and cost of project are evaluated, analysed to make fahs project execution.

Other Abbreviations

Other systems of Networks techniques (used for Computerised Computation System).

PCS — Project Control System (IBM).

PMIS — Project Management Information System.

PMS — Project Management System.

PPBS — Planning Programing and Budgeting System.

RAMPS — Resource Allocation Manpower System
Resource Allocation Multiproject Scheduling.

RPC — Resource Planning and Control.

RPSM — Resource Planning and Scheduling Method.

IMPACT — Implementation, Planning and Control Technique.

CRAM — Contractual Requirements Recording Analysis and Management.

LESS — Least Cost Estimating and Review Technique.

COMET — Computer Operated Management Evaluation Technique.

CPA — Cost Planning and Appraisal.

PEP — Programme Evaluation Procedure.

PLANNET — Planning New Work.

SPERT — Scheduling Performance Evaluation and Review Technique.

QUESTIONS

1. (a) Justify the importance of network techniques in civil engineering works.
   (b) Explain importance of, Control monitor phase in network technique.
   (c) Explain PERT.
2. Prepare a time schedule for a networks, shown below and mark critical path.

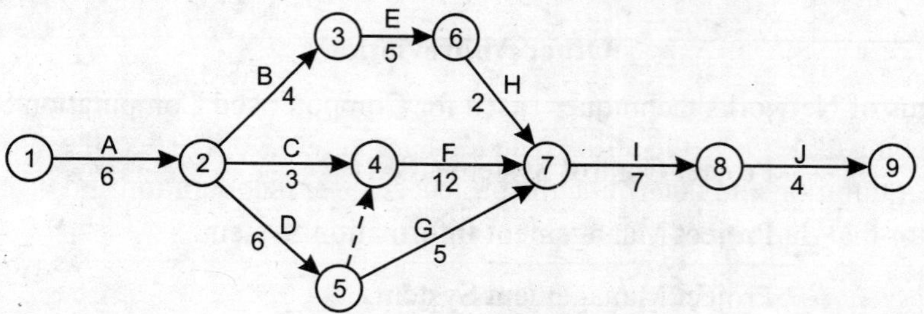

3. (a) Explain the meaning of "Critical Path Method".
   (b) A arrow diagram for a project is give in Fig. 2. Work out the minimum project duration. Mark the critical path on the diagram, giving the earliest and latest event times.

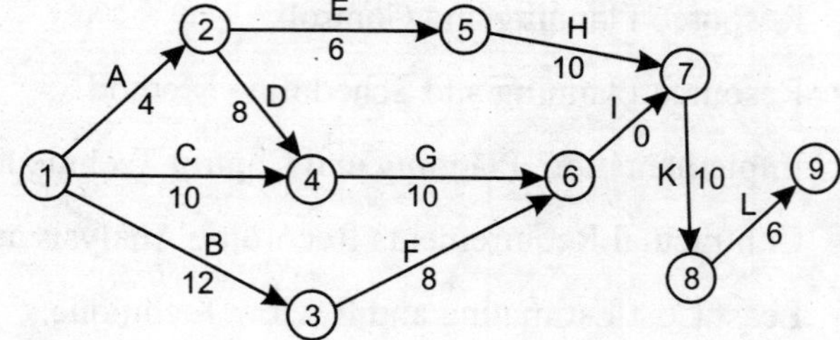

4. (a) State the important stages of a project, in general, where CPM can be applied advantageously.
   (b) Write down the limitation of the PERT technique in Civil Engg. construction.
   (c) Prepare an arrow diagram and calculate necessary data to show critical path.

Activity	*Duration*	*Activities which precede*	*Immediately follow*
A	3	None	B, C
B	5	A	D, E
C	4	A	F, I
D	7	B	G
E	6	B	H
F	11	C	H
G	6	D	J

H	4	E, F	K
I	3	C	K, L
J	6	G	M
K	5	H, I	M
L	7	I	M
M	6	J, K, L	None

5. Following table gives the utility data for a project network. Crash the network to minimum project duration and determine the project cost for that duration and determine the project cost for that duration

S.No.	*Activity*	*Normal*		*Crash*	
		Time (days)	*Cost (Rs.)*	*Time (days)*	*Cost (Rs.)*
1.	0-1	1	5000	1	5000
2.	1-2	3	5000	2	12000
3.	1-3	7	11000	4	17000
4.	2-3	5	10000	3	12000
5.	2-4	8	8500	6	12500
6.	3-4	4	8600	2	16500
7.	4-5	1	5000	1	5000

Indirect cost = Rs. 1200/- per week

6. Taking the indirect cost as Rs. 250/-per day, prepare the optimum schedule for the project network whose utility data are as follows :

S.No.	*Activity*	*Normal*		*Crash*	
		Time (days)	*Cost (Rs.)*	*Time (days)*	*Cost (Rs.)*
1.	1-2	4	600	2	800
2.	1-3	2	500	1	900
3.	2-4	6	1000	3	1750
4.	2-5	4	1200	3	1200
5.	2-5	5	1000	5	1200
6.	3-7	10	2500	5	3500
7.	4-5	5	1300	6	1300

8.	5-6	8	2000	6	2100
9.	5-7	0	0	0	0
10.	6-8	7	2000	7	2000
11.	7-8	8	1600	5	1750

7. The data for a network are as under :

S.No.	*Activity*	*Duration (Days)*
1.	1-2	3
2.	1-3	2
3.	2-3	4
4.	2-4	7
5.	3-4	6
6.	3-5	7
7.	4-6	3
8.	5-8	9
9.	5-7	5
10.	6-8	6
11.	6-9	12
12.	7-9	8
13.	8-9	10

A review of project after 15 reveals the following situation :

(*i*) Activities 1-2, 2-3, 2-4 are completed.

(*ii*) Activities 3-5, and 4-6 are in progress and need 2 and 4 days more respectively.

(*iii*) The revised estimate shows that activity 8-9 will take only 8 days but activity 7-9 will need days.

Formulate a new network after updating and determine the critical path. show the progress of the project on a bar char and indicate the modification based on updating.

8. (a) Briefly mention the principle of any two methods of planning of works (using system approach) and state suitability and limitations of CPM and PERT in Civil Engineering construction.

(*b*) The activities relationships of a highway project is as under :

Activity		Duration in Weeks	Activity precedes	Which follow
(*a*)	Move	1	0 (None)	B
(*b*)	Clear land, part 1	3	A	C, D
(*c*)	Clear land, part 2	5	B	E
(*d*)	Drainage structures, part 1	6	B	E, F
(*e*)	Drainage structures, part 2	6	C, D	G
(*f*)	Earth work, Part 1	6	D	G
(*g*)	Earth work, Part 2	6	E, F	H
(*h*)	Concrete pavement	8	G	I
(*i*)	Clearing	1	H	J
(*j*)	Move out	1	I	0 (None)

Draw arrow diagram and calculate necessary data to show critical path.

9. (a) For the following data draw the network diagram. Show critical path. How will you find the probability of completion of the project in a given project duration.

Duration in weeks

Duration	*Acvity*	*a.*	*m.*	*b*
8	1–2	3	8	13
5	1–3	2	5	8
7	1–4	3	8	7
7	2–5	4	7	10
4	3–6	2	3	10
7	4–6	7	9	11
6	4–8	4	6	8
9	5–7	6	9	12
6	7–9	2	5	14
6	8–4	4	6	8

a = optimum time estimate

m = most likely time estimate

b = pessimistic time estimate

10. (a) Explain the difference between levelling and smoothing of resources.

(b) What is updating of the Network?

(c) The following data refers to a small work. Carry out the resource levelling.

Activity		*Duration*	*Men required per day*
1–2	A	5	10
2–3	B	5	5
2–4	C	6	10
3–5	D	5	6
4–5	E	10	4
5–6	F	6	10

11. (a) Plot the network diagram and mark the critical path for the project of which the precedence relationships are tabulated below. Find the earliest and latest start and finish times:

Activity	***Duration in***	***Activity which immediately***	
	Days	***recedes***	***succeeded***
A	5	–	B
B	15	A	CD
C	25	B	B
D	30	B	E, F
E	30	C, D	G
F	30	D	G
G	30	E, F	H
H	40	G	I
I	5	H	J
J	5	I	–

(b) What are the advantages of critical path method of planning a project?

12. (a) Justify the importance of network technique in civil engineering works,

(b) What are the basic differences between CPM and PERT?

(c) Define EOT, LST, FF, non-critical activities.

13. Explain following :

(i) Construction schedules.

(ii) Financial schedule.

Draw a specimen tabulation of the above for a construction project.

14. Describe in detail how a proposed construction project can be analysed on the basis of a network diagram, bringing out clearly the significance of the concepts of dummy

activities, float times and critical path. Explain also how a time grid diagram is prepared and how this is different from the CPM network.

15. (a) Plot the network diagram and mark the critical path for the project for which the precedence relationships are tabulated below. Find the earliest and latest start and finish times. If the work starts from 30th Jan. draw the time grid diagram

Activity	***Duration in***	***Activity which immediately***	
	(Days)	***Precedes***	***Succeeds***
A	1	–	B
B	16	A	C, D
C	10	B	E
D	12	B	E,F
E	12	C, D	G
F	12	D	G
G	12	E, F	H
H	16	G	I
I	2	H	J
J	2	I	–

QUALITY CONTROL AND INVENTORY

7.1 CONCEPT OF QUALITY

Quality is an essential requirement not only in civil engineering but to any product used by any person. Present consumer is said to be quality conscious, it means todays consumer is ready to pay a little bit more if quality product is available. Today quality control has become an integrated part of an organisation.

Quality in civil engineering construction work is required at all stages of construction *i.e.* right from purchase of land, materials to its ultimate construction and possession. The idea of quality concept can be expressed schematically as shown in Fig. 7.1.

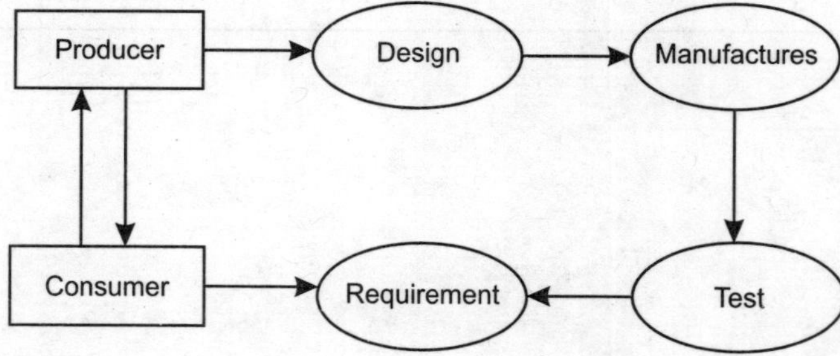

Fig. 7.1. Meaning of Quality

The quality in civil engineering comprises of purchase of good construction materials, their testing as per I.S. norms, alignment and layout, maintaining levels, gradients and slopes as the case may be, erection of formwork and its levelling in case of concreting, checking of bar bending and placing in position, maintaining of plumb in masonry construction, mixing of mortor and concrete, their placement curing and many other works. Negligence in any one of the item may spoil the quality works.

7.2 STAGES OF CONTROL

Quality control in construction is expressed in three stage in general. These stages are (1) selection of material (2) processing of the materials (3) the final product (*i.e.* execution)

1. **Selection of materials.** Right materials must be used in construction. In the case of manufacture of materials like steel, cement, tiles, bricks etc. if certain qualities of the properties are already available then problem of selection is made comparatively easy. In case of materials like stone, aggergate etc. the properties vary from place to place and therefore selection of right type of material depends upon proper determination of the engineering properties of these materials.
2. **Processing of Materials.** The raw materials of construction need processing during construction to obtain the final product *e.g.* concrete or masonry. Quality of final product, therefore depends upon adopting the proper procedure of construction in addition to proper selection of materials. In certain cases ; there are no tests to check that processing is being done properly. It is only the properties of the final product that reveals the inadequacies of processing. It is therefore essential to have good vigilance during this part.
3. **Final Product.** Checking that properties of final product helps in different ways viz.
   1. It indicates that every thing has been done according to specifications and whether quality control has been exercised fully and properly.
   2. Whenever possible it helps in rejecting the product in time and replacing it by a proper one.
   3. It also helps in knowing whether design assumptions are being fulfilled or whether any modifications are necessary in design itself and
   4. It is useful for comparison of work at different places.

The whole programme and execution of quality control operations has, therefore, to be planned keeping all thee aspects in view.

7.3 MEANS OF CONTROL

The mean in general employed for quality control in civil engineering works can be stated as follow:

1. **Testing.** It is required for determining the properties of materials which help in their selection as well as for ascertaining the quality of the final product. In the testing of materials, the adoption of proper method of sampling and procedure of testing are important aspects. There are method to collect samples to be tested to represent a bulk of material properties.

   These materials further be tested as per the standard procedure laid down by I.S.I. relevant IS codes should be referred. The equipments used for testing are also required to be maintained properly and calibrated periodically for the authonticity of test results.
2. **Supervision of the Construction Procedure.** This involves constant and direct primary supervision at the lower level of sub-ordinates, intermittent supervision at middle and senior levels of officers as well as surprise inspection. Persons must have a full knowledge of their work and also their responsibility. They must know what is to be seen and where and how often. They must be fully aware of the adverse effects on the structure because of failure on their part in the matter of supervision or inspection apart from the disciplinary action for personal failure in the execution of duties.

3. **Compilation and Analysis.** This part compilation, analysis and interpretation of test results have to be carried out systematically and properly by using statistical method at the end of every month with due regards for the specifications and design assumptions. The results should be reported regularly to appropriate authorities for further necessary action if required. The data available and collected for analysis purpose must be adequate. The frequency of testing sampling procedure must be specified. Sufficient number of samples as per standard procedure laid down should de drawn if data available are not sufficient. A set of standard specifications available should be used. There should be no ambiguity.

7.4 ORGANISATION FOR CONTROL

A separate cell should be established for quality control. Most of the departments, now a days have a separate department dealing with quality control of construction. The organisation set up for good execution of all these different aspects of the quality control work has to be carefully decided. Every project organisation is principally responsible for quality as well as for quantity of the work. Quantity at the cost of quality can not be thought of. These two aspects are complimentary of each other. The testing work and interpretation of results is a specialized work and therefore can be separated from the day to day work of execution. Supervision and inspection which are necessary for obtaining compliance of the proper construction procedure is very intimately connected with the construction itself. A division of work and responsibilities to obtain proper quality as well as quantity is therefore possible and also desirable upto a certain level of a administration on a project. Moreover, the knowledge that somebody is checking the quality of the work independently should be conductive to more vigilance in construction and should give ultimately a better final product. The various person required to quality control cell or quality control wing varies from a person of the rank of Executive/Deputy Engineer to Research Assistant, Laboratory assistant, Senior clerk, tracer etc. The class of work is as given below :

Class A : Laboratory having concrete and masonry works more than 5 lakh m^3 but less than 15 lakh m^3.

Class B : Laboratory having concrete and masonry works more than 90,000 m3 but less than 5 lakh m^3

Class C : Laboratory having concrete and masonry works less than 90,000 m^3.

For projects having combined soil and concrete/masonry testing laboratory or for the works to be executed in more than one shift or for laboratory involving concrete and masonry more than 15 lakh m^3, staff pattern can be changed. More executive engineers can be deputed under such conditions.

There should be a set relationship between the department of quality control and that of execution. They should not have any relation which will affect the quality of the work. It can never be forgotten that the ultimate aim of the project organisation is to execute a sound and stable work according to specifications at the most economical rate.

7.5 QUALITY CONTROL MANAGEMENT

Quality control is required not only at construction, but during pre-construction phases from planning through design and selection of contractors. Effective management of quality control requires a realistic and workable quality control programme. Quality of any product/service has three basis elements : quality characteristics, quality of design and quality of performance. Like any other management system, the quality control system design must be based on four basic components of quality control programme - planning, monitoring, analysis and managing. Quality control programme must therefore by designed and tailored to meet the requirement of the project during all its phases and rigorously implemented for effective quality assurance.

For effective management of quality control programme, it is essential to design a realistic and workable quality control programme for a particular project. It is also important to clearly understand and decide what is to be controlled at what state.

Basic Element of Quality. The three basic elements of quality are (*a*) quality characteristics (*b*) quality of design (*c*) quality of performance.

These are dealt here with suitable examples

(*a*) **Quality Characteristics.** This refers to one or more properties that define nature of the product for quality control purposes. As an example, quality characteristics that are commonly specified for concrete are

— compressive strength at 28 days
— slump
— type of cement and its grade
— type and size of aggregate
— w/c ratio
— surface finish
— reinforcing steel quality like plain M.S. bars, tor steel, high strength steel etc.
— shuttering pattern etc.

Quality of Design. Quality of design refers, specification referred, design concepts, tolerance and acceptability limits. The engineers/architects has to keep in mind the effect of higher quality standards and tighter tolerances on cost. Depending upon requirement of project, designs and tolerances limits are finilised. Certain examples are quoted here.

(*a*) Specifying use of M20 or M25 concrete for single or double storey building is not needed. At the same time owner should not try to overload columns not designed for that load.

(*b*) Constructions of framed structure (R.C. frames) for one or two storey building is not needed. Load bearing structure using good quality bricks will economies the building, maintaining functional performance.

(*c*) Designers many times do not give bar bending schedule, this sometimes results in mistakes and wastage.

(*d*) Specifications not defined clearly also create confusion which leads to extra cost and delay in construction (*e*) Specifying concrete finish for buried footings is not required. It is unnecessary, hence asking for it by engineers settings too high standard will make structure costly just for no reason.

Quality of performance refers quality control during actual execution ; construction of the project. Once the quality of design has been specified quality of performance is the degree to which physical work conforms to the design standard. It is the duty of site engineer to see that the construction executed strictly as per drawing and specifications laid down. He is required to refer drawing and specifications throughout the process of execution and guide the workers according. This is the most important part of quality control which is mostly neglected and as such quality control is not achieved. Construction must be executed under the strict supervision of site engineer and he should be given liberty to get the work executed as per his will, maintaining quality as well as cost. A few examples are quoted.

1. If form work is of special type and design engineer has suggested to replace form work by new one after a defined number of uses, it is duty of site engineer to check and see that form work is not used after certain reuses as it cannot resist the load of plastic concrete and fails. There are examples of form work failure due to more usage of form work.
2. If design engineer has mentioned type of aggregate, their sieve analysis, site engineer is required to set sieve analysis of aggregate done at site and reject if not confirming to specifications.
3. Type of electric wiring if specified, same must be used.

Depending on tolerance of dimensions of various civil engineering items of constructions, cost of quality control methods can vary considerably. The cost of performance is influenced by :

1. Actual construction methods at the site, including skill of available workers, type of tools and equipments available at site. Quality of raw materials also affect the quality.
2. Supervision and managerial control of engineer at site. Strict supervision and control produce good quality of work.

Knowledge and experience of supervisory staff also affect quality considerably. Skilled and experienced supervision can extract good quality work from unskilled workers.

The owners needs are expressed in the design criteria that guide the engineering and design process which produces technical specification for the project. This is in effect, sets the quality of design.

Quality of Performance. Quality assurance is the basic responsibility of engineer during the design phase and that of the site engineer during execution state. They must ensure that the appropriate quality engineering and quality control measures are adopted. Thus quality of design and their effective performance together determine the quality of final product.

The quality of performances is influenced by a few factors like

1. actual field construction method including the skill of workers, tools and plants available and quality of raw material.
2. Supervision and managerial control of labourers at site.
3. Inspection and quality control procedure applied including knowledge and skill of supervisor and reliability of their methods and tools for measuring quality characteristics specified by the designer.

Quality assurance is the basic responsibility of design engineer during planning and deign phase and that of site engineer and executive manager during construction phase. These persons must ensure that appropriate quality control measures are adopted to guarantee the construction facility of the desired quality within the performances standards specified. Thus, quality of design and quality of performance together determine the quality of final product.

One of the major advantages of this quality control management is that it provides a good fore warning mechanism. Because of constant monitoring and analysis, the construction manager is in a position to know the reasons for time and cost overturn, sub-standard quality if and any anticipated future trends. One can take the remedial measures in time, advise to take certain decisions which may have become necessary due to certain changed conditions to achieve the project objectives of completing the project in the shortest time at a minimum cost and of acceptable quality. All management tools available for project quality control are to be judiciously used to serve a particular purpose.

7.6 STATISTICAL QUALITY CONTROL

Statistics is the study of population of aggregates of individuals, rather than individual and hence this study involves formulation and application of different methods to observe the structure of the population, variation of characteristics in the population by processing the data in a compact from suitable for further mathematical treatments. The principle aim of statistical method is to summarize data, which by their very bulk are difficult to comprehend and which, therefore, need to be analysed by appropriate methods before they can be understood or made to supply useful informations. Statistics therefore can be regarded as

(*i*) the study of population

(*ii*) study of variation and

(*iii*) the study of method of reduction of data

Walter A. Sheward applied the statistical method to the problem of quality control for the first time in 1924. He prepared the first sketch of modern control charts. Statistical quality control is defined as "The method of applying statistical techniques to the collection and analysis of inspection and other data in order to achieve and maintain maximum economy in manufacturing process. It is based on the application of the theory of probability. It helps to control the quality. Statistical quality control is a function to control, to maintain and present a record of quality production.

A few common terms used in statistical quality control are :

(*a*) **Variations.** When the record of an actual measured quality characteristics such as compressive strength expressed in N/mm^2 is made, the quality is said to be expressed by variables. A variable can have all possible values between any two given limits or outside the limits.

(*b*) **Attributes.** When a record is made as to the number of items conforming to any specified requirements, it is said to be recorded by attributes. For example whether the compressive strength of concrete is within specifications or not. If it is as per specifications, It can be called good otherwise bad.

Variables are treated as attributes. It means whether one is interested in actual compressive strength *i.e.* variable or simply interested to know that strength *i.e.* variable or simply interested to know that whether the concrete is good or bad *i.e.* attribute. Hence at certain times the variables are treated as attributes. Thus the measured quality of objects in dimensions is called attribute (good or bad).

(*c*) **Unit.** It is a single member from the group of similar items, objects, individuals or any other thing.

(*d*) **Universe.** It is a common source from which the total collection of units is obtained.

Steps involved in Statistical Analysis. A few steps involved is statistical analysis of any numerical date are :

(*i*) Collection

(*ii*) Presentation

(*iii*) Analysis and

(*iv*) Interpretation.

These have been dealt herewith in detail with a suitable example.

1	16.5	21	9.25	41	15.00
2	21.5	22	17.00	42.	12.50
3	24.5	23	11.00	43	16.50
4	22.0	24	14.50	44	20.25
5	18.5	25	12.00	45	14.00
6	15.0	26	10.00	46	11.75
7	14.5	27	14.50	47	13.75
8	16.5	28	13.25	48	11.50
9	20.5	29	15.00	49	14.00
10	16.00	30	22.00	50	18.50
11	17.00	31	18.00	51	20.50
12	15.00	32	10.50	52	17.25
13	19.00	33	15.50	53	18.00
14	21.25	34	18.50	54	19.00
15	22.00	35	17.50	55	16.50
16	18.50	36	19.00	56	14.50
17	15.50	37	19.25	57	15.50
18	20.50	38	24.00	58	16.00
19.	17.50	39	16.25	59	17.00
20	20.75	40	11.50	60	8.00

The manner in which the data are collected is very important. It is essential that the data should reflect the universe. It is therefore, necessary to ensure that the procedure prescribed for collection of any data is strictly followed. Equipments used to collect the data or for testing purpose must be reliable and uniformity is required to be maintained in either collection of data or for testing purposes. The next step collection of data consists of summerisation and presentation in form a suitable for analysis and interpretation. For example data of compressive strength of concrete in N/mm^2 is collected for 60 concrete cubes, details of which are as given below.

By merely inspection these individual records, it is not possible to get any definite knowledge of the materials as a whole concreting and character measured. It is therefore necessary to put some order in data before deriving a useful informations from them. The first step therefore consists of classifying the observations and to study their resulting distribution. These collected data can be summerised in different ways *i.e.*

(*i*) by preparing frequency table

(*ii*) by calculating various measures of central tendency and dispersion.

This process of summarising data is called as analysis of data.

There are various methods of analysis of data, but the most common method used for concreting is by frequency distribution. This method is also common for other civil engineering data analysis also.

Frequency Table

The statistical data is generally a number of observations of a characteristic. Let us consider data of 60 concrete cubes, the compressive strength of each cube is as given in tabular from. The data of these 60 cubes can be summerised in a frequency table. In this frequency table, the number of items are grouped into classes and number of observations in each class noted. This step consist of range in which observations are spread (*viz* minimum to maximum value) *i.e.* divided into different classes ; each class being of uniform interval say 2.0 units. Then each observation is put in one and only one class depending upon the reading. The smallest and largest value of observations are 9.25 N/mm^2 and 24.5 N/mm^2 respectively. Thus there is no observation below 9.00 N/mm^2 and above 25.00 N/mm^2 and therefore the classes can be formed as 9-11, 11-13, 13-15, 15-17, 17-19, 19-21, 21-23 and 23-25 *i.e.* a total of 8 classes. The first observation is 16.5 N/mm^2 and therefore it falls in the class of 15-17. This way each observation is allotted to its respective class. The number of observations in each class is called as "Frequency", while the difference between the limits of each class is termed as the class-interval. The frequency table from above data is as given below :

S.NO.	*Classes*	*Number of Observations i.e. Frequency*
1	9-11	4
2	11-13	5
3	13-15	13

4	15-17	12
5	17-19	13
6	19-21	6
7	2123	5
8	23-25	2
	Total	**60**

The aforesaid presentation of frequency distribution is in the form of open class interval in the sence that exact terminal values in the class interval are repeated in preceding or following intervals. However, it is always advisable to make the class intervals mutually exclusive as 9 to 10.9, 11 to 12.9, 13 to 14.9, 15 to 16.9 17 to 18.9, 19 to 20.9, 21 to 22.9, 23 to 24.9. In practice, however, the presentation in either of the two forms is used because mid point of the moderately large class interval is unaffected for general calculations, in both such cases.

The above table is now comparatively easy to analyse and to interprete the results than that of earlier data. If number of data is very large, it becomes more easy to interprete. In this summeried data, value of individual unit is lost and therefore the observations are now available for a particular class, however it is assumed that the value of particular unit belongs to that class *i.e.* the value of that particular unit is equal to the mid value of that particular class. This loss of information will be more when number of classes are less *i.e.* when the size of class interval is made large and the *vica versa*. Thus in preparing a frequency table, the choice of number of classes to be adopted and class interval play a very important role. The choice is normally made by taking into consideration the type of data being grouped. In choosing a class interval, certain important considerations should be made. These are :

1. The class interval should be of uniform width and of such size that the characteristic features of the distribution are displayed.
2. The range of the classes should cover the entire range of the data and the classes must be continuous.
3. As a general rule, the number of classes should be about 15 and never more than 30 and not less than 6. The following class intervals are generally used in a few civil engineering data.
   Concrete – 1.5 N/mm^2
   Mortar – 1.0 N/mm^2
4. It is convenient to make the mild point of a class a whole number. The class interval can be determined approximately by use of following formulae,

If $\qquad I =$ Range

$\qquad N =$ Number of tests

then

$$I = \frac{\text{Difference of Maximum and Minimum Values}}{1 + 3.222 \log_{10} N}$$

One important advantages of the frequency table is that, it enables to easily locate the central value about which other values are scattered. However, location of exact centre value is difficult. Therefore, for detailed analysis one has to adopt difference measures of central tendency and dispersion.

Measure of Central Tendency

It is most common method used. The various terms used are mode, mean and median. These are explained in brief. "Mode" is the value of the variable which occurs most frequency or is repeated the largest number of time. "Median" is the value which is located in the middle of a series and divides the series into two equal halves, half the number of observations lying above it and half below.

"Mean" is defined in three ways *i.e.* Arithmetic mean, geometric mean and harmonic mean and are used as measure of central tendency. The arithmetic mean is not commonly used in statistic and therefore by mean, it is generally assumed to be arithmetic mean, unless it is specially stated. The arithmetic mean *i.e.* mean is the sum of individual item in the series divided by the number of items in the series and in mathematical notations it is expressed as

$$\text{Arithmetic mean} = \bar{x} = \frac{\sum_{i=1}^{n} x_i}{n}$$

When $\sum_{i=1}^{n}$ stands for summation of n different values of x *i.e.* $x_1, x_2, x_3 .... x_n$ and n is the number of observations. When the data is in a particular series is summerised in a frequency table, the arithmetic mean x_r is given as

$$x_r = \frac{\sum_{t=1}^{n} f_i x_i}{n}$$

where f_i is the frequency of x_i and $n = \sum_{t=1}^{n} f_i$ and r is the number of classes adopted in frequency table.

Dispersion

Any average fails to give a complete description of the frequency distribution. The degree to which the items included in the original distribution depart or vary from the central value (*i.e.* arithmetical mean) is also necessary. This variation from the central value is called the dispersion. The average by itself is of little significance unless the degree of variation in the given distribution is known. If the variation is so great that there is no pronounced central tendency, an average has no significance. With a decrease in the degree of variation, an average becomes increasingly significant. Whether a single frequency distribution is being described or comparison is being made with other distribution, the measure of central tendency if supplemented by the measure of variation, it becomes more meaningful. Different measures to obtain the degree of variation or dispersion are described below.

(*i*) **Range.** The range distribution is the difference between the largest and smallest observations and gives us some idea of the amount of variability present.

(*ii*) **Mean Deviation.** The mean deviation of a series of observations is the arithmetic mean of their deviations from an average value (arithmetic mean). In the process of summation and averaging, the algebraic signs of the deviations are disregarded. The value of the mean deviation is least when the deviations are measured from the mean.

(*iii*) **Variance and Standard Deviation.** The form of frequency distribution curve showing the percentage of areas included in one, two and three standard deviations - measured both ways plus and minus about the arithmetic mean is given below :

(*i*) The 'variance' is defined as the mean of the squares of differences of individual observations from the arithmetic mean. It is calculated by the formula

$$\text{variance,} \quad v = \frac{\sum_{t=1}^{n}(x_i - \bar{x})^2}{n}$$

This formula can be written, taking into consideration for loss of information due to grouping.

$$v = \frac{\sum_{i=1}^{n}(x_i - \bar{x})^2}{n} + S$$

where $S = \frac{I^2}{12}$ where I = magnitude of class interval

(*ii*) Standard deviation is the square root of variance and is more commonly used in practice and denoted as σ

$$\therefore \quad \sigma = \sqrt{v^2} = \sqrt{\frac{\sum(x_i - \bar{x})^2}{n}}$$

(*iv*) **Coefficient of Variation.** The above measures of dispersion are not absolute being expressed in units in which the observations are recorded. A measure of variation which is independent of the unit of measurement and therefore useful for comparison between different population is provided by the standard deviation expressed as a percentage of the mean. This measure is known as the coefficient of variation and is given as

$$= \frac{}{\bar{x}} \times 100$$

The calculations of various measures of central tendency and dispersion considered above are explained below with the example of concrete 1 : 2 : 4.

Example 1. *For ungrouped data :*

S.No.	*Compressive Strength, N/mm²*	*Deviation* $\bar{x} - x_i$	*Square of Deviation* $(\bar{x} - x_i)^2$
1.	17.5	+ 3.18	10.1124
2.	19.5	+ 1.18	1.3924
3.	21.5	– 0.82	0.6724
4.	19.5	+ 1.18	1.3924
5.	22.5	– 1.82	3.3124
6.	20.75	– 0.07	0.0049
7.	26.67	– 5.99	35.88
8.	24.00	– 3.32	11.022
9.	23.00	– 2.3	5.3824
10.	25.75	– 5.07	25.70
11.	8.50	+ 12.18	148.35
12.	19.00	+ 1.68	2.8224
Total	$\Sigma = 248.17$		$\Sigma = 246.0437$

$$\therefore \text{ Mean } \quad \bar{x} = \frac{248.17}{12} = 20.68$$

$$\therefore \quad \text{Standard deviation} = \sigma = \sqrt{\frac{246.0437}{12}} = 4.528 \text{ N / mm}^2$$

$$\text{and coefficient of variation} = \frac{4.528}{20.68} \times 100 = 21.895 \ \%$$

Example 2. For grouped data of values quoted earlier of compressive strength of 60 concrete cubes:

Range of compressive strength N / mm²	*f*	*d*	*f × d*	*f × d²*
9-11	4	– 3	– 12	36
11-13	5	– 2	– 10	20
13-15	13	– 1	– 13	13
15-17	12	0	—	—
17-19	13	0	—	—
19-21	6	+ 1	6	6
21-23	5	+ 2	10	20

23-25	2	+ 3	6	18
$n =$	$\sum = 60$		$\sum fd = -13$	$\sum fd^2 = 113$

$$E_1 = \frac{\sum fd}{n} = \frac{-13}{60} = -0.2166$$

and $\quad {}^2_1 = 0.0469$

$$E_2 = \frac{\sum fd^2}{n} = \frac{113}{60} = 1.8833$$

$\bar{\ } =$ average value $= A + iE$

where A = value of assumed mean (having maximum time)

i = range of interval

$\therefore \quad A = 17, i = 2.0$

$$\therefore \quad \bar{\ } = 17 + 2 \times (-0.2166)$$

$$= 17 - 0.4332$$

$$= 16.5668 \text{ N / mm}^2$$

$$\therefore \quad \text{Standard deviation} = \sigma = \sqrt{\ _2 - \ ^2_1 - \frac{1}{120}}$$

where $\frac{1}{120}$ is sheppards correction for grouping

$$= 2\sqrt{1.8833 - 0.0469 - 0.00833}$$

$$= 2.704$$

and coefficient of variation $= \frac{2.704}{16.5668} \times 100$

$$= 16.32 \text{ percent}$$

However this result vary considerably. Further assumption of mean affect the results of standard deviation and coefficient of variation and particular when number of variables is even, assumption of average value becomes rather confusing.

7.7 INSPECTION

It is an art of comparing raw materials or finished product with that of established standards. It is an instrument of quality improvement. Inspection gives more importance to the quality of past output whereas quality control gives importance to the future product. Following are the important methods of inspection

1. Destructive inspection
2. Efficiency inspection

3. First piece inspection
4. Key operation inspection
5. Percentage inspection
6. Pilot piece inspection
7. Pre-assembly inspection
8. Trial run inspection
9. Working inspection
10. Functional inspection or inspection of assembly for proper functioning.

These inspections are required to determine the fitness of the article made. The main objective of inspection are to detect and isolate faulty material before it is used, to move or dismantle defective parts of defective construction, to bring these defects to the knowledge of concerned person to avoid its repetition, to detect sources of weakness of defects in final product and to get assured that the defective piece does not reach to consumer. It is essential to promote reputation for quality, thus minimising customers complaints.

The inspection is therefore carried out of in following steps :

(*a*) Incoming material inspection

(*b*) Process inspection

(*c*) Metrology inspection *i.e.* measurements of dimensions

(*d*) Equipment and tool control

Various methods of inspection, given earlier are used at required stages. Non destructive testing methods have also developed and mainly used to test the final product without damaging it.

7.8 SAMPLING AND SAMPLING TECHNIQUES

This word sampling in not new. We do sampling in day to day practice also. It is used to select certain items. For example when one goes to purchase any thing from market, he surveys, he collects sample if possible and then choose one of those samples collected. This sample taken represents the quality of the lot. If the sample is found good, it is assumed that total lot is good or vica versa. The same method can also be used in civil engineering or any of other field of engineering. In case bricks, samples of bricks are taken and tested in laboratory. If the results confirm to IS requirements or owners requirement, the lot is assumed good and the whole lot or part of it is purchased as required.

It means sampling inspection is conducted to evaluate the quality of a lot of that product, through the careful examination of a small number of units of products drawn from the lot. The purpose of sampling is to test that appropriate material as per specifications laid down have been used. In other word constructions has been executed as per norms laid down in contract deed and materials confirming to requirements are used. The main purpose of acceptance sampling or sampling inspection is to distinguish between good and bad and to classify the lots according to their acceptability or non-acceptability.

Sampling techniques are usually classified into four categories :

1. Single sampling plan
2. Double sampling plan
3. Multiple sampling plan
4. Sequential sampling plan

Single sampling plan is a type of sampling inspection in which the decision to accept or to reject a lot is based on the inspection of a single sample. A plan of this type requires following measures to be specified.

The lot size (the number of items in the lot) = N

The sample size (the number of sample size) = n

The acceptance number (the number of defective items allowable in the sample) = C

Rejected number = r

It means test samples n from lot N and if sample contains C or less than C defective samples accept lot N and if number of defective sample is equal r or more than r, reject lot N. This system of sampling therefore calls for decision on acceptance or rejections of a lot on the basis of the evidence of one sample from that lot.

Double sampling plan involves the possibility of putting the decision on the lot until the second sample is taken. The decision of acceptance is based on the results of the two samples combined. It gives second chance to eliminate doubtful lots.

If number of items in lot N

first sample size and second sample sizes are n_1 and n_2

first acceptance and second acceptance number are C_1 and C_2

First rejection number = r_1

Second rejection number for both samples combined = r_2

Then number of samples in combined = $n_1 + n_2 = n$

Inspect sample n_1 first and decision is taken as shown below :

1. If sample contains C_1 or less than C_1 defective items accept the lot N.
2. If sample contains defective items between C_1 and r_1. Inspect a second lot of n_2.
3. If the first and second sample combined *i.e.* $n_1 + n_2$ combined
   (*a*) Contains defective pieces less than C_2 accept the inspection lot N.
   (*b*) Contains defective items equal to more than C_2, reject the inspection lot N.
4. If sample n_1 contains more than or equal to r_1 defect items, reject lot N.

Therefore there are four possibilities for acceptance or rejection of a lot in double plan method of sampling

(*i*) acceptance after first sample

(*ii*) rejection after first sample

(*iii*) acceptance after seconds sample

(*iv*) rejection after second sample.

Inspection of first sample of 50 items ($n = 50$)

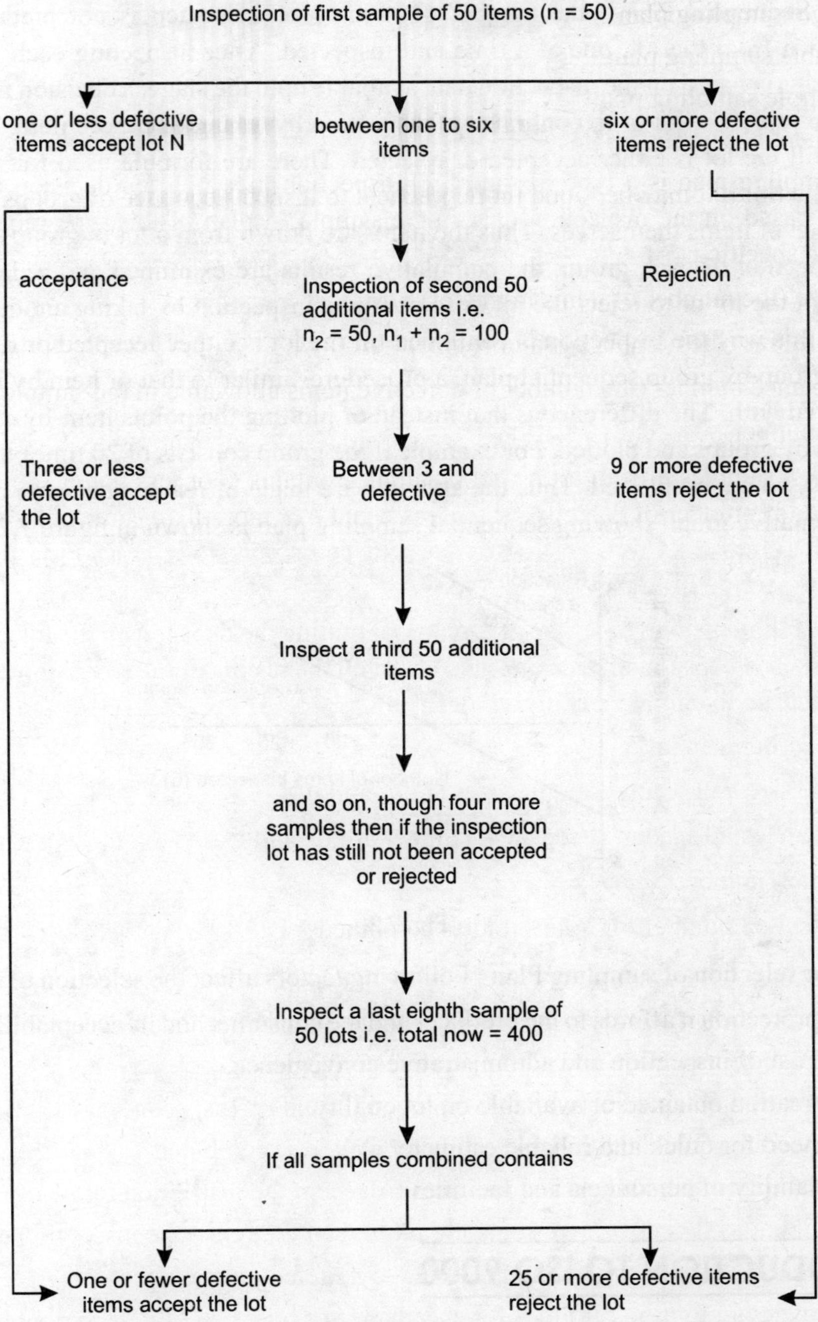

In multiple sampling plans, more than two samples are taken before decision to accept or reject the lot is reached. This is an extension of the double sampling plan and is specified by the lot size N. A schematic representation is as shown :

There are two type of sequential sampling plans namely :

(*a*) item sequential plans

(*b*) group by group sequential

In "Item Sequential Plan", the number of item to be inspected is not predetermined. The items are drawn from the lot, one at a time and inspected. After inspecting each item and using appropriate criteria based on the inspection data available upto the stage, a decision is made whether to accept or to reject the lot or to continue the inspection by taking one more item. This procedure is continued till the lot is either accepted or rejected. There are formula used for acceptance and rejection. Some times it may be found more practical to inspect sequence of groups of items rather than a sequence of items themselves. Thus the items are drawn from a lot in groups and inspected. After the inspection of each group, the cumulative results are examined and a decision is made either to accept the lot or to reject the lot or to continue inspection by taking another group of the same size. In this way the inspection is continued till the lot is either accepted or rejected. For the operation of group by group sequential plan, a procedure similar to that of item by item sequential plan is followed with. The difference is that instead of plotting the points item-by-item, the results are calculated by groups and plotted. For example if the group consists of 20 times then 40, the first 40, the first 60, so on are plotted. Thus the attempts are made of reach a decision at less frequent intervals. A tentative graph showing sequential sampling plan is shown in figure 7.2.

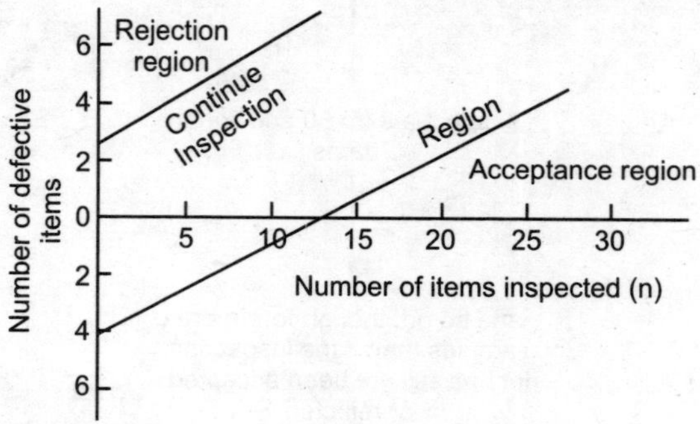

Fig. 7.2.

Criteria for selection of sampling Plan : Following factors affect the selection of sampling plan:

1. The protection it affords to the producer and the consumer and its acceptability to producer.
2. The cost of inspection and administrative convenience.
3. Information obtained or available on lot quality.
4. The need for quick and reliable estimate.
5. Availability of personnels and facilities.

7.9 INTRODUCTION TO ISO 9000

The International Organisation for Standards (ISO) is a body of international level experts who have developed a set of guide lines (known as ISO 9000) to help in the development of healthy, perfect and economical system with respect to various products/services of a company/organisation. The guide lines show the way to quality reliability and a continual survival. The ISO codes depending upon level of assurance specify standards about all following elements

— Management responsibility

— Quality system principles

— Auditing of quality system
— Economics-quality costs
— Quality in marketing
— Quality in specifications and design
— Quality in procurement (Purchasing)
— Materials control and traceability
— Inspection and testing
— Inspection measuring and test equipments
— Non performance control
— Corrective action
— Handling, storage, packaging and delivery
— After sale service
— Quality documentation and record
— Personal (training)
— Product safety and liability
— Use of statistical methods

The Bureau of Indian Standards (BIS) *i.e.* Government of India published IS-14000 series incorporating ISO 9000 series in totally for Indian users.

The need of introducing ISO-9000 due to changing scenario on quality front. Today customer is highly quality conscious and choosy. Buyer market has emerged with increased competition investment. Expectations about services, assurance and improvement keep rising. In order to achieve these objective, it was felt necessary to define a profile of well defined procedures, instructions, activities and records for each product and services of a company or organisation ISO 9000 is a outcome of these requirements.

Quality has been defined by ISO as seen total of characteristics and features of product or service having bearing on its ability to satisfy all stated and implied needs. In the developed countries, application of ISO 9000 quality has already been treated in following services.

1. Purchasing, contracting, management, distribution
2. Technical consultancy, photography, test process.
3. **Administration.** personal, office, management computing
4. **Professional.** Architecture, engineering survey, instrument, security, project management, quality management, consultancy, training and education.
5. **Financial.** Accounting, banking, insurance, personal property, services
6. **Trading.** Wholesale marketing, retail stockiest, distributors, packaging
7. **Scientific.** Research and development, studies and decision.
8. **Utilities.** Cleaning, waste management, water supply, ground maintenance, electric supply, gas and energy supply, fire policy and other policy services.
9. Hospitality services, catering, hotels, tourism, entertainment, radio, television, leisure

10. **Communication.** Airport, rail, road, sea, transport, telecommunication, postal, data management etc.
11. **Health Services.** Medial staff, doctors, dentist, clinics, hospitals, ambulances, laboratories, optician etc.
12. Maintenance, electricity, machinery vehicles, heating system, air conditioning, building, computers etc.

The series of ISO 9000 and its equivalent Indian Standard and the applicability of each is given below.

International standards	*Equivalent Indian Std.*	*Applicability*
ISO 9000	IS-14000	Gives knowledge required during the selection of proper ISO code.
ISO 9001	IS-14001	Gives a set of guide lines that help manufacture.
		Assurance of customers about production and
ISO 9002	IS-14002	installation qualities.
ISO 9003	IS-14003	Final inspection and testing.
ISO 9004(1)	IS-14004(1)	Principles and procedure to be adopted by organisation wishing to establish for itself quality management system.
ISO 8402	IS-1404(2)	Elements to quality management to services.
ISO 8402	IS-13999	Gives knowledge required during the selection of proper ISO and IS code.

Elements of ISO-9000 Quality Management System Relevant to public services/public administration. Characteristics of service could be measurable (quantitatively) or comparable (qualitative). For example - facilities, capacity, number of personnel, quantity of materials, waiting time, delivery time, processing time, hygiene, safety, reliability, security, responsiveness, accessibility courtesy, comfort, aesthetics, environment, competence, dependability, accuracy, completeness, state of art, capability and effectiveness. The control over these services is achieved by checking their performance to relevant standards stipulated in the contract or order. Involvement of customers, consumers, masses in management responsibility, quality system structure, personnel and material resources take care that the customer of public services is always at the focus. Quality policy, responsibility and commitment at the higher level of management percolates to each functional level through documented by :

— Grade services to be provided

— Corporate policy for maintaining organisation image/reputation for quality in service.

— Objectives for quality of service

— Approach to be adopted

— Role of organisation personnel at each level.

Quality objectives are identified, established and documented ensuring

— Consumers satisfaction consistent with professional ethics standards

— Continuous appraisal of services

— Due consideration to society, environment, efficiency etc.

These quality objectives are achieved through

— well defined needs of consumers to avoid their dis-satisfaction.

— preventive action and control to avoid their dis-satisfaction.

— optimizing costs for requisite quality of services.

— continuous review of services requirements and achievements.

— prevention of adverse effect on society and environment.

Authorities achieve this quality responsibility through well defined and established effective control, education and implementation of service quality at each stage, consistent with means and methods. Management review plays an important role in it. Motivation of personnels, Training and development, proper communication, materials handling, proper documentation system. quality audit, interact with customers and consumers etc. help to achieve quality control and quality management.

The social responsibility is the prime functions of every public organisation. As required in ISO 9000/IS 14000, series, it has to be satisfied that all the relevant laws, statues, rules, regulations, codes, environment/health/safety factors as well as energy and material conservation aspect have been taken due care of while designing the services specifications and the service control specifications. It is thus seen that ISO 9000/IS 14000 series of quality system standards take care of all aspects relating to effective and efficient management system. The overall responsibility and commitment to quality management is to be ensured at topmost level.

7.10 INVENTORY MANAGEMENT

Construction activity accounts for more than 60 percent of national investment in the development outlays. Considering the total national outlay which has been spent for any development work, whether it is for power, steel, health, care, building construction etc. the major portion of the expenditure goes into construction activity. It is of course surprising to note that of this construction activity where major portion of expenditure goes, more than sixty percent of this goes into material expenditure, rather any activity either in construction or in industry cannot be carried out without material supplies. Before undertaking any construction work either in civil engineering or any other production work anywhere in any industry ; it is essential that material must be available on hand and further there should be an assurance of its continuous supply as and when required to keep up the schedule.

Material management is the concept that integrates all the activities of planning, scheduling and controlling, right from the design till final production including delivery to the consumer. It undertakes the function of planning, organising, integrating and measuring the volume and flow of material. Proper and scientific management of material is required not only to optimize the period of construction, but also to economics the cost of construction.

It is further important to note that

1. Management are going to represent a larger and larger proportion of the total cost of what we make and
2. Problem of materials procurement and control are going to become more difficult and complicated in the further time.

Aims and Functions

As such proper and efficient use of material and its management are essential today. The basic aim and functions of materials management are :

1. Low cost and higher inventory turnover.
2. Continuity of supply of material as and when required maintaining quality and cost.
3. Proper and valid relations with supplier.
4. Maintaining proper record of issue and supply and use record of materials.
5. To keep knowledge of latest materials developed, their quality and use.
6. Standardization of dimensions to have simplified designs and specifications.
7. Forecasting of change in trend, designs at market and replanning according to future designs.

The functions involved in material management is illustrated in the form of a flow diagram in Fig. 7.3.

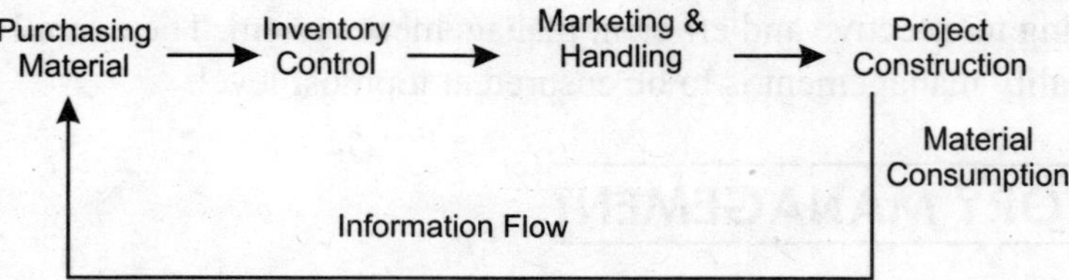

Fig. 7.3. Flow and function involved in material management

There are three basic key aspects of material management function namely.

1. Purchasing
2. Material handling and control
3. Inventory

1. Purchasing function. It is essential for management involved in purchasing function to have efficient control over purchases and expenditure fund and inventory flow of material. Purchasing must be done at reasonable price. Many persons believe in lowest price quoted. If all the suppliers are valid and reliable maintaining the quality, it is always better to purchase material from lowest bidder. But then there should be proper competition in bidders. In absence of this purchasing materials at lowest bidder may not maintain the quality work. Specification laid down must be in details and well specified having no ambiguity.

Purchasing is a service function that supports the activities of other operations. In turn it receives assistance from other operating units. Effective functioning requires a steady and reliable flow of information between concerned departments. The relationship of the purchasing structure of other parts of the system of project is illustrated in Fig. 7.4

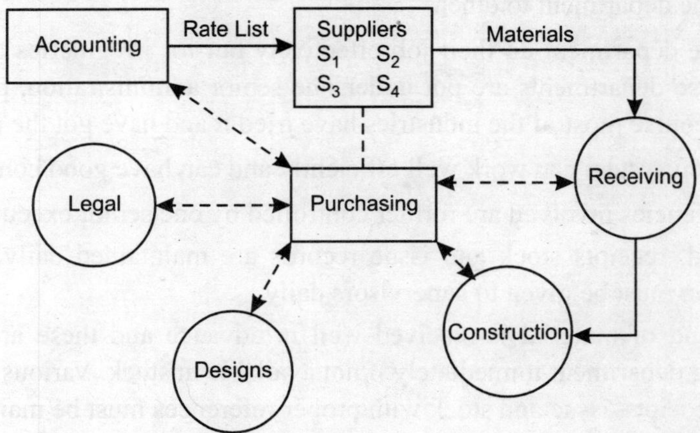

Fig. 7.4. Information and material flow in the purchasing system

The basic objective of purchase committee is to purchase right material at right time in right quantity and at reliable right rate. As far as possible is should be purchased from reliable supplier or through reliable agency. The purchasing is therefore required to maintain the list of materials required their specification and quality. The committee is also required to maintain a list of regular supplier of these various materials and their market background. It is also responsibility of the committee to see that once the material has been purchased, payments are made.

Methods of Purchasing: There are various methods of purchasing adopted each depending upto the condition, time and urgency. Almost all the materials are of course purchased from the market. The methods are :

1. Inviting tenders
2. Rate contract purchase.

In first case, tenders are invited for purchase of materials from various supplies. The rates for various items either are as given or may be quoted by supplier. The order is placed where bid is accepted. Hence it is essential to quote the quantity and quality of material in detail. In rate contract purchase, rates are invited for supply of material for a certain contact period. Once the rate is accepted for a specific period of supply it is binding on the purchaser as well supplier to stick to the rates quoted whatever may be the market rate. This method though popular today, but many supplier now-a-days have tendency to not to accept such offer due to drastic fluctuations in market price, further the rates presently are following the increasing unknown trend, no supplier likes to take risk.

2. Material handling and control. Material handling includes the movement of material from receiving and through manufacturing or in process storing to finished good storage ready to dispatch. Transportation of material from one place to other consumes not only time but consusmes a major portion of project budget. In India, the transportation cost of materials is very high, the reason being zig-zag and poorly maintained roads. Straight road not only reduces the fuel consumption, they also reduce the time of travel.

Presently, the various department of same industry like finance, purchase, transport store, issue, control, sales and various other are managed by different agencies with little and late flow of information from one department to other.

Further all these department do their job effectively but for irregularies throw responsibility on others. If all these departments are put under one senior administration, probably things will become smooth of course most of the industries have tried it and have got the result also.

Material handling system can work well efficiently and can have good control if

1. Various agencies involved are further controlled by one senior executive.
2. Use record, receipts stock and issue records are maintained daily and properly. This information must be given to supervisors daily.
3. The demand of material is received well in advance and these are communicated to purchasing department immediately if not available in stock. Various standard proformas used for receipts, issue and stock with proper references must be maintained.
3. **Inventory.** Efficient material management much more depends on inventory. Inventory, in the project context, is an idle resource, which acts as a cushion between supply and demand. Industries inventory enables the firm to meet the orders and also helps in keeping up the manufacturing cycle. There should be a proper control on inventories. It increases the profit.

In civil engineering projects, many times the works are held up due to non-availability of material. This increase the project duration, hence the administrative cost also increases. This either reduce the profit of the contractor or increase the cost of the project.

7.11 INVENTORY ANALYSIS

Various types of models are available for this purpose and suitability of model to be used can only be decided on the basis of the problem under consideration. Conflicting objectives of inventory analysis are illustrated in Fig. 7.5.

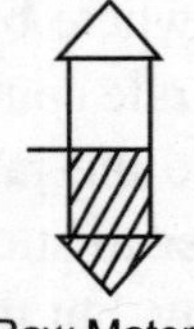

Raw Material.

Purchasing. Allowing larger inventories means bigger saving from discount on larger purchase quantities.

Finance. A large inventory means more money is invested on materials, hence can not earn interest.

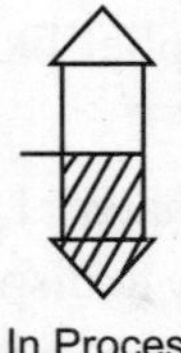

In Process

Construction. Large inventories allows to decide for successive steps in project and allow larger flexibility.

Controller. Larger inventories are subject to more breakage and increase the cost of handling and congestion.

Fig. 7.5. Conflicting objectives of material management.

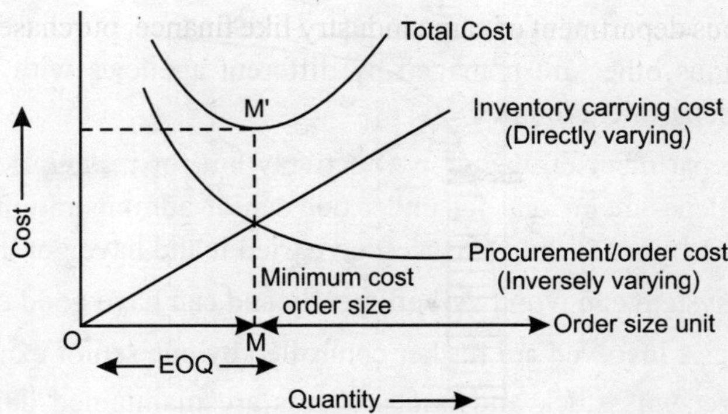

Fig. 7.6. Cost-order size-order size relationship.

The point to be noted at this stage is, analysis of inventory costs.

— costs that vary directly with size of an order and

— costs that vary inversely with order quality cost, order size relationship is shown in Fig. 7.6.

The total cost is obtained by adding inventory carrying cost and the order cost. The cost per period is minimum at M'. Hence OM represents the economic order quantity (EOQ). This can be calculated using following formula.

$$EOQ = \sqrt{\frac{2 \times \begin{array}{l}\text{Annual Consumption}\\ \text{of raw material}\\ \text{item } i.e. \text{ item/yr.}\end{array} \times \begin{array}{l}\text{Procurement cost}\\ \text{(per quantity)/order}\\ \text{Rs./order}\end{array}}{\begin{array}{c}\text{Unit cost} \times \text{Inventory carrying rate or I.C. rate}\\ \text{(Purchase cost per unit)/Rs./unit}\end{array}}}$$

$$= \sqrt{\frac{2S \times C_o}{Cu \times I}}$$

where I = carrying cost percentage of purchase cost.

S = Annual consumption

C_o = Ordering cost (Rs per order)

Cu = Unit Cost in Rs.

Therefore the design of an inventory system starts with the determination of economic order economic order quantity *i.e.* the size of an order that minimise the total inventory cost. It is evaluated quantitatively.

7.12 ABC ANALYSIS

Large organisation are required to maintain stocks of different materials. The division of inventory into three classes according to monetary usage in known as ABC analysis. It is illustrated graphically.

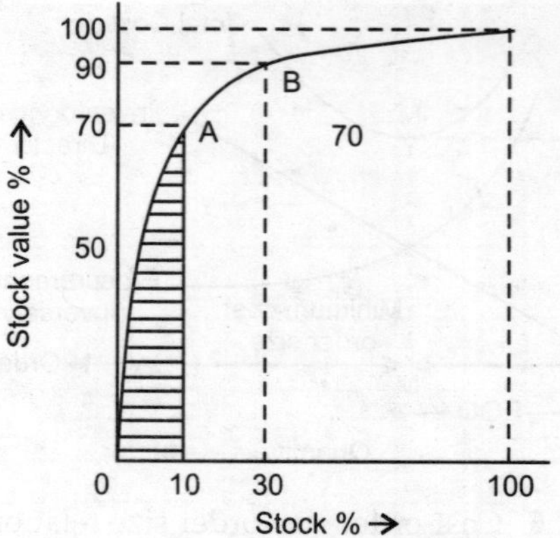

Fig. 7.7. ABC Analysis

The above Fig. 7.7 indicates a graph between the percentage of total number of items held in stock and the percentage of total stock value. It can be seen that number of items held in stock A is only 10% whilst the same of B is 20% and C is 70 percent but the cost contribution is just revere *i.e.* 70% of the stock value is for item in stock A, 20% for B and 10% for C. It shows that item stock A must be given top priority of supervision whilst C should be given minimum items B are medium valued and lies between A and C.

Thus ABC method of analysis helps us in determining which inventory items deserve most attention. This is essential because the time and record keeping activities require to control inventories cost an organisation money.

Example on ABC analysis

Following are data available from a record of store :

Item No. in store Record	*Anual usage units*	*Unit cost in Rs.*	*Annual usage in Rs.*
101	28000	1.00	3000
102	300000	1.20	40000
103	3000	0.90	350
104	105000	1.40	6000
105	5000	0.80	200
106	160000	0.70	24000
107	15000	1.50	1000
108	7000	1.10	8000

Prepare ABC ranking and its summary.

Solution. Table given below shows the percentage investment of total on that amount and their ranking

Item No in store record	*Annual usage units*	*Units Cost Rs.*	*Annual usage in Rs.*	*Annual amounts Spent*	*Investment percentage of total Annual*	*Ranking*
101	28000	1.00	3000	3000	3.45	5
102	300000	1.20	40000	48000	55.20	1
103	3000	0.90	350	315	0.36	7
104	105000	1.40	6000	8400	9.66	4
105	5000	0.80	200	160	0.18	8
106	160000	0.70	24000	16800	19.31	2
107	15000	1.50	1000	1500	1.72	6
108	7000	1.10	8000	8800	10.12	3

Ranking wise distribution and their cumulative percentage is given in tabular form below

Item No. in Store record	*Ranking*	*Percentage Investment Annual*	*Cumulative Percentage*	*Category*
102	1	55.20	55.20	A
106	2	19.31	74.31	A
108	3	10.12	84.63	B
104	4	9.66	94.29	B
101	5	3.45	97.74	C
107	6	1.72	99.46	C
103	7	0.36	99.82	C
105	8	0.18	100.00	C

Summary of ABC analysis is given

Class	*Item*	*Percentage of item*	*Cost Percentage Rs.*
A	102	two out of eight *i.e.*	74
	106	25%	
B	108 & 104	25%	19.78
C	remaining 4 items	50%	5.71

7.13 MATERIAL REQUIREMENT PLANNING (MRP) SYSTEM

Material requirement planning system (MRP system) is a technique for determining when to order dependent demand items and have to reschedule order to adjust changing requirements.

Before 1960's there were no satisfactory methods available for handling the inventory of dependent items. Inventory system was often patterned after order points, and was either misapplied or broken down into a maze of informal methods when it come to handling dependent items. There was no feasible method of keeping accurate records of the thousand of inventory items that went into project, so management relied upon the safety stocks of the order point model to keep them out too much scheduling trouble. Unfortunately, they did not always achieve that objective, but they did always make a healthy contribution to the inventory carrying and storage cost.

Thus the manual and informal control system is use before 1960, could not adequately cope with the multiple of items and complex activities that took place in an average sized project.

During the 1960's the computer opened the door to MRP, an inventory ordering system that could at one time keep up-to-date records of status of thousand of items in inventory. If also facilitated rapid rescheduling of jobs when planned order receipts got out of phase with planned requirements. MRP is not highly theoretical, it is an sophisticated inventory management technique. It is just something that works to handle the every day problem of priority and capacity.

Key features of MRP system are:

1. The time phasing of requirements
2. Planned order release
3. Generation of level requirements and
4. The rescheduling capability provided.

At each week passes, planned orders are (manually) released and any change in the requirement and due dates are entered into the computer database as they occur most (firms). The management wing of MRP then 'run' their MRP progress once a week (over week end). The computer automatically explodes the material requirements to all lower level components (sub-components) and point out the necessary planned order release dates. These are of course, based upon the individual order quantities and lead times which have previously been specified for each item. The order quantity may, by the way, represent EOQ, simply a lot size that is convient from project construction stand point.

The features of MRP system have come out largely due to availability of computers, which are not intimately link inventory management with scheduling. This enables management to reuse or eliminate the safety stock that were formerly used to overcome the uncertainties of trying to mesh to distinct activities.

Market Research

It is essential for persons involved in material management is to have latest knowledge of material. They must have the ideas of changing trend. The cost of material depends much more on demand and supply. It is also affected by development of new materials. For example the cost of electronic equipments are continuously decreasing, it is presumed that this will come to standstill in due course and then will start increasing again. Other example is cost of doors and windows. The teak wood doors and windows have become very costly newly developed timber door of well seasoned sapier wood door cost less, steel doors also less, hence the present trend is to use steel doors or plywood doors. Hence it is essential to material management to do market research and to have latest know how of material development and market cost, resources available. Now a days plastic doors have also became common.

7.14 INVENTORY MANAGEMENT

Inventory Management is carried out to determine the frequency of replacement of consumable articles and fast wearing parts. The basic purpose of preparation of articles and their management is to ensure that the same is available when required and that the cost of procurement is kept minimum. This is achieved by procuring only such qualities as would be used during the execution of the project when demanded.

The reasons of holding inventions are :

1. To create a safety between input and output. The output and input rates varies many time and inventory meet the demand when output rate is greater than input rate.
2. To ensure against delay in deliveries : When an order is placed for fresh stock it may be not be available immediately. There occurs some delay in the supply and this is called "Material lead time". To meet the demand during this lead time, it is always essential to hold stock.
3. To allow for a possible increase in output if required. Reserve of stock of raw materials is necessary to meet the sudden increase in production rate.
4. Materials and components if purchased in large qualities, they always cost less due to large discount and lower overall transportation cost. Thus there are advantages some times if large quantities of material are stored.
5. It also ensures against scarcety of materials in the market. It happen when there are fluctuations in the availability of certain materials and there exists a likelihood that material might become scarce. Hence a reserve stock of raw material is always needed to run the production smoothly.
6. As far as possible materials must be purchased when prices are low.

Objective of Inventory Control

1. There should not be excessive inventions this may be lead to wastage or un-necessary stock holding.
2. There should not be shortage of inventory, otherwise may lead of stopping of production.
3. There should be acceptable customer service.

The problem of inventory control is very complex because it affects many factors like credit terms, transportation, purchasing, material, market situation etc. The inventory control functions are illustrated in Fig. 7.8.

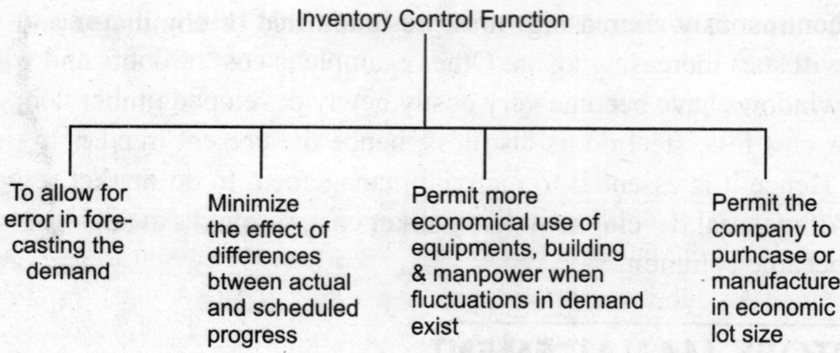

Fig. 7.8. The inventory control functions.

Steps in Inventory Control

1. Planning the material and operation of purchasing etc. and its effective acquisition.
2. Adequate physical verification of material time to time.
3. Proper storage, not be damaged during storage.
4. Security of stock.
5. Maintenance of stock, receipt and issue record of inventory and posting information sincerely and regularly.
6. Fixing responsibility of performance.

Management is interested in maintaining supply of stock which provides acceptable customer service. The inventory consumption saw tooth figure where it is assumed that there is uniform and constant usage rate. Also the replenishment is instantaneous as shown in Fig. 7.9. If replenishment taken 12 days then order is placed as shown in Fig. 7.10.

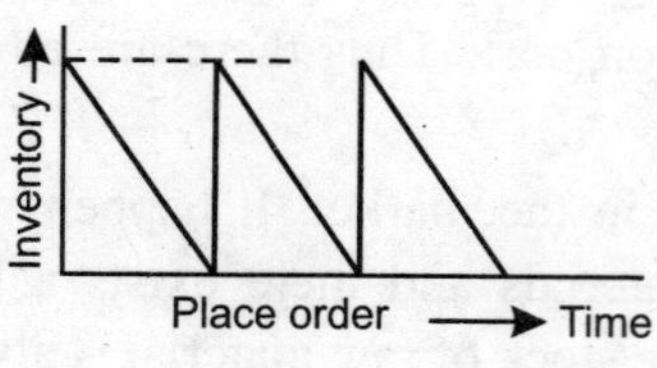

Fig. 7.9.

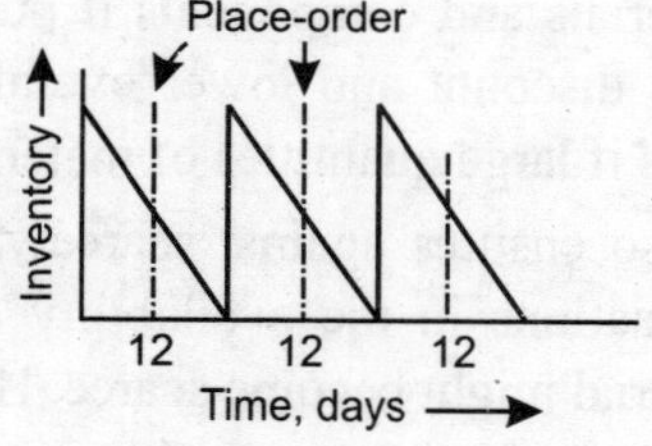

Fig. 7.10.

Here are manager has to face two basic problems :

1. Economic lot size or ordering lot size.
2. Re-order point.

7.15 INVENTORY-COST

Inventory cost consists of three types of cost.

1. Processing or ordering cost.
2. Carrying cost.
3. Stock out cost.

The third major cost is stock out cost must be minimised or eliminated and it is possible by logical planning.

The processing or ordering cost consist of

(*a*) Cost of processing the requisition

(*b*) Cost of following up and expediting orders

(*c*) Cost of receiving the material (including inspection and labour)

(*d*) Cost of processing the vendor's invoice and

(*e*) Cost of material delivered to stock room and piling

The carrying cost consist of :

(*a*) Loss due to no return on investment on inventory

(*b*) Space requirement

(*c*) Handling during storage and issue

(*d*) Insurance, deterioration, damages and taxes.

The cost of ordering and processing and carrying out cost for a stocked item are calculated a per the formulae given here, If

C_o = Cost per order in Rs./order

C_u = Unit cost *i.e.* purchase cost per unit in Rs./unit

S = Annual requirement of the items in number of item/year

Q = Ordering quantity or order in units per order

I = carrying cost percentage or purchase cost (also called inventory interest rate)

It is further assumed that issue rate or consumption rate is constant and the stock reduces to zero value finally just before new order is received.

Therefore average inventory $= \dfrac{O+Q}{2} = \dfrac{Q}{2}$

∴ Ordering cost = Cost per order × Number of order placed per year

$$= C_o \times \frac{S}{Q}$$

$$= \frac{C_o \times S}{Q} \qquad \text{...(1)}$$

Annual Carrying = Average inventory × Purchase cost per unit × Inventory interest rate

$$= \frac{Q}{2} \times C_u \times I$$

$$= \frac{Q\,C_u I}{2} \qquad \text{...(2)}$$

∴ Total inventory cost/year = Annual order cost + Annual carrying cost

$$= C_o \frac{S}{Q} + \frac{Q\,C_u I}{2} \qquad \text{...(3)}$$

The graphical representation of these equations is shown in Fig. 7.11.

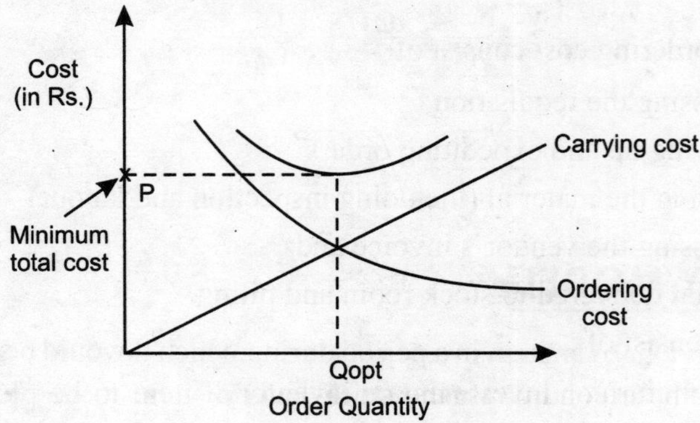

Fig. 7.11.

It can be seen from this Fig. 7.11. that there exists a particular value of Q (optimum) for which the total cost is minimum (C_r). This is called Economic Order Quantity.

This can be derived analytically

Total Cost $$C_r = \frac{C_o S}{Q} + \frac{C_u I Q}{2}$$

for C_r to be minimum

$$\frac{d\,C_r}{d\,Q} \text{ should be positive according to minima principle.}$$

$$\frac{d\,Cr}{dQ} = \frac{-C_o S}{Q^2} + \frac{C_u I}{2} = 0$$

$$\therefore \quad \frac{C_o S}{Q^2} = \frac{C_u I}{2}$$

or $$Q_{\text{optimum}} = \sqrt{\frac{2\,C_o S}{C_u I}} \text{ and}$$

further $$\frac{d^2\,Cr}{d^2\,Q} = \frac{2\,C_o S}{Q^3} \text{ is a + ve quantity.}$$

Hence

$$Q_{\text{optimum}} = \sqrt{\frac{2\,C_o S}{C_u I}} = EOQ$$

i.e. $$EOQ = \sqrt{\frac{2\,C_o S}{C_u I}}$$

further

$$\therefore \quad \text{Total cost} = C_r$$

$$\therefore \quad C_r = \frac{C_o S}{Q} + \frac{C_u I Q}{2}$$

putting value of Q_{opt}.

$$= \frac{C_oS}{\sqrt{\frac{2\,CoS}{cuI}}} + \frac{cuI}{2}\sqrt{\frac{2\,C_oS}{C_uI}}$$

$$= \sqrt{\frac{C_oS}{2/C_uI}} + \sqrt{\frac{C_uI\,C_oS}{2}} = \sqrt{2\,Co\,Cu\,IS}$$

7.16 INVENTORY MODEL

This represents the behaviour of an item in a period during which it would be issued and replenished. Obviously to represent these conditions, the stock level of item to be plotted against time. It is obvious that stock level varies with time.

It is a common practice of designing an inventory model based on data derived from the item's history. A well designed model works out successfully in practice. If this inventory designed model adopted in practice, inventory cost should be minimum and stock-outs should reduce to a minimum . There are three types of models generally used. There are :

1. The Wilson's model or Fixed - Reorder quantity model
2. Repenishment model and
3. Two Bin model.

1. **The Wilson model or Fix-Reorder Quantity model** is one as illustrated in Fig. 7.12. As the model appears an inventory cycle dimensions remain constant

As the model shows each time an order is placed for order quantity *Q* which is equal to *EOQ* for the item. The demand for the item is assumed constant, the item is issued at a constant rate, whatever be the cycle. Therefore the lead time for each order is the same. As order is placed in such a manner that the supply from the vendor reaches the stock exactly when the stock level falls to zero i.e. lead time ends when the stocks level become zero. Here *EOQ* or *Q* (Optimum) is calculated by collecting previous data. On the basis of varying daily demand, an average value of daily demand can be calculated. This will be used as the slope of the issue or demand line.

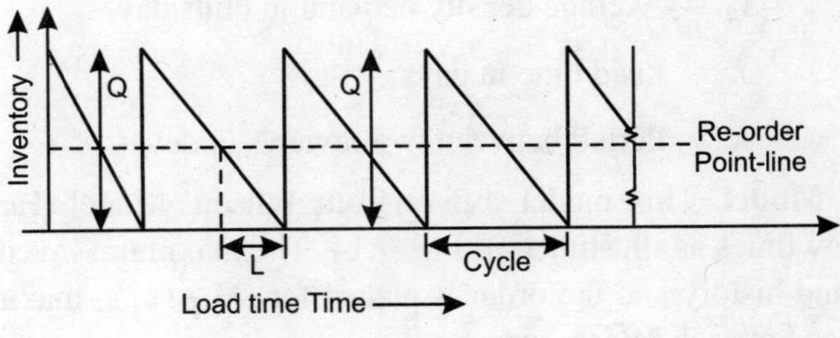

Fig. 7.12.

While a model of this type is very useful to start with, but faces a few problem on implementation. The problem arises are :

1. The demand or issue is presumed as average. If due to some reason demand is found more than the average, particularly during lead time, a stockout becomes imminent.

2. Lead time has been assumed to be constant for every cycle. If lead time in a cycle becomes greater than the average, a stock out occurs.

Therefore to modify it, a buffer stock or safety stock is introduced into model, to be put to use. The stockout can be negated or reduced depending upon the size of the buffer stock. Introduction of buffer stock make the model as illustrated in Fig. 7.13.

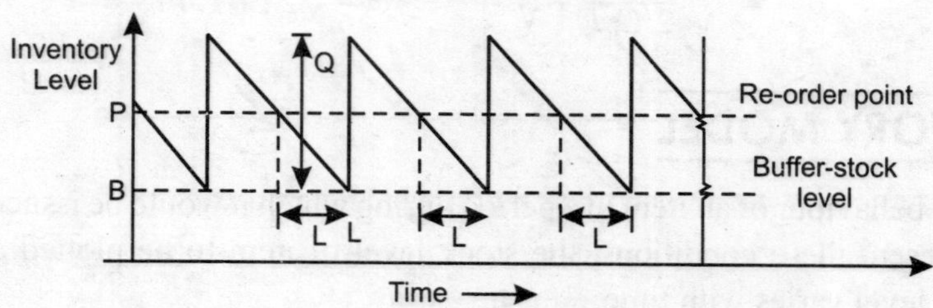

Fig. 7.13.

This model works in a more realistic manner. Minor changes may occur during implementation, but the works successfully. The values of P, Q and B are fixed. Whenever stocks level falls to the P value, an order of the size Q is placed. During lead time L, the total demand would be $L \times S_d$. Hence P level would be $P = L \times S_d$ for instance where no buffer stock is kept. When buffer stocks are present the P level would become $L \times S_d + B$, where B is the Buffer-stock level.

This Wilson's model is used extensively in practice for regularly issued items. It necessitates the maintenance of daily records without which it would be difficult to know when an item has reached the P level. Obviously the controlling work involved by using the Wilson model is quite rigorours and hence more costly. It is generally used for items of A class or high B class.

The re-order point is units is calculated by formulae.

$$P = B + S_d\left(L + \frac{R}{2.0}\right)$$

where P = Re-order point in units

B = Buffer stock in units

S_d = Average density demand in units.day

L = Lead time in days

R = Period between two reviews in days

2. **Replacement Model.** This model differs from Wilson' Model. Here re-order quantity changes at each review times as illustrated in Fig. 7.14. The maximum inventory level is fixed in the basis of the demand history and the order is placed for (M - I) *i.e.* maximum inventory level minus inventory on hand at each review time.

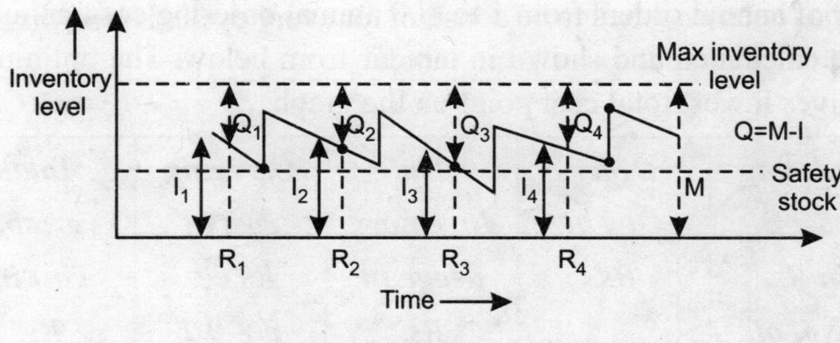

Fig. 7.14.

This system is generally used for *B* class and lower *A* class items.

3. **Two Bin Model.** This is the simplest inventory model in which an order is placed when inventory on hand falls to half the maximum inventory level and the order quantity will by M/2 units each time. This illustrated in Fig. 7.15.

Here there is fixed recorder quantity *i.e.* $Q = M/2$ when $I = P = M/2$ where M is the size of a bin. This model some times become un economical as *EOQ* is not determined. Inventory cost is also high as inventory on hand is quite large, hence mainly suitable for class *C* items, where severe control is not needed due to low investment involved. It is used generally for nut-bolts or drawing papers etc. No estimated Buffer is provided.

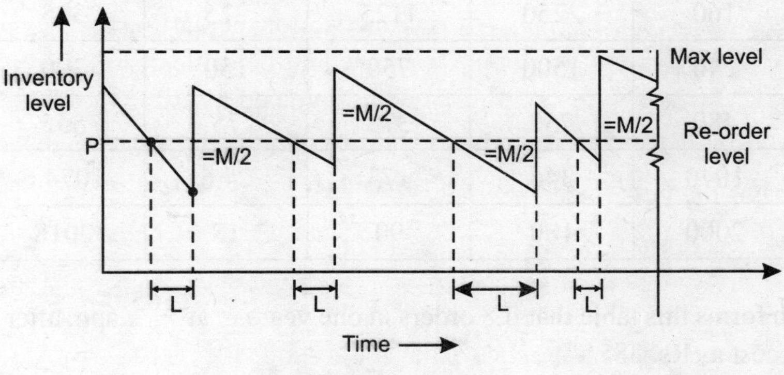

Fig. 7.15.

Example. *Following data are available for an item*

1. Annual usage in Rs. 18000/-
2. Lead time = 1 month constant
3. Re-order point in given by lead time × monthly usage

$$1 \times \frac{18000}{12} = 1500/-$$

4. Ordering cost per order = Rs. 20/-
5. Annual carrying cost for inventory = 20% of average inventory investment.

Find the total inventory cost by total variable cost curve and by *EOQ* formula.

Solution.

Method 1 :

Optimum solution by total variable cost curve.

Varying number of annual orders from 1 to 100 annual ordering cost, annual carrying cost and annual total cost are calculated and shown in tabular from below. The optimum number of order will be that which gives lowest total cost point on the graph.

Annual number of Orders *a*	*Ordering cost B in Rs. (No. of orders × 20) =* *a × 20*	*Order Qty. in Rs.*	*An Inventory usage in Rs. Average* $d = \frac{0+Q}{2}$	*Carrying cost in Rs e = d × 0.2*	*Total variable cost in Rs. f = b + e*	*Remarks*
1	20	18000	9000	1800	1820	Once in year
2	40	9000	4500	900	940	Six monthly
4	80	4500	2250	450	530	Quarterly
8	160	2250	1125	225	385	–
12	240	1500	750	150	390	Monthly
24	480	750	375	75	555	Fortnight
52	1040	346	173	24.6	1074.6	Weekly
100	2000	180	90	18	2018	–

It can be seen forms this table that if 8 orders in one year *i.e.* at $\frac{365}{8}$ app. after 46 days is placed we get optimum cost as Rs. 385/-

∴ Number of order = 8

Ordering cost = Rs. 160/-

Carrying cost = Rs. 225/-

Total cost = 160 + 225/ = Rs. 385/-

These reading can be plotted on graph and results can be held from graph more accurately.

By *EOQ* formula

$$Q_{\text{optimum}} = \sqrt{\frac{2\,C_o\,S}{C_u I}}$$

C_o = ordering cost in Rs. per order = Rs. 20/-

S = total annual demand in units = 18000 C_u

I = annual carrying cost for inventory = 0.2

$$C_u = \text{unit cost of item in Rs. per unit}$$

$$S \times C_u = 18000$$

$$\therefore \quad S = \frac{18000}{C_u}$$

$$\therefore \quad Q_{\text{optimum}} = \sqrt{\frac{2 \times 20 \times 18000}{C_u \times C_u \times 0.2}}$$

$$= \frac{100}{C_u}\sqrt{\frac{72}{0.2}} = \frac{18.98 \times 100}{C_u} = \frac{1898}{C_u}.$$

$$\therefore \quad Q_{opt} \times C_u = 1898$$

$$\therefore \quad \text{Number of orders} = \frac{\textit{Annual usage in Rs.}}{Q_{opt} \times C_u}$$

$$= \frac{18000}{1898} \therefore (Q_{op} \times C_u = 1898)$$

$$= 9.5$$

For this total ordering cost

$$= 9.5 \times 20 = 190/-$$

$$\text{Average Inventory} = \frac{\frac{18000}{9.5} + 0}{2} \text{ i.e. an Inventory usage}$$

$$\text{An Inventory usage} = \left\{\frac{18000}{9.5} + 0\right\} /2$$

$$= 947$$

$$\therefore \text{Annual inventory carrying cost} = 947 \times 0.2$$

$$= 189$$

$$\therefore \quad \text{Total inventory cost} = 190 + 189 = \text{Rs. } 379$$

The difference analytically and graphically is nominal.

7.17 BUFFER-STOCKS

Buffer stock is a stock maintained and utilised to take account of certain uncertainties which may arose due to :

(*i*) fluctuation in consumption rate

(*ii*) production delay

(*iii*) variation in lead time.

If buffer stock, also termed as safety stock is not provided, stockout may occur

The cost associated with stock outs may either be :

1. Loss of customer good will if it occurs in a shop. This is also an indirect cost.
2. Cost associated with company not having material when needed. It must be remembered

that the consequence of this may be a shortfall or delay in production. To control cost of this nature it is essential to maintain some positive minimum inventory in the form of safety stocks. The following condition should exist at the time of placing an order for an item.

Inventory on hand + order quality = safety stock + forecast demand overlead time

The amount of safety stock thus depends on

1. Accuracy of forecast of demand and
2. Lead time

To determine the safety stock level, the following informations are required.

(*a*) Permissible frequency or remaining out of stock

(*b*) Usage and production rates history

(*c*) Lead time

If policy of never missing out of stock is made, it will prove very costly because an overstock has to be carried and hence it becomes essential to set a level of assurance. Higher the level of assurance, high will be the level of safety stock and higher will be inventory cost.

There are three methods to determine safety stock, applicable under three different conditions

Methods A. For new items or for items where records are not available. It is further presumed that there is a constant lead time and variable demand.

This method is applicable when no information and data regarding inventory is available and only average demand is known for a few days experiences. Poison distribution has been found to describe sales or issue in many situations. This method is again based on standard deviation σ. This buffer stock is calculated by multiplying σ to lead time and limits of probability assumed.

Method B. When information is available, the buffer stock is calculated in a manner similar to method A. Here, standard deviation is calculated from available data. Also frequency of different demand values is calculated in general by multiplying the difference of maximum demand and mean demand by lead time. This give buffer stock level.

Method 3. Lead time and demand both are variable. When lead time and demand both are variable, the bufferstock is calculated by 'Monte Carlo Simulation. For this data demand and lead time should be known. With this data, demand during lead time can be simulated and safety stock can be found for various risk levels of stock out. By selecting the derived risk level, the corresponding safety stock can be found out.

These method are not essentially, required to be dealt in detail as their use in much less to civil works hence not covered.

7.18 LEAD TIME

It is the time that elapses between the voicing of a need for anything and the time taken to satisfy the need. If one order is placed today and after a month it is fulfilled, then one month period is called as lead time. Hence it is obvious that the order must have been placed a month earlier. This is reason why it is essential to maintain inventories in stock. These three points *i.e.* lead time, re-order-point and replenishment period are the almost same. It is very essential to analyse the lead time, as it determines the amount of material to be kept in reserve. As it increases, the stock also increases. Therefore, it is further essential to reduce the lead time.

Lead time consists of

1. Time to process the requisition and to place the order.
2. Time required to deliver the order.

7.19 QUALITY CIRCLES

Organisations everywhere are now in search of excellence in their diverse activities. The pressure of circumstances and newly emerged challenges in the socio-economic scenerio are prompting them to do so. Things are happening and changing much faster. Only those companies who will keep up pace with time and change will only stay alive. Competition is increasing; new technologies are developing.

The concept of quality circles has been interpreted in many ways in India and abroad. The most commonly accepted definition in keeping with the essence of the philosophy as it originated in Japan is as follows :

"Quality circle is a small group of employees in the same wok area or doing a similar type of work who voluntarily meet regularly for about an hour every week to identify, analyse and resolve work related problems, leading to improvement in their total performance and enrichment of their work life".

It is necessary to comprehend the significance of every part of this definition to ensure that it is implemented in the right spirit and that the fundamental difference between quality circles and other techniques that are available and used are clearly appreciated.

The important parts of this definitions are:

(*a*) **Small Group of Employees.** A group of 8 to 10 employees is considered to be adequate as in case of small group less than 5; or if a group is large that is more than 15; the efficiency is lost. More people can not take proper decision in less time and less people rarely gather at a time. As such 8 to 10 persons in a group is considered adequate and can works more efficiently.

(*b*) In the same work area or doing similar type of work, A quality circle is a homogeneous group and not inter department or interdisciplinary one. People of that group activity must be on the same wavelength. Persons discussing any matter should be intelligible to each. This is only possible if the composition of circle includes employees working in the same work area or enganged in a similar type of work. The level of employees i.e. designation of employees may be different but they should have been engaged in common activity. In case of small organisation persons of some organisation can form one circle, though their activities may be different but related to each other. For example, in a small workshop, turner, driller, electrician and unskilled persons can form one quality circle.

(*c*) **Voluntarily.** It is voluntary circle where no individual gets any benefit. As such employees will have to come forward to join the quality circle for the welfare of the organisation, employees and employer and the owner. They will have to keep their egos aside. Those who are interested be allowed to join the circle. No one should be forced to join else he will not work sincerely.

(*d*) Meet regularly for about one hour every week. Unless meetings are arranged; topics are discussed; alternate solutions are found and implemented; the purpose of quality circles can not be achieved. It is therefore essential that members of quality circles should meet atleast once in a week for one hour. The number of meetings can be varied as per requirements but regular meetings are essential with no prejudiced mind.

(*e*) To identify, analyse and resolve work related problems. It is a fact that employees who work in any organisation know the problems and have some solution to resolve that problem. Many of them also know the effect of one problem/solution on other problem. Therefore the persons working in quality circle are required to identify problem in detail; they are required to analyse it and then to reach to some possible and fruitful resolution. Many times it happens that people know problem \; they also analyse it but do not reach to a common solution and as such problem remains as it with no resolution.

One important aspect that should be borne in mind and emphasised before volunteers are invited to enroll themselves as circle members, is that only "work related problems" come under the preview of quality circles and not other issues such as grievances or demands or any other such problems are tackled by trade unions or grievance committees or other similar bodies. However work related problem may be a matter of ambiguity, however otherwise work related problem in itself is clear.

(*f*) Leading to improvement in their total performance, quality, productivity, cost reduction, safety etc. are factors if discussed, resolved, improves the total performance which helps to the whole organisation. The benefit to the organisation include enrichment of the worklife of their employees apart from attitudinal changes, cohesive team culture etc. The quality of worklife (QWL) of any individual is said to be good only if he is enjoying the work that he is engaged in, is happy in the environment and finds his workplace meaningful. Poor quality of worklife is attributed by

— High absenteeism and turnover

— Poor quality product

— Low productivity

— Human relation conflicts

— Non-involvement and apathy.

As such it can be said that if quality circles are formed and implemented, these helps in improvement of the enterprise, respect humanity and build a happy workplace and enables employees to utilise their intrinsic wisdom and creativity in the works that they are engaged in. It also helps employees for their self-development.

The impact of quality circles are :

1. Improvement in human relations and work area morale.
2. Promotion of participative culture.
3. Enhancement of the job interest.
4. More effective team works.
5. Reduces defects and improves quality.
6. Improves house keeping, cost effectiveness, safety and other such related areas.

7. It improves productivity qualitatively as well as quantitatively.
8. Enhance understanding, problem solving capacity and encourage employees on attitude of problem prevention.
9. It improves communication between employees, employees and employer, employees and their superiors.

7.20 PROCEDURE OF RECEIVING MATERIAL AT SITE

Once the order is placed and supplied the material through trucks at site, how the truck proceeds, from one section to another, what formalities are completed and all other details are illustrated Fig. 7.16.

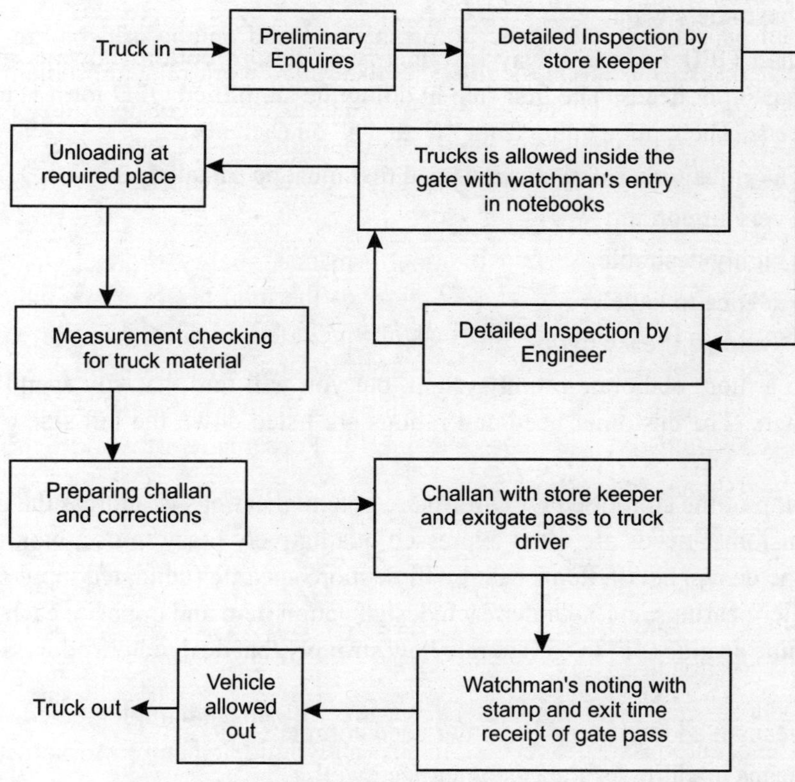

Fig. 7.16. Schemetic-Procedure of receipt of Material

7.21 QUALITY FUNCTION DEVELOPMENT (QFD)

What a customer really needs is often not truly understood during the design or change of a product, process, or service. A simplified QFD, if done carefully, will minimize issues arising from this lack of understanding.

QFD originally stood for quality function deployment. Years ago, when quality departments were generally much larger than now, quality engineers were 'deployed' to the customers to rigorously probe the customer's need and then create a series of forms that transitioned those customer needs into a set of actions for the supplier to take. The simplified QFD attempts to accomplish the same task in a condensed manner.

The simplified QFD is usually used to Define or Improve steps of the DMAIC (Define, Measure, Analyse, Improve and Control process. A simplified QFD is a Six Sigma tool that does not require any statistics. But it is usually necessary to do a simplified QFD to understand what actions are needed to address a problem or implement a project. The specific actions that are identified in the OFD, or in any of the other qualitative tools, are often what trigger the application of the statistically based Six Sigma tools.

Many product, process and service issues are caused by not incorporating from customers and/or suppliers of components and raw materials early in a design or process change. Often the manufacturer decision makers just assume that they and the people from whom they source already know what the customers want.

The simplified QFD form ·is a way of quantifying design options, always measuring these option against customer needs. The first step in doing the simplified QFD form is to make a list of the customer needs. Then, put a value from "1" to "5" on each need :

- "5" is a critical or a safety need, a need that must be satisfied.
- "4" is very important.
- "3" is highly desirable.
- "2" is a "nice to have".
- "1" is "wanted if easy to do."

One can do a more elaborate rating system, but you will find that you spend too much time assigning numbers. The customer need and ratings are listed down the left side of the simplified QFD form.

Across the top of the simplified QFD form are potential actions to address the customer needs. Note that the customer needs are often expressed qualitatively (easy to use, won't rust, long life, etc.), whereas the design action items listed will be more specific (tabulated input screen, stainless steel, sealed roller bearings, etc.). Under each design action item and opposite each customer need, you will determine a value ("1" to "5") to rate how strongly that design item addresses the customer need :

- "5" means it addresses the customer need completely.
- "4" means it addresses the customer need well.
- "3" means it addresses the customer need somewhat.
- "2" means it addresses the customer need a little.
- "1" means it addresses the customer need very little.
- "0" or blank means it does not affect customer need.

A negative number means it is detrimental to that customer need. (A negative number is not that unusual, since a solution to one need sometimes hurts another need !)

Put the rating in the upper half of the block beneath the design item and opposite the need. Then, you multiply the design rating times the value assigned to the corresponding customer need value. Enter this result into the lower half of the square under the design action item rating. These values will have a possible high of 25.

Once all the design items are rated against every customer need, add up the values in the lower half of the boxes under each design item and enter the sums into the "Totals" row at the bottom of the sheet. The solutions with the highest values are usually the preferred design solutions to address the customer needs.

Once these totals are reviewed, someone may feel that something is away and want to go back and review some ratings or design solutions. This second (or third) review is extremely valuable. Also, the customer "5" ratings should be discussed one at a time to make sure that they are being addressed sufficiently.

In the simplified QFD form, near the bottom, design action items are grouped when only one of the options can be done. The priorities within a group are only among the items in that group.

The simplified QFD form should be completed for all new designs or for process modifications. The time and cost involved in having the required meetings will be more than offset by making the correct decisions up front, rather than having to make multiple changes later.

Traditional QFDs

A traditional QFD, is likely to be one of the following.

The first is a QFD consisting of four forms. The first form is the "House of Quality". This form covers product planning and competitor benchmarking. The second form is Part Deployment, which shows key part characteristics. The third form shown Critical-to-Customer Process Operations. The fourth is Production Planning.

Two other QFD are the "Matrix of Matrices" QFD, consisting of 30 matrices, and the "Designer's Dozen," consisting of 12 QFD matrices.

License Plate Holder Option for Luxury Automobiles Ratings **5 = Highest** **1 = Lowest for negative number**		**Design Items**									
		Metal Cast Rim	**Plastic Cast Complete**	**Stamped Steel Rim**	**All Holes Already In**	**Optional Punched Holes.**	**Gold/Silver Plating**	**Hex/ Slotted Plastic Screws**	**Hex/Slotted Platted steel screws**	**Plastic Lens, Separate**	**Tempered Glass Lens, Separate**
Customer Needs	**Rations**										
Embossed Name	5	5	5	3	0	0	0	0	0	0	0
		25	25	15	0	0	0	0	0	0	0
Place for Dealer Name	5	5	5	5	0	0	0	0	0	0	0
		25	25	25	0	0	0	0	0	0	0

Must Hold All State Plates	5	5	5	5	5	5	0	0	0	0	0
		25	25	25	25	25	0	0	0	0	0
Solid Feel	3	5	3	2	0	0	0	1	4	2	5
		15	9	6	0	0	0	3	12	6	15
Gold/Silver Option	2	5	5	3	0	0	5	5	5	0	0
		10	10	6	0	0	10	10	10	0	0
Easy to Install	4	3	5	3	5	3	0	5	3	3	2
		12	20	12	20	12	0	20	12	12	8
Corrosion Resistance	4	3	5	2	0	0	3	5	3	4	5
		12	20	8	0	0	12	20	12	16	20
Light Weight (for MPG)	1	1	4	5	0	0	0	5	2	5	1
		1	4	5	0	0	0	5	2	5	1
Luxury Look	4	5	2	1	2	4	5	1	5	2	5
		20	8	4	8	16	20	4	20	8	20
No Sharp Corners	5	4	5	2	4	2	0	5	3	5	2
		20	25	10	20	10	0	25	15	25	10
Transparent Lens	4	0	3	0	0	0	0	0	0	3	5
		0	12	0	0	0	0	0	0	12	20
Last 10 Years	5	4	4	3	0	0	3	5	4	3	5
		20	20	15	0	0	15	25	20	15	25
Low Cost	2	1	5	4	5	2	1	5	3	4	2
		2	10	8	10	4	2	10	6	8	4
Keep Appearance 10 Years	4	4	3	2	0	0	2	5	3	2	5
		16	12	8	0	0	8	20	12	12	20
Groupings											
Totals		203	225	147	83	67	67	142	121	119	143
Priorities			1		3		4	2		NA	NA

Needless to say, these other QFDs take much more time and effort than the simplified QFD. Meetings to complete the traditional QFDs generally take as least four times as long as the meetings required for the simplified QFD on an equivalent project.

Are the traditional QFDs worth the extra effort versus the simplified QFD? Perhaps so on very large and very complex programs. However, the benefit of being able to use the simplified QFD on every project or change, which is not realistic with the more complex QFDs, gives it a decided advantage. The customers' inputs are needed on all levels of projects !

The major benefit of any QFD comes from getting the input of everyone affected. If the form is too complex or the meeting to do the form is too long, people lose focus, their eyes begin to blur, and the quality of the input diminishes. The simplified QFD is designed to get needed input with the minimum of hassle ! Simplified QFD meetings are often only or two or three hours long.

In summary, we can say that,

- The simplified QFD is usually used in to Define or Improve step of the DMAIC process.
- Many product, process, and service issues are caused by not incorporating inputs from customers and/or suppliers of components and raw materials early in a design. Often the manufacturer just assumes that what the customers really want is already known.
- The use of a simplified QFD can minimize a lot of issues before a design or modification is implemented.
- Everyone affected by the project such as operators, suppliers, users, engineers, maintenance staff, and customers must participate in generating the simplified QFD form.
- The simplified QFD should be done as early in a project as possible.
- Revisit the simplified QFD's priority results several times to make sure they truly reflect the group's intent.
- The cost of doing a simplified QFD will be more than offset by the benefits of a superior design with fewer modifications required.
- The simplified QFD should be used on every new product or product/process change.
- There are more complex and detailed QFDs that way be worth considering for very large and complex programs. However, the effort and people required for these QFDs usually preclude their being used on all designs and changes. The simplified QFD is extremely practical, since it can be used on all designs and changes.

Case Study :

1. Kitchen Trolley: In construction projects, customer needs to decide about installing a Kitchen Trolley in the Kitchen. It is quite difficult and time consuming task to decide which facilities are needed and how to make a final decision about it.

We can use following QFD for this purpose.

Kitchen Trolley Options For Apartments Ratings 5 = Highest 1 = Lowest		**Design Items**							
Customer Needs	***Ratings***	**Most Components**	**Hex/Slotted Screws**	**Optional Punched holes**	**All holes already in**	**Customization option**	**Small footprint**	**Variable in Size**	**Standard items**
Easy to install	1								
Solid feel	2								

Colour Resistance	1								
Corrosion Resistance	5								
Luxury look	4								
No sharp corners	4								
Low cost	2								
Space saving	3								
Grouping Totals Priorities									
Best Quality	4								
Space Saving	2								
Scope for future modification	2								
Grouping Totals Priorities									

7.22 FAILURE MODES AND EFFECTS ANALYSIS (FEMA)

On any project there can be collateral damage to areas outside the project. A simplified FEMA (failure modes and effects analysis) will reduce this likelihood. A simplified FEMA will generate savings largely through cost available and it is usually used in the Define or Improve steps of the DMAIC process.

Note that the simplified FEMA format is very similar to that used for the simplified QFD. This is intentional, since the goal is to use both on every new product or change. Since many of the same people will be involved in both the QFD and the FEMA, the commonality of both forms simplifies the task.

A simplified FEMA is a method to review things that can go wrong even if a proposed project task, or modification is completed as expected. Often a project generates so much support and enthusiasm that it lacks a healthy amount of skeptics, especially in regard to any effects that the project may have on things not directly related to the project. Everyone is working on the details of getting the project going and little effort is spent on looking at ramifications beyond the specific task !

The simplified FEMA form is a way of giving a project a critical look before it is implemented; it often saves a lot of cost and embarrassment. In doing a simplified FEMA, it is assumed that all inherent components of the direct project will be done correctly. (They should have been covered in regular project reviews). The emphasis in a simplified FEMA is to identify affected components or issues downstream or tangentially related processes in which issues may arise because of the project.

Just as in the simplified QFD, the critical step is getting everyone together who has anything to do with the project, especially those having to deal with the effects of the project. These people could be machine, operators, customers, or even suppliers. The proper group of participants will vary per project.

The left side of the simplified FEMA form is a list of things that could possibly go wrong, assuming that the project is completed as planned. The first task of the meeting is to generate this list of concerns. On this list could be unforeseen issues on other parts of the process, safety issues, environmental concerns, negative effects on existing similar products, or even employee problems. These will be rated in importance :

- "5" is a safety or critical concern.
- "4" is a very important concern.
- "3" is a medium concern.
- "2" is a minor concern.
- "1" is a matter for discussion to see if it is an issue.

Across the top of the simplified FEMA is a list of solutions already in place to address the concerns or additional solutions that have been identified in the meeting . Below each solution and opposite the concern, each response item is to be rated on how well it addresses the concern :

- "5" means it addresses the concern completely.
- "4" means it addresses the concern well.
- "3" means it addresses the concern satisfactorily.
- "2" means it addresses the concern somewhat.
- "1" means it addresses the concern very little.
- "0" or a blank mean it does not affect the concern.

A negative number means the solution actually makes the concern worse.

Enter this value in the upper half of the block, beneath the solution item and opposite the concern. After these ratings are complete, multiply each rating times the concern value on the left. Enter this product in the lower half of each box. All all the values in the lower half of the boxes in each column and enter the sum in the Totals row indicated near the bottom of the form. These are then prioritized, with the highest value being the 1 consideration for implementation.

As in the simplified QFD, these summations are only a point of reference. It is appropriate to reexamine the concerns and ratings.

In summary we can say that,

- A simplified FEMA is usually used in the Define or Improve steps of DMAIC.
- Much effort goes into making sure the specific details of a project, process, or service are correct. However, areas not inherently tied to the project are often ignored.
- A simplified FEMA emphasizes identifying concerns in other affected areas and prioritizing potential solutions to these concerns.

- Everyone affected by the proposed project, process or service should participate in the simplified FEMA.
- Revisit the results of the simplified FEMA several time to make sure that they truly reflect the group's intent.
- The cost of doing a simplified FEMA will be more than offset by the costs avoided on the project's potential negative effects on other areas.
- Traditional FEMA are more complex, but may be justified on extremely large and complex programs. However, they are unlikely to be used on every program or change, which is the value of the simplified FMEA.

1. **Construction is not as per design or customer needs**

Let us study the case of critical problem of not making the construction as per design or need. There are several reasons we can explore for the problem. But FEMA is mostly useful for exploring the analysis of the effect caused by the event. Following FEMA form we can use for it.

FEMA : Note done as per design or needs **Ratings :** **5 = Highest** **1 = Lowest**		**Design Items**				
Concerns	**Ratings**	**Modify the budget**	**Make a over budget**	**Optional Punched holes for handling the problems**	**Use low cost material**	**Optional Punched holes delete some part**
May cause conflict in execution	1					
May stop construction	2					
Modification in the plan while execution is going on	1					
May have to dismantle some work	5					
May produce low quality	5					
May produce low utility	3					
Financial Loss	5					
Totals Priorities						

QUESTIONS

1. How the quality of any project is checked?
2. What do you mean by inventory, how it is maintained?
3. Describe the expression for economic order quantity (EOQ) for the material

4. What do you mean by term 'Concept of quality?'
5. Name various stages to control the quality of work.
6. Explain various means of control of quality of work in construction.
7. What statistical quality control? Explain various steps involved in statistical quality control.
8. Explain in detail the sampling and sampling technique.
9. Write short notes on:
   (a) ABC Analysis
   (b) System thinking
   (c) Inventory management
   (d) Buffer stock

Chapter

FORMS FOR CONCRETE STRUCTURE

8.1 INTRODUCTION

Every kind of reinforced concrete construction requires first of all the building of mould in which to pour the concrete. These moulds are called as 'form work' or 'Shutt ering'. It is an essential part of the concrete works and hence due regard for form work design must be given.

When concrete is placed in a mould (form work), it is in plastic state. Hence, it is necessary to use forms to confine and support it until it is rigid and self-supporting. Due to initial plasticity of concrete, it can be cast to any desired shape provided forms can be built to confirm with that shape.

While selecting and designing form work for concrete work, following must be considered :

1. Strength.
2. Economy and Functional Requirement.

Different types of concrete components will need different types of forms as per their functional requirement. Monumental work where beauty is primary criteria, timber forms will not be useful. Circular columns can be caste if plywood or steel forms are used. Hence, therefore, according to functional requirements forms are used.

In initial stage form work has to resist all the vertical load of plastic concrete or the hydrostatic pressure as may be the field conditions. Hence forms must be strong enough to carry the dead load of concrete plus live load applied during concreting. Forms should be rigid enough to retain the shape without undue deformation. Economy consistent with strength is main requirement. Economy in treatment of concrete surface depends upon tightness with which the forms are built.

Factors which should be taken into consideration for designing form work and to achieve economy in its use are as given below :

1. To reduce number of irregular shapes so that they can be used many a times.
2. Forms should be designed to provide adequate but not excessive strength. Unreasonable limitations on size should not be used if they have got adequate strength to resist the load.
3. These should be removed as soon as it is safe to do so, the early removal of forms will permit to reuse it early.
4. The form should have modular size; further the planner should have modular size of rooms to that unnecessary cutting of forms will be less.

5. Most economical material, with more reuse value should be used.
6. Not more than needed nails should be used. As far as possible nut-bolts should be used.
7. The joints in the form work should be rigid so that the bulging, twisting or sagging due to superimposed load is as small as possible. Excessive deformation will disfigure the surfaces of concrete hence more finishing material will be required to make it plane.
8. The form work should be easily removable. Concrete is liable to stick to the form hence before concrete is placed a layer of lubrication should be applied to form work. It will avoid the possibility of damage of form work as well as of concrete while removing.

8.2 FUNCTIONS OF MEMBERS FOR FORMWORK

The 'Sheating' as shown in Fig. 8.1 forms water-tight enclosure and provides a smooth surface against which the poured concrete can rest until it has hardened. Playwood is used on curved surface instead of plain timber boards of sheating. They are also called as form panels.

The 'Shoe' acts as a plate to hold the studs at equal spacing and as a straight edge to the form. The 'Stud' forms a solid backing for the sheating and holds in it alignment.

The 'Wales or walers' hold the panels of the form in a straight line and provide rigidity and bracing. The 'Spreaders' hold the forms at the proper distance apart until the concrete has been poured.

The 'Wires' hold the panels against spreaders thus keeping the forms spereading apart under the pressure of the semi-liquid concrete. The wires are twisted to draw and hold panels tightly against the spreaders.

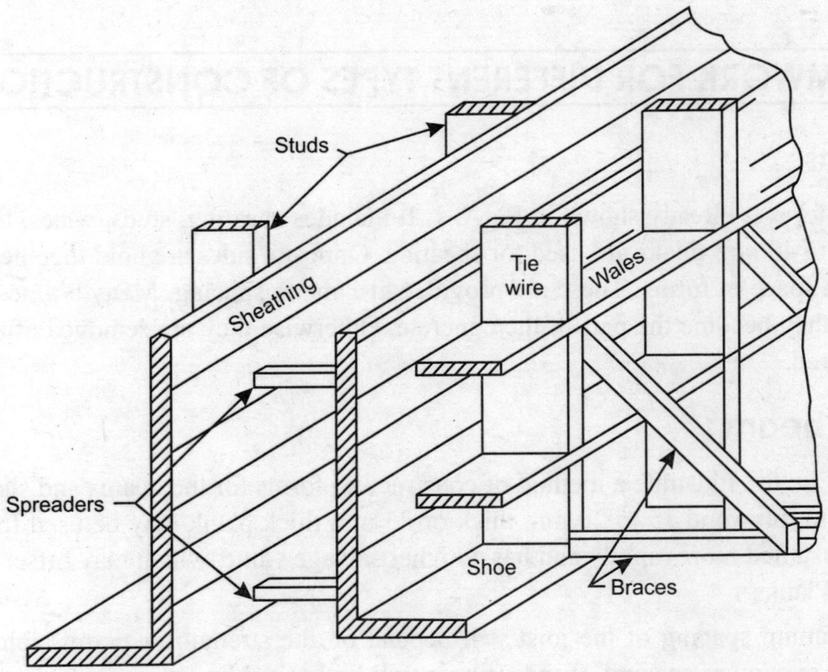

Fig. 8.1. Forms for a concrete wall

The 'Braces' hold the form in an upright or solid position and prevent the form moving under the pressure of the wet concrete. The 'Stakes' give nailing support for braces.

'Props' are the vertical supports to 'Stringer' or 'Joist' on which sheating rests. In case of slabs and beams sheating rests on joists which then rests on stringer as shown in Fig. 8.2. Props are also termed as 'Shores', 'Scab' are used for nailing stringer to prop incase of slabs and 'ledgers' are used to nail side sheating of beams with horizontal one of slab as shown.

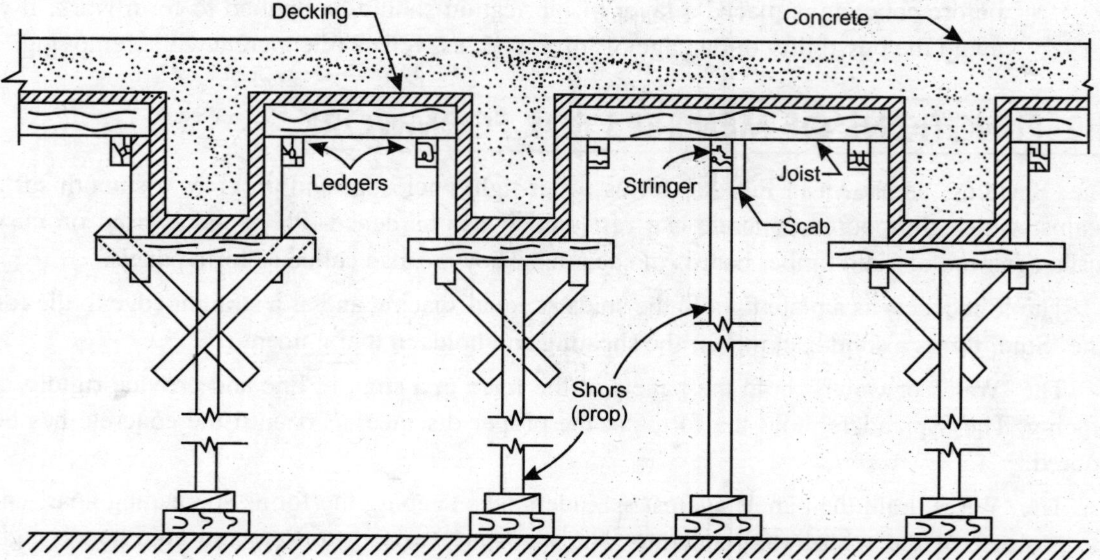

Fig. 8.2. Wood forms for beam and slab concrete floor

8.3 FROMWORK FOR DIFFERENT TYPES OF CONSTRUCTION WORKS

Wall Forms :

Typical wall form is already shown in Fig. 8.1. It includes sheating, studs, wales, ties and braces. Planks of 25 to 40 mm thicks are used for sheating. Opposite sides are hold together by the ties to have uniform space in forms. These are provided at 1 metre spacing. Many a times steel rods are used as ties, they become the part of the concrete. Otherwise they are removed after the concrete has been poured.

Slab and beam :

Fig. 8.2 and Fig. 8.3 illustrate a method of constructing forms for the beams and slab for this type of floor system, plywood 15 to 20 mm thick or 25 mm thick plank may be used for the decking. Plywood is installed more rapidly and has a higher salvage value, which may offset its higher cost compared to Planks.

The maximum spacing of the joist will depend on the strength or permissible deflection of decking. The maximum spacing of the stringer will be limited by the strength or the permissible deflection of the joists. The maximum spacing of shores under the stringers will be limited by the strength of the permissible deflection of the stingers or the capacity of the shores.

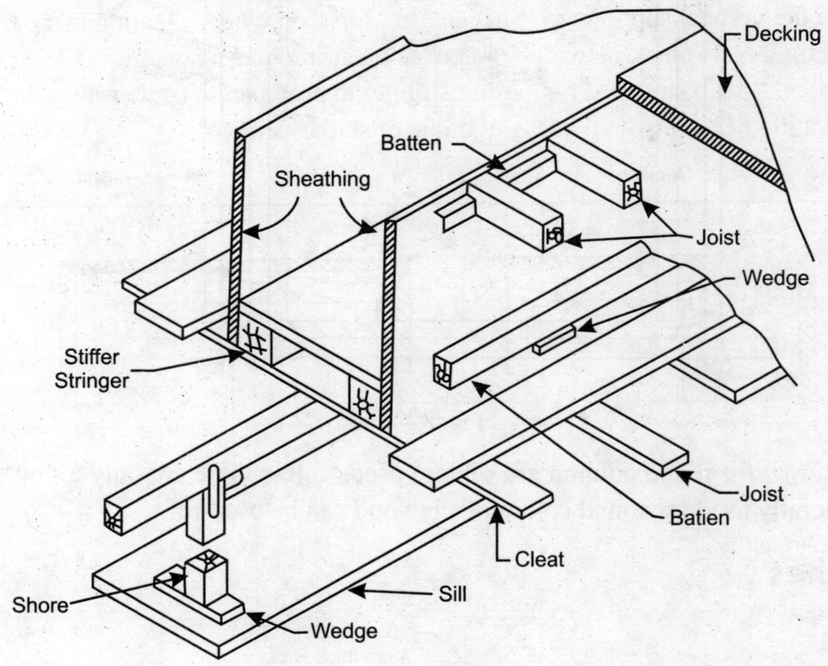

Fig. 8.3. Wood forms for beam and slab concrete floor

The later type forms was developed by contractor. The assembly was designed to be fastened together by wedges, with no erecion nailing required. The primary objectives are to reduce the damage to the forms in order to obtain more uses and to reduce the amount of labour required to erect and remove the forms.

Column Forms :

Forms for columns usually are made of vertical planks with normal thickness of 25 mm plywood. The component parts are prefabricated, then assembled in the location where they will be used. Adjustable steel clamps generally are used to resist the pressure from the concrete as shown.

Forms shall be designed to resist the high pressure resulting from quick filling. If forms are filled in 30 minutes or less, concrete may exert the full hydrostatic pressure based on a weight of approximately 2400 kg/m^3.

Column forms require 0.06 to 0.09 cubic metres of timber per square metre of contact and an average value can be used for estimation. The larger the column less the timber required per square meter. It is usual practice to prepare form of all columns on the lowest floor and to use the same forms again on the upper floors.

Column forms of more than the ordinary height upto 4m. should be braced or stiffened at intermediate points to give them sufficient rigidity. Yokes are generally used to stiffen them.

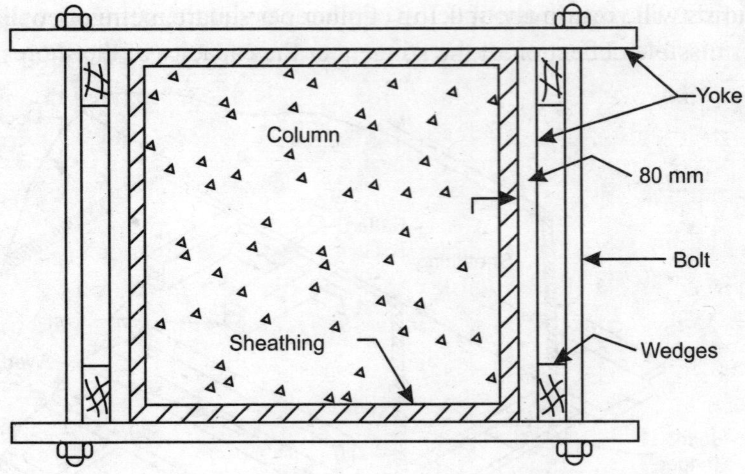

Fig. 8.4. Plan

Wooden forms for round column are seldom used, unless there are only a few columns. Steel forms are generally used for round columns. Plywood can be used for it.

Small Arches :

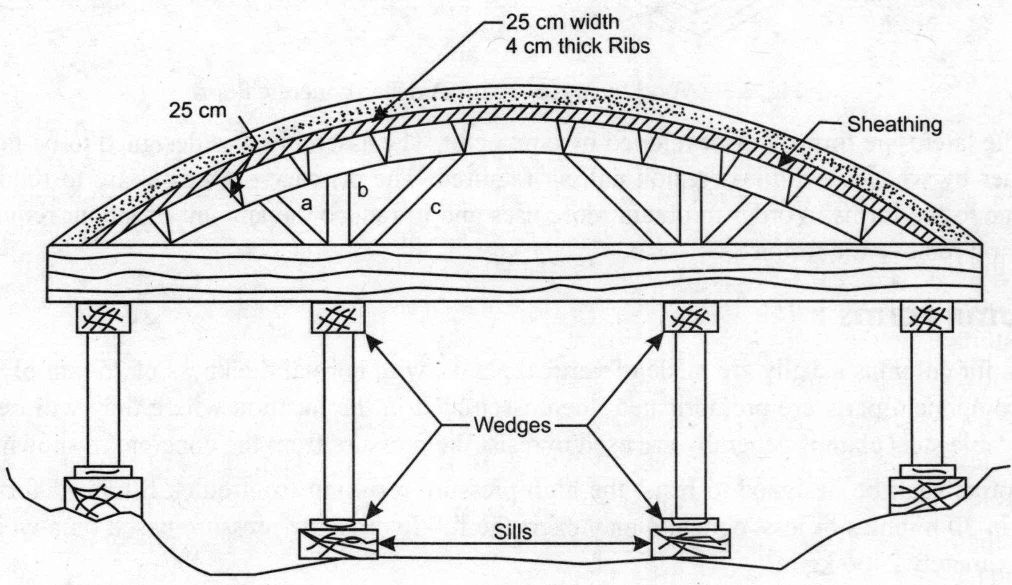

Fig. 8.5. Arches

One small arch form is shown in Fig. 8.5. These can be built in two ways. Either joist themselves can be cut to the required radius and supported by straight ledgers, or the joists can be straight ledgers cut to the radius. The first method shown would be better for a short curve and small area.

In power house construction, there is often the problem of building forms for a slab that is flat at one end and curved at the other, the curvature gradually increasing from straight line to a maximum. This can be done by using straight joists resting on a horizontal ledger at one and a ledger at the other.

Curved slab forms will require about 0.1 m^3 timber per square metre of contact area.

The tie beam is fixed through wedgoes to the prop and load is transferred to the ground. On this vertical posts as shown a, b, c members are fixed and on them are fixed double ribs made from planks 25 cm wide × 4 cm thick on which sheathing is laid as shown in Fig. 8.5.

Circular water tanks :

All structures having high thin walls, whether for storing water, grain, cement, sand etc, are built by similar methods. The same general principles apply whether the structure is square, rectangular or round or whether it consists of one unit or a number of similar units connected together.

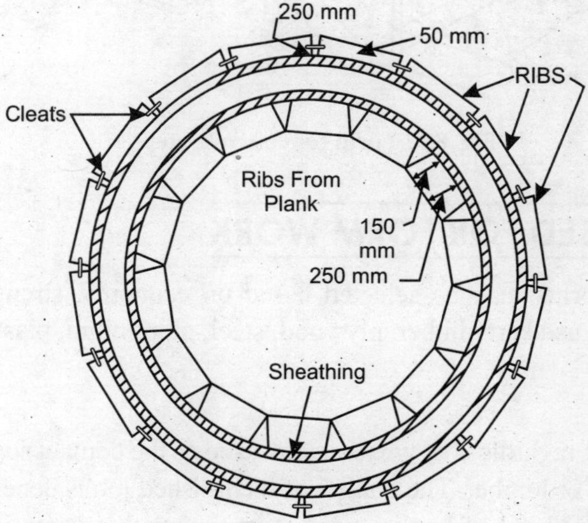

Fig. 8.6. Circular water tank

If the tank or bin is not over 3 to 4 m high and the diameter is not over 10 metre it is usual to build forms for the whole structure rather than to use moving forms. Very high structure such as silos, storage bins, chimneys etc. or other structure with large diameter such as oil storage tank are built with moving forms.

As it is desired to have no construction joints as in water tank and stand pipes, so that there will be less possibility of leakage, Continuous sliding forms are used or complete tank is cast at one stretch.

In the overhead tanks forms consist of vertical sheahing fixed to ribs which are cut to circular shape. These are made from 4 cm think plank at about 25 cm. wide. The ribs have radial joints and are joined by means of cleats. The ribs form complete circles and are under hydrostatic pressure, they develop only the hoop stresses. The inner ribs are in compression, the outer ones in tension. The spacing of ribs vertically is determined by the thickness of sheating which spans between the ribs. The joints in outer ribs must be strong enough to bear the hoop tension. To keep uniform thickness of the wall all around, few spacers are inserted at regular intervals between the outer and inner forms.

Forms for staicase is aslo shown in Fig 8.7.

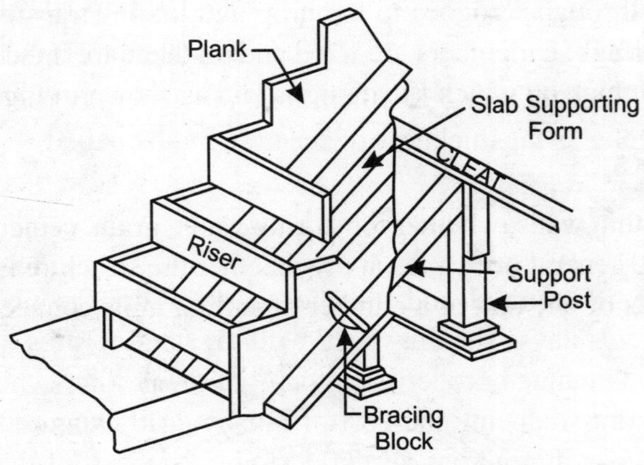

Fig. 8.7. Form for concrete steps

8.4 MATERIALS USED FOR FORM WORK

The materials used for forms may be selected based on economy, strength and necessity. The materials most commonly used are timber, plywood, steel, aluminium, plaster of paris etc.

Timber :

Practically all form works, regardless of what may be used as the contact surface with the concrete, involves the use of timber or lumber. The quality of the finished job is dependent to a considerable extent upon the quality and kind of lumber used. Any timber that is straight, Structurally sound and strong and property seasoned can be used for form work. Partly seasoned timber is the best for formwork as if it is to dry, it will tend to swell from absorption of moisture either from atmosphere or from concrete while green timber will tend to dry and shrink in hot weather causing bins and ridges on the concrete. Timber used for from work should be free from all type of defects like knots, wanes etc. Soft wood should be preferred for form work as it is being soft, easier to work. Hard wood is used only for caps and wedges, further it is also difficult to wrap.

Timber planks used for form work should be of uniform thickness. It is very necessary with sheathing otherwise it will require great labour while planning for joints. Similarly joints and studs too should be of uniform dimensins other wise considerable trouble will be caused during fitting.

Fir is mostly used for an form work; but is also depends on availability and cost. In the East, fir and hemlocks are generally used for studs and wales whilst on South and West very little hemlock is used. Thickness of timber will depend on the available supply and the loads to be carried. Following are the general sizes of different form works:

(*a*) Sheathing — 25 to 50 mm thick

(*b*) Beam and column sides — 25 to 50 thick

(*c*) Beam bottoms — 40 to 50 mm thick

(*d*) Joist and ledgers — 7.5 × 10cm to 15 × 15cm.

(*e*) Posts or props shores 7.5 × 10cm to 15 × 15 cm.

(*f*) Column yokes 5 × 10cm to 10 × 10 cm.

(*g*) Stud and Walings 5 × 10cm to 15 × 10 cm.

Common wire cut steel nails of size 5 to 10 cm long are generally used. A common rule is to use a nail at least twice as long as the thickness of sheathing to be nailed.

Plywood

Where a very good finish is required and also for curved surfaces plywood formworks are becoming very common now a days. The drawback with plywood form is its high initial cost. But once it is constructed and maintained properly can be used many times, they will be cheaper and will give very good finish thus reducing the cost of finishing. Plywood designed for form work is made of high grade. Douglas fir and has all the desirable qualities of that timber for such use. Plywood of thickness 6 to 20 mm are used for form work. Phenolic resin surfaced plywood which is quite smooth, hard and water resistant is also used form work. Fibre board is used as a form liner where smooth surface entirely free of grain making are desired. It is made of shredded wood chip compressed under very heavy pressure at a high temperature. The board is treated to minimize absorption.

Steel

Steel forms have become very popular now a days. In most of the construction works steel forms are only used. The reason is that these forms are quite strong and can be used many times hence are economical. Further they give very good finish hence finishing cost is less. The only drawback is its high initial cost.

These forms usually comprise of standard flat or curved panel units made of steel sheets reinforced with wooden sections or angle sections and are clipped together by keys, wedges or other similar devices. Horizontal and vertical stiffners are provided by angle iron or wooden section which are clipped to the moulded panels.

Flat panels can be made of various size but common sizes are 60 cm × 60 cm, 120 cm × 60 cm or 120 cm × 120 cm and 120 cm × 30 cm etc.

The use of steel forms has increased tremendously and now can be obtained for any type of structure and of any shape and has replaced wooden forms almost entirely for most structure elements. Erection of steel forms is too simple just a mechanical job hence no very skilled labour is required. They are manufacturing in factory. Circular shapes can be given properly only by using steel forms. Black iron sheets are generally used for from work as galvanized metal may stick to the concrete leaving it rough even if forms are well oiled before concrete is pourded.

Aluminium

Aluminium forms have not been tried in India to a lar0ge extent yet. But the increasing cost of steel may economize the use of aluminium form work in near future.

Whasoever available material may be used for form work, the two criteria must be kept in mind. They are economy and strength. Economy is in the sense of cost and life (Reuse) and strength in terms of load resistance.

Special Materials

Ornamental details in architectural concrete involving floral designs, interlaced or pierced tracery human forms, warped and intricately curved surfaces are generally formed in plaster (Plaster of Paris) moulds. Such moulds are called as waste moulds because they are broken while stripped off and hence can be used only once. Thee moulds are made of casting plaster containing jute fibre and are reinforced and braced to prevent breakage. These mould are made of a shape that of finished concrete. Waste moulds are usually given two coats of white shells to make them water proof absorbent. Before concrete placed, waste moulds must be greased to faciliate stripping.

8.5 ERECTION OILING AND STRIPPING

Forms must be erected by a skilled person who has knowledge of erection of it. All forms should be so maintained as to eliminate the formation of joints due to shrinkage of the lumber. These should be so constructed and erected that the portions, where finishing is required, may be removed without disturbing the portions of forms to remain. There should be no play between form structures. Double headed nails should be used, they make stripping easier.

Timber forms must be soaked in water before concrete is placed. This makes the joint tight and prevent absorption of water from concrete hence facilitate stripping. Plywood fibre board forms must be oiled before concrete is placed. It is done generally with the help of brush. Linsed oil is used for oiling the form work. Steel forms should also be brushed with oil for esay stripping.

Forms should be stripped off carefully. Careless stripping may damage the concrete. Forms recessed in concrete require special care in stripping. After forms are stripped, all materials must be thoroughly cleaned of hardened concrete. All nails must be pulled from sheating, boards etc. Cleaned sheathing should be sorted in size and be stored in neat piles.

Stripping Time

Forms should not be struck, until the concrete has reached a strength al least twice the stress to which the concrete may be subjected at the time of removal of form work. The strength referred to is that of concrete using same cement and aggregate, with the same proportions and cured under conditions of temperature and moisture similar to those existing on the work. Where possible, the form work should be left longer as it would assist the curing.

In normal circumstances and where ordinary portland cement (OPC) is used, forms may generally be removed after the expiry of the following periods.

Table 8.1.

	Types of surface work	*Time of stripping of forms*
1.	Walls columns vertical faces of all structural memebres.	24 to 48 hours as may be decided by the engineer in charge.
2.	Slabs (props left under).	3 day
3.	Beam soffits (props left under).	7 days.
4.	Removal of props under slab	
	(*a*) Spanning upto 4.5 m.	7 days.
	(*b*) Spanning over 4.5 m.	14 days.
5.	Removal of props under beams and arches.	
	(*a*) Spanning upto 6 m.	14 days
	(*b*) Spanning over 6 m.	21 days

For other cements, stripping time may be modified.

Where the shape of the element is such that form work has restrint angles, the form should be removed as soon as possible after the concrete has set to avoid shrinkage cracking occuring due to restraint imposed.

8.6 SLIP FORM CONSTRUCTION

Concrete has been in use for construction of civil engineering structures for last many decades. It has been always been preferred to many other materials. Though it can replace steel at various places, however it is still a versatile material of construction. Formwork is very essential in concrete construction. Plastic concrete is poured in form and allowed to remain there till it hardens and attains strength. The formwork cannot be removed till it has becomes hard to resist load likely to come over it. This prevents earlier removal of form and likely reuse of form is delayed. This further slows down the progress work consequently increases time required to complete the work. Slip form is the alternative process of concreting where construction is such that from can be lifted vertically upward and can be rolled (slide) horizontally side way. Slip form construction is a type construction in which forms are raised while concrete is in transferring stage from plastic to hardened concrete.

It is a type of construction in which forms are raised while concrete is in plastic state. By placing the concrete continuously for an entire structure, construction joints are eliminated which avoids the possibility of leakage through it. Silos, bins, bridge piers, tall rectangular wall panels etc. which have uniform cross-section throughout can be caste using slip forms. It reduces the joints further the cost of form work also goes down as they are used continuously. The construction is also very rapid. These forms are lifted with the help of jacks. The forms can be lifted at a rate of 20 to 25 cm per hour. If they are lifted rapidly, concrete at bottom will bulge due to very much in plastic stage whilst if they are lifted very slowly the concrete will stick to the form. Great care is required to handle slip forms in actual use.

The 275 m high chimney of thermal power station Mumbai was constructed using slip form construction technique in 1987-88 by HCC for Tata Electrical company of Mumbai. The average speed of construction was approximately 3 m/day. The process of slip form construction is continuous, once starts, it is not desirable to stop the work till it gets completed. However in actual practice it is rather difficult. Therefore a few stoppages during construction are unavoidable. These stoppages can be planned before start of construction work and therefore called as Foreseen Stoppages for example Hoist installation, winch replacement erection of working platform etc. However once construction starts; at same stage, it becomes essential to stop the work due to failure or breakage of some part of formwork or of equipment machineries involved, climatic factors, electrical failures, accidents, labour strikes etc. Such stoppages are called as Unforeseen Stoppage. However where there are stoppages, precautions are very essential. Joint between old and new work must be taken care of.

This type of slip form construction is very common in construction of chimneys, shafts of overhead tanks, concrete towers etc.

Types of Slip Form

Slip form construction is a construction method in which concrete is poured into a continuously moving form.

- The slip form primary consist of shuttering panels which envelop the structure along its design the shape and then it removed upwards continuously living below the structure in its final form.
- Slip forming is used for tall structures (such as bridges, Towers, buildings and dams), as well as horizontal structures such as Roadways.
- Slip forming enables continuous, non-interrupted, cast-in-place "flawless" (i.e. no joints) concrete structures which have superior performance characteristics to piecewise construction using discrete form elements.
- Clip forming relies on the Quick setting properties of concrete.
- Concrete needs to be workable enough to be placed into the form and consolidated (by vibration), yet quick setting enough to emerge from the form with strength.
- This strength is needed because the freshly set concrete must not only permit the form to "slip" upwards but also support the freshly poured concrete above it.

In vertical slip forming the concrete form may be surrounded by a platform on which workers stand, placing steel reinforcing rods into the concrete and ensuring a smooth pour. Together, the concrete form and working platform are raised by means of hydraulic jack. Generally, the slip form rises at a rate which permits the concrete to harden by the time it emerges from the bottom of the form.

In horizontal slip forming for pavement and traffic separation walls, concrete is laid down, vibrated, worked and settled in place while the form itself slowly moves ahead. This method was initially devised and utilised in interstate highway construction.

Components of slip forming :

1. Centre point
2. Starter/ kicher
3. Form
4. Yokes
5. Normal bracket
6. Central control system
7. Jack holder
8. Yoke leg clamp
9. Panel clamp
10. Water cleat plates
11. Working platform
12. Lifting equipment
13. Pump
14. Jack

Advantages of Slip Form Work

- High speed of erection (works' execution speed increases) and as a result, rapid completion of project - considerable savings of time;
- Rapid completion of the elevator machine room constructing process, time saving to quick installation of Elevators, installation of water and electricity supply;
- Minimal use of a crane, as formwork is lifted by hydraulic Jacks;
- Optimal quantity of scaffolding and work platforms as the slip formwork system pre supposes the use of its own platforms for implementation of works. This helps to maintain cleanliness at the construction site, because there is no redundant material to be stored;
- Uniformity of wall sections with the project drawings, smooth wall surfaces are achieved by the use of slip formwork, which cannot be said about other formwork systems;

Reduced labour costs due to the intensity of the working process.

QUESTIONS

1. What are the essentials requirements of a good form work.
2. Draw a neat sketch of form work for construction of a
   (a) Staircase
   (b) Beam and slab
   (c) Water tank
3. Name various materials used for form work and give merits and demerits of each.
4. Write a detailed note on Slip Form Construction.
5. Write a note on Erection-Oiling- Striping of form work.
6. What are the various loads required to be taken in consideration while designing a form work.
7. What is slip form construction? Discuss its advantages.
8. Explain different types of slip forms.

Chapter

9 PREFABRICATED STRUCTURES

9.1 INTRODUCTION

A large section of the urban population of India dwells in inadequate and sub-standard houses even under conditions of overcrowding. The population of country has run passed a figure of to 125 crores, the shortage of housing threatens to the worse than even before unless urgent steps are taken to tackle this situation in more dynamic way. The present housing shortage in the country is reckoned about more than 100 millions units. The task before us is, therefore, to construct so many houses and of constructing them as expeditioulsy and economically as possible bearing in mind certain minimum standards of accommodation, hygiene etc.

One may well ask how we propose to meet this situation with shortage of building materials and labour and further with tremendous increase in the cost of matierals. The answer lies in evolving new building materials and more overmass construction techniques which must bring about an appreciable saving both in the cost and time of construction. Hence on the basis of recommendations of expert committee appointed by the Ministry of Works and Housing of Government of India, housing factories in big cities like Ahmedabad, Poona, Mumbai, Delhi, Chennai etc. had been established for prefabricated construction of different units of a house and then to assemble them to form a house.

Such large prefab plant is the real answer to the housing problem in urban as well as in rural areas also. Large number of factories should be alloted, further sufficient publicity is needed for such factory made houses. Standardisation of dimensions be made to suit the need ; further planners should be advised to keep the dimensions as per standard size of different prefab, precast units available. This will increase the speed of construction and will economise the cost.

9.2 MODULAR CO-ORDINATION

Standardization and dimensional co-ordination have long been recognised as essential to efficient and economical construction. It is, therefore, necessary to plan houses on a modular basis, so that the dimensions of the structure will conform to the standard dimensions of modular product, without the necessity of cutting and patching at site. The basis module designated as 1 *m* which increasingly being adopted in this country, for hollow blocks and other small precast units, it is 10 cm. Thus if size of all the spaces are made multiple of 10 cm, there would be no need to cut and patch blocks. Thus all the building components should be standardized in modular co-ordination. Thus there can be a large number of choices and the speed and economy in construction can be achieved. Hence standardization has a very important role in modular co-ordination. Building components can be standardized in multiple module or submodule and can be cast in factory and can be utilised in any buildings, thus economizing the cost and increasing the speed of construction. Doors; windows, slabs, beams, wall panels, sanitary fitments, furniture etc: all should be standardized in modular co-ordination.

It is important to keep in mind that modular coordination has been created as a natural and necessary complement to standardization. Consequently, the very purpose of modular co-ordination is exactly the same as that of standardization namely to promote industrial mass production. The object-of-modular co-ordination is, therefore, first to allow advanced standardization and industrialization of the production of building components and secondly to accelerate the removal of their production from the building site, with its unfavourable working conditions, to factories, for the production of building parts, where conditions are more suitable for rational production. That is why modular co-ordination is basically an industrial tool for standardization.

9.3 MODULAR PLANNING RECOMMENDATIONS

In concrete as in other masonry construction, it is essential to design walls to conform to modular dimensions, then the full advantage can be taken of full and half length units. Besides a minimum cutting, fitting of units by mason mean speedy construction and neater appearance of the finished wall.

It is all essential that door and window frames conform to modular dimensions so as to fit in with full and half size units. All horizontal dimensions should be multiplies of nominal half length units whilst all vertical dimensions be multiple of nominal full height units. Thus with the nominal $40 \times 20 \times 20$ cm^3 blocks both horizontal and vertical dimensions should be designed to be in multiples of 20 cm.

The relationship has proceeded on the principles of establishing a simple dimensional relationship by means of basic unit which governs the dimensions of buildig components. The principle of co-ordination is called modular co-ordination and the basic unit as module.

The word "module" has come from the Latin "modulus" meaning a small dimension. A module is a unit of measurement it may be any number of a unit of a measurement like centimeter, metre etc. and may be applied to the width, depth and height of building component or a whole building. The module forms the basis of a three dimensional rectilinear frame which when used in building, ensures that. standard material and components can be used together and they will in turn fit into general design layout.

The basic module used in building industry for the purpose of dimensional co-ordinator is 10 cm. In addition a planning module or multiple are also used by multiplying modules such as 1 m, 2 m etc. but generally adopted equal to 100 cm. Basic module can be further sub-divided as M/2 or M/4 called as sub-module.

Hence all the buildings should be so dimensioned that walls, floors, openings, placing of equipments, pipe connections etc. are coordinated by a modular square grid laid over-plans, elevations and section. Horizontally, in the case of walls whose nominal thickness is a multiple of the module the modular grid is laid so as to coincide with the modular surface of the unfinished wall. In the case of walls, whose nominal thickness is not a multiple of module, the grid be laid as to coincide with the central lines of the walls. Partition wall be designed to fit in centre line basis so as not to disturb the grid arrangement. Doors and windows and other openings should be located so as to fit into modular surface system. In the vertical plan, the modular grid is laid as to coincide with the modular surface of the structural floors, finishing not being taken into consideration. The height of the storey measured from floor to floor should be modular. In the placing of doors and windows, a clearance of 4 cm is provided for floor finishing and height of shutters is adjusted accordingly.

All building components used in the skeleton frame, fixture and installation should be so dimensioned and shaped that they should fit into modular dimensions with a minimum of wasteful cutting and fitting. In arriving at the work dimensions of components, appropriate clearances necessary for making joints, for assembly and for fitting into in jointing material should be allowed.

The building height is restricted to be not less than 2.6 m, further increase if made in multiple of 2 modules of 10 cm each, *i.e.* further heights can be 2.8, 3.0 etc. upto 3.8 m. for further height, it should be increased in multiplies of 4 modules, *i.e.* 40 cm.

Assembly. Any competent or group of components having a modular dimension which are to be assembled with other components should be placed within its own modular areas as defined by lines or places of modular grid and not penetrate into the area of the adjoining components. A modular component should be placed in the centre of its modular area or space. In locating a component between its modular planes, considerations should be given to the joints required between two adjoining components and to the accuracy with which the component can be placed in position. For every competent a space is required for joining and the minimum and maximum and limits for joint thickness should be defined. The position of the component should be governed by the limit. Generally the limit is equal to half the minimum joint thickness.

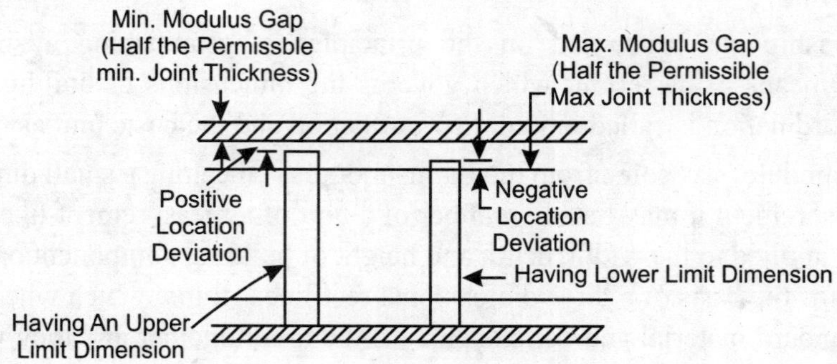

Fig. 9.1.

In assembly pre-fabricated elements, to achieve a stable structure is an important problem. Structural stiffness, insulation properties, water tightness, durability etc depend on a good execution of joints. Joints between different structural elements transfer forces. Hence good continuity has to be obtained between successive segments otherwise unnecessary stresses may be developed.

9.4 MODULAR TOLERANCE

Tolerances in building industry are employed to have limited variation in the length, shape, angle and position. These are primarily meant to ensure constructionally correct, economically profitable and aesthetically satisfying assembly and installation of pre-fabricated constructional elements.

It is not possible to specify the size of a component without deviations. The difference in the size is caused due to various factors such as carelessness on the part of workers, discrepancies in the method of manufacture, imperfect tools used, shrinkage of materials etc.

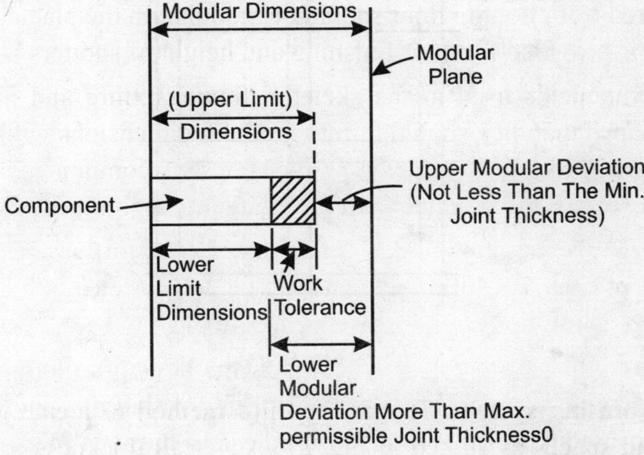

Fig. 9.2.

System of modular co-ordination is to a great extent governed on the location of joints. There are three methods of locating joints in modular co-ordination.

1. Joints at grid intersection.
2. Tartan joints
3. Modular jointing system (M.J.S).

1. **Joints at grid intersection.** It is important to have joints at grid line intersections. In grid the intersections are suitable place for locating the joints as shown Fig. 9.3.

   If narrow panels are used, these type of joints possess high degree of flexibility.

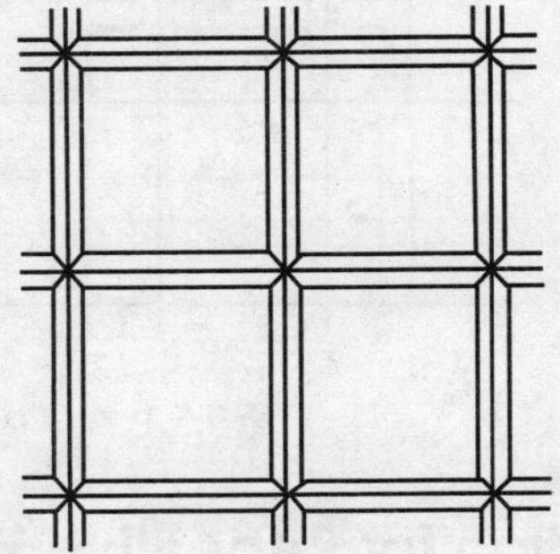

Fig. 9.3.

The drawback with this system is, if the components are located centrally between the grid lines, in certain situations the length of some of the components will not be modular, so long as the thickness of the components are not equal to the grid sire

2. **Tartan joints.** This is another method to locate the joints in which joints are located at the intersection of the pairs of grid lines called Tartan grid as shown in Fig. 9.4. If all the joints are identical then all the components will have same thickness. In this there will be no non-modular component. The diadvantage of this system is that all the components should have the same thickness.

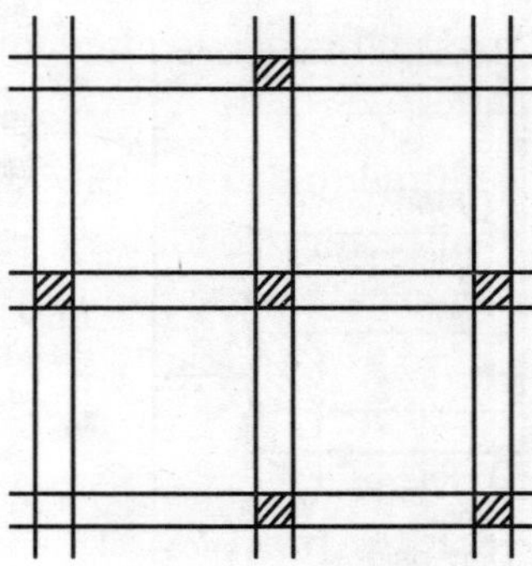

Fig. 9.4.

3. **Modular jointing system.** This is the third method to locate joints at centres of grid lines or grid spaces as shown in Fig 9.5. Any wall thickness upto the grid size can be accommodated in this case. The square joints of the Tartan grid are replaced by corner components in this case. In this system components will be modular. In this system in all types of junctions (end to end, *L*, *T* or +) the same face to face joint details can be adopted. Among the different methods of locating joints, modular joints system and system of locating the components such that these centre lines coincide with grid lines, are the most suitable ones of modular planning.

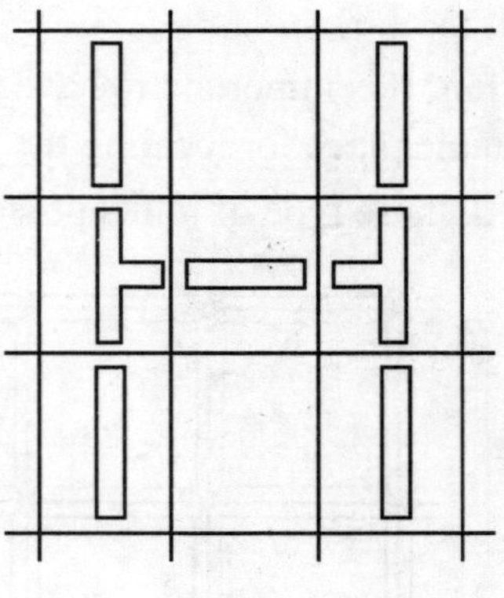

Fig. 9.5.

9.4.1 Modular Construction for Repetitive Works

Modular construction is a type of construction in which each and every element is of size one/two/ to multi module or half module. The size of a structure also in terms of module. Therefore in case of repetitive construction, if all elements are in terms of module, construction becomes easy and fast. In case of construction of large number of houses or flats of identical in nature, this makes construction very fast. Weather delays are no longer affected with modular construction either, so contractors are able to complete daily build goals and easily avoid setbacks that are common with on-site construction.

Modular construction is a term used to describe the use of factory produced pre-engineered building units that are delivered to site and assembled as large volumetric components or as substantial elements of a building. However it can be used for conventional construction of repetitive nature and large in quantity. Because construction teams are able to work in controlled environment using repetitive techniques and convenient equipment like large scale jigs, modular construction results and faster build times.

Volumetric construction (often referred to as modular construction) involves the production of three-dimensional units in controlled factory conditions, prior to transportation to site. Three dimensional (3D) volumetric modular construction is a modern method of building by which large, often room sized models are manufactured in a factory, then transported to site and craned into place. Permanent modular construction (PMC) buildings are manufactured in a controlled setting and can be constructed of wood, steel or concrete. Modular components are typically constructed in doors on assembly lines. Modules' construction may take as little as 10 days but more often one to three months.

As number of units to be constructed (in case of conventional construction) or fabricated in factory and assembled at site (in case of precast construction) is large, labour involved in construction is not required to be guided again and again. The labour being familiars of work and work being identical in nature, its progress is very fast. In case of precast houses a project of approximate hundred houses can be completed in a few weeks. However in conventional construction, more gangs will have to assign different jobs and one gang will follow other as per progress and such large projects can be completed in a few months. To check progress of project of different items in a group of house construction, line of balance (LOB) (refer to chapter management techniques with CPM and PERT) method can be used. This LOB will indicate number of items have been completed in various houses on a particular day, giving complete idea of project to higher authorities.

9.5 PRE-FABRICATION

The reduction of building cost through modernization of building methods and production of pre-fabricated components in factories have received very large attention in all parts of the world. The design has received great deal of attention of architects, engineers and material specialists. One such aspect of design which has been investigated with substantial success is the rationalization of dimensions of building materials and building components. It can be achieved by standardization of

1. Components of pre-fabricated elements.
2. Building types (plan, layout and spans)
3. Services.
4. Construction technique.

Controlled dimensions by modular co-ordination leads to standardization of components. Reduction in variety and size which is essentially for economy can be held by reduction of preferred dimensions. A series of preferred dimensions should have

1. Complete range from thickness of a small component to the total length of building.
2. Should have a constant increment in a particular range.
3. Any dimension should be a certain percentage (7 to 13%) higher than the immediate lower dimension.
4. A single service may not be efficient to cover from small to large dimensions.
5. Functional requirements of space should be the most important criteria in deciding preferred dimensions.
6. Selection should also be based on its effect on cost of structure.
7. Weight of different components should not be too much different so that they can be handled by the same equipment.
8. Production technology and construction techniques to be employed should be considered.

Too few sizes may result in spaces bigger than needed from functional requirements. Increase in size of flexibility of components will make them less versatile so that only fewer cases for them can be found. Components which may be standardized are :

1. Bricks and building blocks of various modular sizes.
2. Doors, windows, ventilators etc.
3. Built in cup-board, cabinets etc.
4. Beams, columns, floor stab and wall components.
5. Partitions roof trusses.
6. Stair-case, lintels and sun-shades.

Further to aid it, standard building types *i.e.* planning of space, typification of repetitive designs should be made. Further their design should be based on modular co-ordination.

Pre-fabrication is an advanced and upto-date method of steel and reinforced concrete construction. Prefabrication means that the structure is disunited into members and these are precast either in factories, built and equipped particularly for this purpose or in temporary plants established on the site. Then the precast reinforced concrete members are shipped to the place where they are to be used, here they are hoisted, set in their final places and assembled to form a complete structure.

The literary meaning of pre-fabrication means to manufacture parts or to fabricate components or assemble parts prior to erection or installation to form a structure.

In India full pre-fabrication has not been adopted so far. In full pre-fabrication, all the parts of a building are pre-fabricated in factories or workshops and then these parts are transported through trucks or other methods to the site of construction and then with the help of heavy machineries buildings are erected. In full prefabrication, heavy machineries, power, good transportation and erection facilities are required.

In case of lacking in these, partial pre-fabrication is the only choice. In this load bearing structure or heavy structures are cast in site while doors, windows, wall panels, roofings etc. pre-fabricated in factories and then transferred to site where they are put or erected in buildings.

Transportation is a major problem. To transfer heavy units, heavy trucks are required, further road conditions should be good otherwise due to jerks, there is every possibility of showing cracks in pre-fabricated elements.

The speed of construction in full pre-fabricated system is certainly more, further it is economical also.

Degree of Prefabrication. It is the ratio, in percent of the estimated cost of prefab structures and parts (including prefab units of sanitary, electric etc.) to the total cost of the project. Higher the ratio, more is the prefab component.

9.6 ADVANTAGE OF PRE-FABRICATION

Following are the advantages of pre-fabrication system :

1. **Speed.** There is about 40% saving in time for a single storey construction and about 25 to 35% in multistory buildings. In case of pre-engineered buildings i.e. steel structure, time saving is very large.
2. Reduction in cost about 30%.
3. Proper utilization of scare materials like cement.
4. High productivity due to better organisation and supervision.
5. Good technical control and quality control hence more strength.
6. Good resistance against weather conditions.
7. Clean and dry work at site.
8. Further 'there is an improvement in working conditions for labours as they get a permanent job. hence unemployment problem to certain extent is solved.

The disadvantages of pre-fabrication system are not more but still certain demerits are these :

1. They are required to be made more strong and hence heavily reinforced. For handling stresses are likely to occur hence extra precaution is necessary.
2. Large initial cost.
3. Care in joining of pre-fabrication components require more skill.

9.7 COMPARISON BETWEEN PRE-FABRICATION AND MONOLITHIC CONSTRUCTION METHOD

1. Partial or total saving of materials for scaffold in prefabrication system.
2. Possibility of far more accurate and better workmanship is possible in pre-fabrication system whilst good quality control in monolithic construction can be achieved only by employing skilled labour.
3. The work in pre-fabrication can be organised similarly to manufacturing with a very high degree of mechanization, stages arising from standardization and serial production can be utilised, the work on the site can be restricted to no more than the assembling of the precast members.

4. Working time shortens in pre-fabrication works.
5. Requirement of skilled carpenter and scaffold hands are hardly required for prefab work, hence interruption in concreting can be omitted.
6. Fewer expansion joints are required.
7. Pre-fabrication system promotes the wide spread use of prestressed structures hence allows for the establishment of lighter, more beautiful and slender structures having a better space effect. Further it reduces dead load of structure also.

9.8 PRE-FABRICATION SYSTEMS

Prefabrication system can be defined as:

1. On site pre-fabrication, *i.e.* site fabrication
2. Off site pre-fabrication, *i.e.* plant fabrication.

"Off site pre-fabrication" system gives permanent work to labour hence have better control. In this almost all the components are fabricated in factory and then assembled at site of construction. A plant can be economical only when more than 1000 housing units are prefabricated every year,

On site pre-fabrication

When using this kind of fabrication for concrete structures, the reinforced concrete members are produced on the site, chiefly in the open air, and most members of smaller size are precast in temporary covered sheds established for this purpose. Difficulties occurring in construction works, in general, cannot be avoided here either. Each new building site entails, in most cases, the employment of new labourers and the use of different building materials, the properties of which are often not sufficiently known. Mechanization can not be of such a high degree as in a permanent plant, because of the temporary nature of site works, its duration being generally short, at the most one or two years. So a high degree of mechanization similar to that applied in a permanent workshop would not be profitable.

Laboratories on sites are, in general, not so well equipped as in a permanent factory for building members. The works must be carried out in the open air and this means particularly in winter a serious draw back, though it is true that the greater part of the members can also be produced in appropriate seasons under favourable weather conditions.

A very great advantage of site pre-fabrication in comparison with pre-fabrication plants is the omission of the need for shipping the cost members over great distance. The large member *i.e.* the main structural elements are pre-fabricated at the site in general, in such a way that their setting in place requires only a vertical hoist without any horizontal motion. Members of smaller size (roofing and flooring, windows) are produced, as a rule, in the vicinity of the construction in progress or on the site itself in a temporary smaller pre-fabrication plant established for this particular purpose.

As the large members are not transported their dimensions and weight are not limited by transportability but only by the possibility of hoisting. Consequently, the members can be larger than when pre-fabricated in factories. The number of members is. therefore, smaller, so advantageously diminishing the number of hoisting and joints.

Plant Pre-fabrication

This kind of pre-fabrication is done in permanent plants of factories established particularly for this purpose. Its advantage is that the work can be performed in covered rooms irrespective of the hardships of weather and outer temperature. With a constant team of workmen, the work itself can be organised factory life. The plant can be furnished to the highest degree with mechanization and automation. The permanent laboratory allows continual control, and so the materials to be used always have similar properties. Owing to these advantageous conditions, prefabrication plants or factories produce in serial manufacturing, in general cheap and reliable structures of good quality and in good quantity.

But a drawback of plant fabrication is that the members must be transported to places where they are to be used. The transportation cost of a precast member to the site costs in general is about 10 to 15 of the total cost required for production and assembly of the same. For the sake of transportability the dimensions of the members must be held within certain limits, so increasing the number of joints in the structures. The above mentioned limitations of the dimensions of the members entails a certain restraint on design, as well as on the further development of pre-fabrication.

Plant pre-fabrication is appropriate for mass production, chiefly for the manufacture of standardized members.

Field of Applications

The following gives some information on the question to which method, *i.e.* whether monolithic, pre-fabricated on site or pre-fabricated in plants should be used provided there is a possibility of choice.

"Monolithic structures" are suitable, in general for all foundation work. Foundations require as a rule, large quantities of concrete and little shuttering. The transport of concrete and the concreting are easier than transport and placing of heavy foundation bodies, therefore, the prefabrication of foundation bodies may not be economical. The monolithic method of construction is suitable in general for all underground construction work and for structures requiring considerable quantities of water-proof concrete like tanks, barrages etc.

In architectural construction the use of monolithic structures is usually expedient for smaller buildings where owing to the few identical members, preparation for prefabrication would not be economical. The application of monolithic method is small, resulting in insufficient utilization of the boards of the moulds, furthermore, where there is no adequate place for pre- fabrication on the site or where the transport and assembly of hoisting machines or the operation of the latter would cause difficulties. Finally monolithic construction is generally expedient where sliding shuttering can be applied *e.g.* silo construction and also for structures with particularly large cross-sections, heavy loads.

The need for monolithic construction depends also, of course, on the availability of modern means and requisites used in shuttering and scaffolding.

"Plant pre-fabrication" is usually appropriate for the production of standardized smaller members. These can be manufactured in serial production, eventually kept in stock. These are mainly members necessary in housing as R.C. joist, floor boards, roof trusses, lintels, posts, prestressed slippers, transmission lines, pressure pipes, wall shafts etc. Members to be used for industrial buildings are store houses can also be manufactured in pre-fabrication plants *e.g.* purlins, roofing members, shells, girders, arches etc.

"Site pre-fabrication" in general is suitable for structures having many recurrent members like industrial halls, ware house etc. In the case of structures having many large and heavy members, manufacturing of these members in factory and then their transportation and erection may be feasible if it may not be economical. If large number of units of identical nature are to be manufactured or to be constructed like trusses, girders, arched structures, roofing members, wall panels etc. it will be economical to fabricate them at site. The greater the height and greater the number of uniform recurrent members, more economic is its construction by using site fabrication. Site pre-fabrication is also suitable for industrial buildings, intermediate floors etc. It is also suitable for post-tensioned structures.

9.9 CLASSIFICATION OF PRE-FABRICATION SYSTEM

Pre-fabrication system can be classified.

On the basis of structural schemes

(*a*) **Load bearing construction,** *i.e.* load bearing walls

(*b*) **Framed Construction :** either rigid frames or diaphragm based construction.

Load is transferred on beams, columns and walls are of partition only

(*c*) **Tube Construction :** either for very tall structures or steel structures.

Classification according to Construction

Size and type of components used as they affect the construction on :

(*i*) Production of elements and their method of construction.

(*ii*) Transportation and handling.

(*iii*) Speed and choice of material.

(*iv*) On overall economy of construction.

Hence accordingly, we have

(*a*) **Large block construction.** Block of size 30 × 20 ×15, 30 × 20 × 10 cm^3 are constructed in general. Blocks of very large size 1 m × 1 m × 0.50 are also constructed, if required.

(*b*) **Medium panel construction.** Generally 1 m wide and of full height size panels are constructed. These panels are used for construction of walls.

(*c*) **Large panel construction.** Constructed full size wall panels like 3 × 3 square metre panels or 3 × 4 sq. m panels and so on.

(*d*) **Special blocks.** Consist of full panel construction unit such as a bathroom, water-closet, drawing room etc. These different special full size blocks are then joined together to produce full unit such as house, dispensary etc. This system is also called as Box system.

9.10 CLASSIFICATION BASED ON CHARACTER OF ELEMENTS

(*a*) **Hollow element.** Economical and reduces self-weight.

(*b*) **Solid element.** For load bearing requirements. Useful where hollow elements may not be able to take load.

(*c*) **Light weight blocks.** For non-load bearing elements, materials used are light weight concrete, fibre-board, jute boards etc. These are used for partition walls, decoration purpose, sound proofing etc.

(*d*) **Sandwitch elements.** Two separate panels are joined together to increase strength or to have sound insulation.

9.11 DISUNITING OF STRUCTURES

To minimise the difficulties connected with transportation, erection, placing of structures. For this, members of different sizes are fabricated and then they are united/joined to develop a whole structure. The structural members to be precast can be divided into namely :

1. **Load-carrying.** Structural system.
2. **Surface forming.** Structural system.

The system used for dis-uniting of structures are dealt further.

System consisting of linear members disunited at joints

Disuniting at joints gives linear members. This facilitates both, *i.e.* for manufacturing and assembling. In this case auxiliary scaffolding are not necessary and the hoisting process is very simple. The disadvantage of this system is that the joints are at the corners, *i.e.* the point of maximum bending moments which leads to the forming of joints difficult. Therefore, the joints must be over-dimensioned.

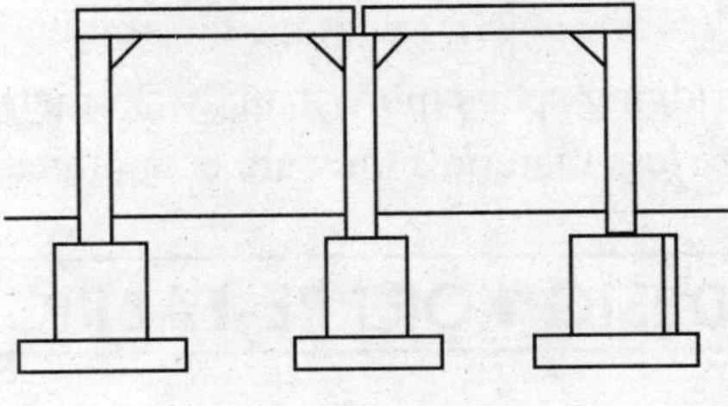

Fig. 9.6.

If moment resisting joints are replaced by hinge joints, the complicated construction of rigid corners can be omitted. Fig. 9.6 shows structure composed of straight numbers joined by rigid or hinge like joints.

System for the pre-fabrication of entire rigid frame

In this case the number of joints are reduced and the members are precast as entire frames. The hoisting of these frames are difficult and complicated then hoisting of straight members and requires careful pre-fabrication.

Therefore, this method is mostly adopted in site prefabrications. Entire frames are precast in a horizontal position on the ground close to their final location. The advantage of entire fabricated frames is the small number of joints needed and the possibility of rapid hoisting works. The application of this method can be recommended for the construction of long hall consisting of a great number of uniform frames.

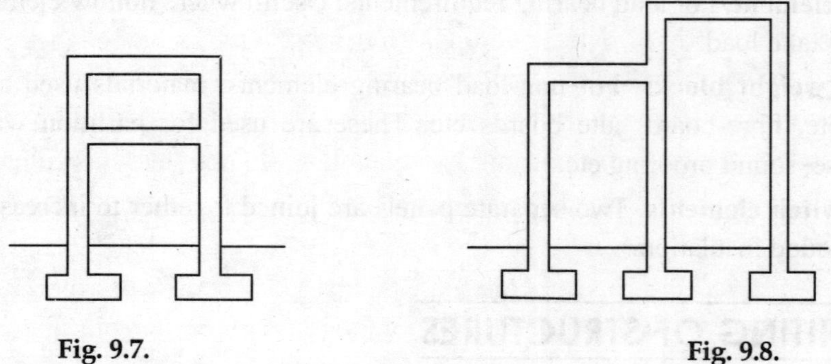

Fig. 9.7. Fig. 9.8.

Many a times when *L, T* shaped members are used, method of their erection is known as Lamda method. In this method the members of a structure for precasting are divided at points of minimum moments. This method came into force due to the difficulties met when carrying out a moment bearing junction of a place where the moment is greatest.

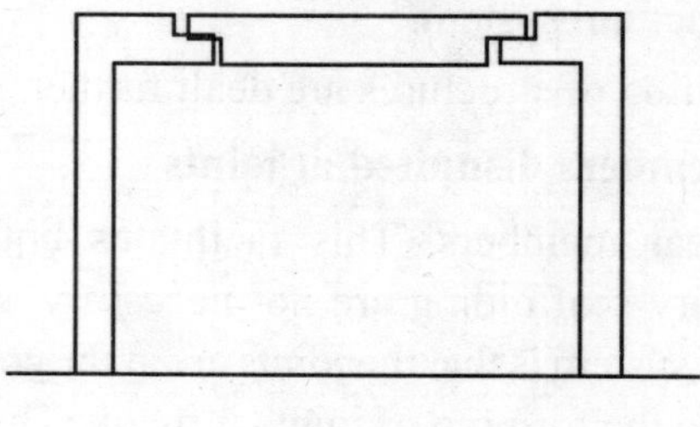

Fig. 9.9.

Arches pre-fabricated

Arches are normally used for bridging spans more than 20-25 metres. The production, placing are more difficult but as they require less materials they are economical for long span structures.

9.12 PRINCIPLES OF DESIGN OF PRE-FABRICATED COMPONENTS

The basic idea of pre-fabricated construction is to split up buildings and other structure into components or elements which can be industrially manufactured and which have shapes that enable them to be combined with one another and can also used for other buildings. The design must be proceeded by analysis, the primary object is to ascertain what structures or structural components are necessary and what requirements, these must satisfy. The evaluation of the analysis co-ordinates the information and information collected makes an assessment of the individual data and compress them with one another. Next is the design of the prototype evolves into a product suitable for serial manufactures. There is no special theory for prefabricated structure. Of course there are little difficulties associated with use of prefabricated structural components.

Handling and erection stresses. Large and sudden stresses are always liable to occur in the course of handling and at erection.

Handling stresses are stresses developed in precast component in the course of transport and erection. These are due to the following reasons :

(*a*) Components are transported in different positions from those in which they are installed in the structure. For example, column is a vertical compression member when erected but during its transport and erection, it functions as flexurally loaded beam.

(*b*) In the structure like slender beams, laterally unsupported beams are braced by other members but during their transport, erection, these do not have such supports.

(*c*) Dynamic stresses are liable to occur during erection and transport, if they are not handled properly.

Hence while doing analysis of structure, stresses associated with transport and erection are also required to be taken into consideration. Temporary conditions arising during construction can be very dangerous and should be carefully investigated. Methods employed for erection should be such that they eliminate surplus stresses arising when the members are hoisted and placed. This aim should be achieved in such a way that no additional reinforcement is required or minimum reinforcement required for that.

9.13 PRODUCTION OF MEMBERS

The concrete being in plastic state, needs form to give a shape. These forms are retained till the cement concrete has partially hardened. All the production technique of pre-cast involves these sequences.

The principles underlying the manufacture of concrete components are :

1. Speediest possible production.
2. Improved quality.
3. Economy using least amount of labour and minimizing wastage of material.

The above aims can be accomplished by using mechanised manufacture which reduce the human error and ensure uniform and good quality of work.

The manufacture procedure comprise three basic stages, *i.e.*

1. Preparation of raw material
2. Making the units
3. Curing and storage.

Manufacturing of pre-cast member can be classified to the following sequence :

1. Production of large members at hoisting places.
2. Production of smaller members
   (*i*) in temporary plants at site.
   (*ii*) in permanent plants.

The manufacturing of heavy members or large members are generally performed under their final places in such a position that the final positioning requires only vertical hoisting. When determining the places for the production of members, following conditions should be satisfied :

1. Sufficient space must be available for their casting as well as for hoisting and erection.

2. Further more adequate room should be provided for the assembly of the lifting equipment.
3. To store raw material for casting and these blocks, adequate space is needed.

Smaller member can be cast in permanent plants or in temporary sheds near the site. The production near the site is only feasible when project is very large and requires sufficient precast units to make it economical. If permanent plant is already available within an economical distance from the given project site the choice between the available permanent factory and the setup to be nearly established at site is to be governed among other factors, by the cost of the new setup against the cost of transportation of the element from the permanent factory. The factors govern the choice are :

1. desired rate of output
2. transportation facilities from the permanent factory to the project site.

9.14 MANUFACTURING METHOD

With respect to the position of moulded elements, following are methods adopted for their manufacture :

1. Stationary production
2. Mobile production

In stationary production the components harden at the point where is moulded while the workmen move to successive moulds. In the mobile production method,the moulds are moved along with the elements from one working position to the next till the elements are hardened.

The methods are further classified as

1. Horizontal moulding
2. Vertical moulding.

Each of the above further classified according to the use of moulds as discussed in subsequent paragraphs.

Horizontal moulding in stationary production is the most universal method. In its simplest form they comprise only edge or side forms placed over a hard base. It needs larger production area. This system is usually adopted for

1. External wall panels,
2. Floor panels,
3. Beams, columns, cored slabs, through units, ribbed slab etc.
4. Steps of stair case,
5. Pile casting.

Horizontal moulding system may be classified as fixed horizontal mould and tilting horizontal moulding system. Both these methods are used for casting large panel elements. Fixed horizontal moulds are less expansive to tilting horizontal mould. The basic difference between these two is that in fixed horizontal moulding system, moulding and demoulding are done in horizontal position whilest in tilting horizontal moulds, moulding is done in horizontal position but demoulding is carried out when moulds are tilted to an almost vertical position. Tilting horizontal moulds are more complex and costlier but they are still better as extra reinforcement is not required, since the demoulding is carried out at an almost vertical position. One edge of the mould base is kept hinged and the tilting operation is usually accomplished either by hydraulic jacking or by tilting the other edge with the aid of lifting equipment. Such moulds are, therefore, intended only for casting wall panels.

Vertical moulding is the other possibility in the stationary production method. Individual moulds or a battery of moulds can be used for such a casting. Internal wall and floor panels are often cast vertically using battery moulds. Vertical moulding gives smooth surface on both the sides hence suited where smooth surfaces on both sides are required. Steps internal wall panels etc are generally produced by vertical moulding system.

Advantages of vertical moulding are :

1. Large number of units can be produced in small space and the process is also very fast.
2. Surface finishing is not required much.

There are many ways of vertical moulding. Single mould is employed for casting volumetric elements as sanitary units, stair cases etc.

Vertical battery costing technique is the most industrialised method of casting wall panels.

Production Technique for Pre cast Elements. The production technique of pre cast elements depends upon number of units to be produced. Based on it there are three techniques of pre cast elements production. These are :

(*a*) Job production
(*b*) Batch production and
(*c*) Flow line production.

As the name indicates, when small number of units are required to be produced, such as small structural frames, job production is done. Here units are cast daily from specific mould and carried to the site. Semi skilled persons can do this type of production. A gang of men carries such type of job. These gangs are expert and do such jobs very efficiently.

Batch production is adopted when components are required to be manufactured large in number. Skilled labour is employed for such jobs. Structural frames, bridge decks etc. are manufactured in batch production. It is a type of semi-mechanized manufacturing unit. When the quantum of work is very large in number and of repetitive type, flow line production system is used. It is a type of total mechanized production and therefore initial investment is large. Railway sleepers are generally produced by this system. This system also needs transporting facilities as large number of units are manufactured: they are required to be transported at required site, as very large storage of these units is difficult.

Pre Cast Technique. Wet casting is adopted for small number of units as in case of job production technique. Moulds are used, they are arranged in casting yard and wet concrete is poured in these moulds.

Flat casting is used when thin and flat panel units are required to be cast such as roofing/ flooring panels, tiles etc. While concreting these moulds, care is required to be taken to avoid displacement of reinforcing bars put in mould. Gang casting is a modified process of wet casting. Number of moulds is assembled and concreting is done to a group of moulds, called gang casting.

Battery casting is a mechanized form of gang casting. Battery moulds are used for large panels. Units are cast in the batteries of cells varying in number from two to twelve. It is a continuous process of casting, hardening, curing and removal of battery mould. Large number of units can be cost efficiently and effectively by this method.

9.15 IMPORTANCE OF JOINTS IN PRECAST CONSTRUCTION

The complexity of joints in precast concrete construction present challenges to the designer. Prefabricated construction is done in three stages.

1. Casting of the components—Slab, beam, column on construction site or factory.
2. Erection of precast components.
3. Assembling components in proper position, alignment, through proper joints.

The joints between precast components plays vital role in the sterility of the building. The uncertainty prevails in the performance and behaviour of joints in adverse conditions like earthquakes, quarry site vibration etc. Thus the study of joints is an area of research in industrialize construction. The failure of the joint can lead to changes of stress pattern which may lead to collapse of the structure. The actual performance of the structure joint can be evaluated only through actual performance of the standing structures. Though techniques have developed to make Joints strong however there is scope its improvements.

Principal requirements of connections (joints) :

1. Connections must be adequate structurally.
2. Means of connection between adjacent members must allow safe and speedy erection regardless of minor inherent inaccuracies in precast elements and in-situ concrete.
3. Physical details of joints and connections must permit economic manufacturing.

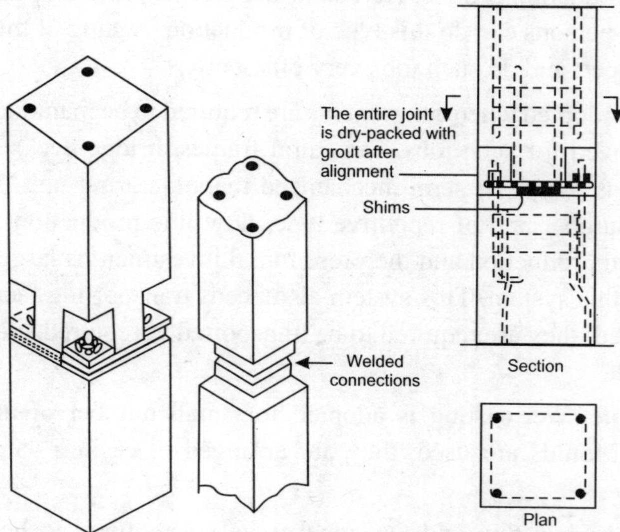

Fig. 9.10. Section through a column-to-column connection using a proprietary sleeve that is cast into the column. Projecting reinforcing bars from the lower column section are inserted in the sleeves, where injected grout makes the connection.

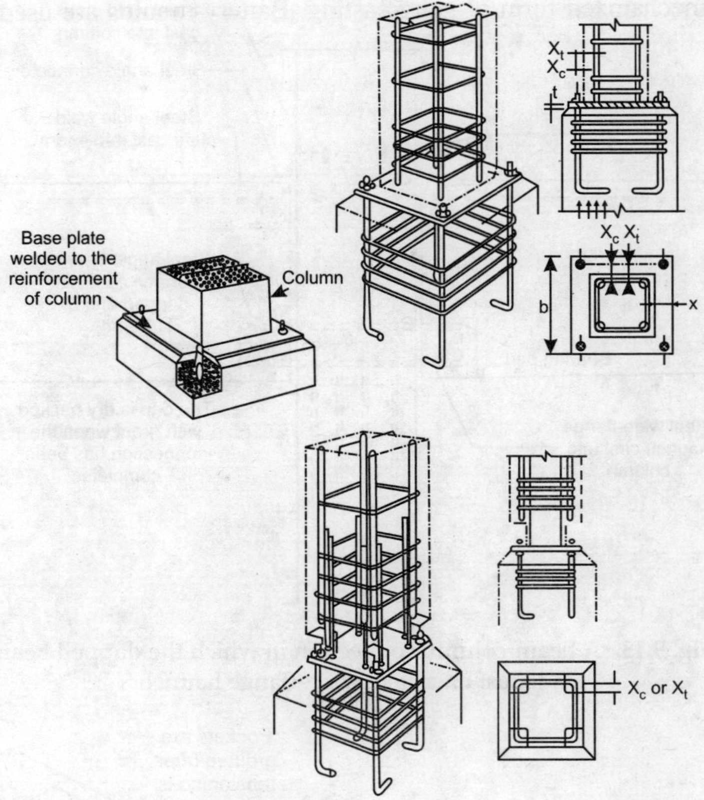

Fig. 9.11. Some typical base details for precast concrete columns. The open corners of the lower details are dry packed with grout after the column is algined and bolted; this protects the metal parts of the connections from fire and corrosion.

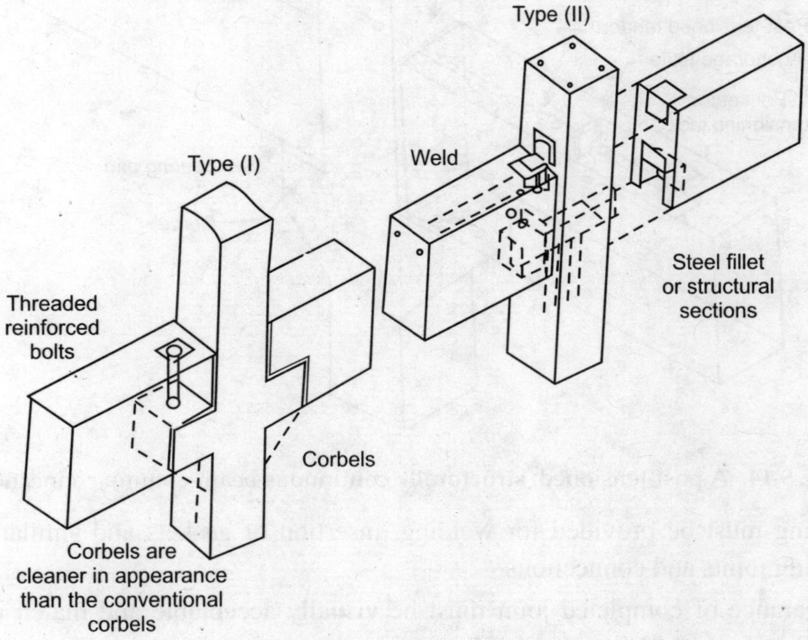

Fig. 9.12. Beam-column connection [Type II]

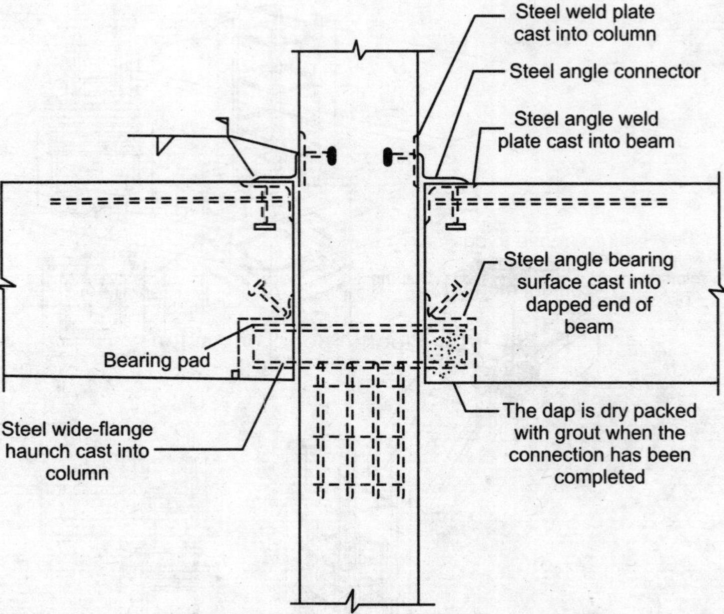

Fig. 9.13. A beam-column connection in which the dapped bean ends rest on a steel wide-flange haunch.

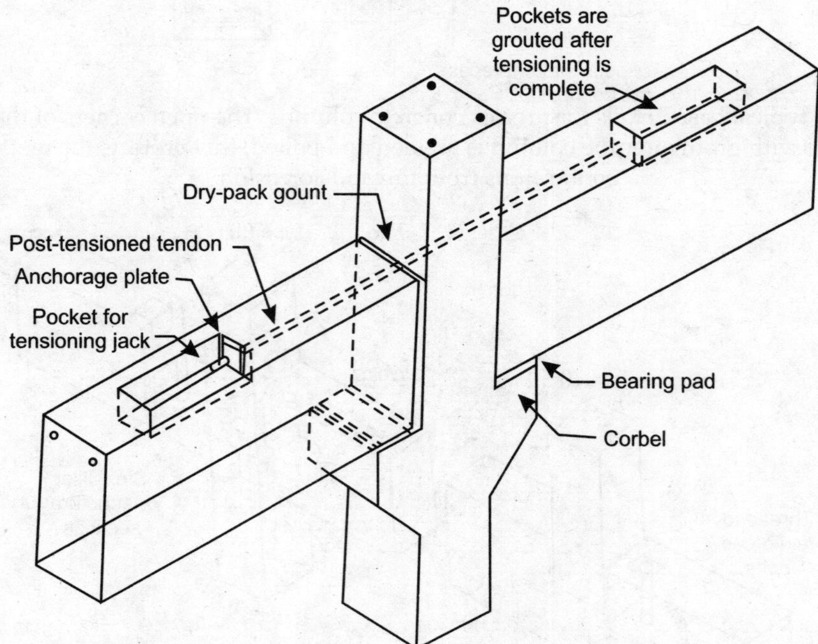

Fig. 9.14. A post-tensioned, structurally continuous beam-column connection.

4. Spacing must be provided for welding, insertion of gaskets and similar operations in forming joints and connections
5. Appearance of completed joint must be visually acceptable and match in scale of the completed structure.

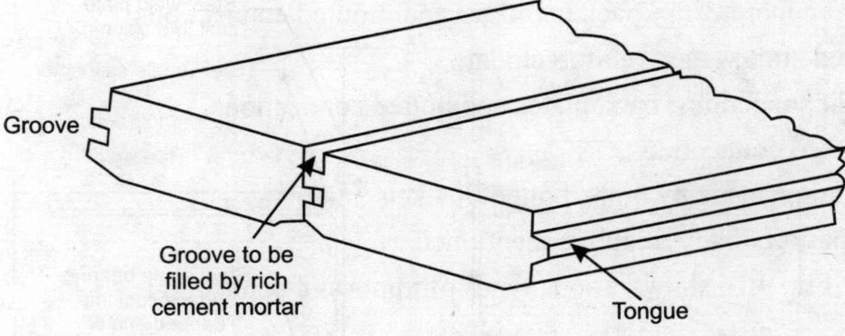

Fig. 9.15. Floor to floor connection.

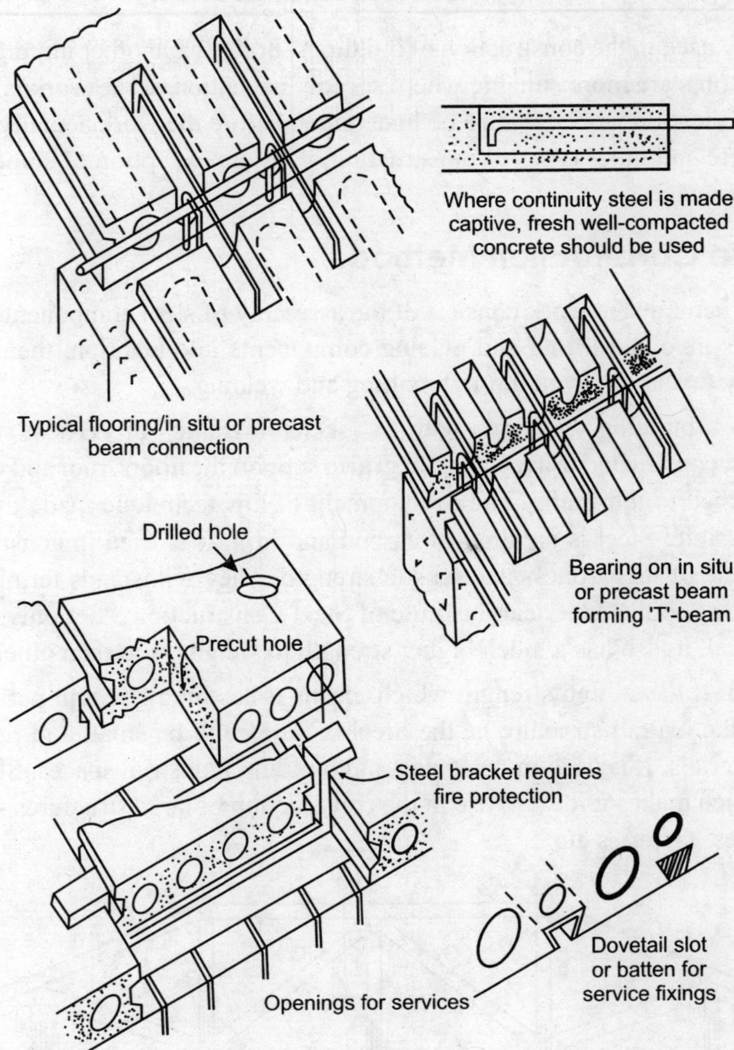

Fig. 9.16. Connection of precast floor.

The principal means of connection between precast elements and between precast and in-situ concrete include.

(*a*) Wet cast concrete, dry packed mortar and grouted connections.

(*b*) Threaded and swaged reinforcement,

(*c*) Welded connections, bars, plates and bolted connections.

(*d*) Prestressed connections.

(*e*) Connections made by bolts, bolted brackets and plates.

(*f*) Combination of any of above mentioned.

Fig. 9.10 To Fig. 9.16 show various types of joints and connections.

9.16 FABRICATION AND ERECTION OF STEEL STRUCTURES

Steel is extensively used in the construction of buildings, bridges and other important structure. The steel framed structures are more suitable where speedly execution of the work is essential. In such cases some pre-fabrication work can also be undertaken before the work actually starts on the site. Reinforced concrete and other types of construction, with the exception of some precastwork, do not posses this advantage.

Steel Structure Construction Method

The erection of structural steelwork consists of the assembly of steel components into a frame on-site. The processes involved lifting and placing components into position, then connecting them together. Generally this is achieved through bolting and welding.

Steel frame is a building technique with a " skeleton frame" of vertical steel columns and horizontal I-beams, constructed in a rectangular grid to support the floors, roof and walls of a building which are all attached to the frame. The development of this technique made the construction of the skyscraper possible. Steel is an alloy of carbon and iron. It is used in construction and other applications because of its hardness and tensile strength. They withstands termites, fire, rust and rot. In fact, according to the American Institute of Steel Construction, "not only is structural steel are stronger material, it also has a much higher strength to weight ratio than other materials."

Steel is tensile. It has a high strength which means it has high strength per unit mass. So no matter how large the overall structure is, the Steel sections will be small and lightweight, unlike other building materials. It is used in designing and building industrial spaces. Structural steel has a high strength which makes it ideal to use in the construction of large structures such as buildings, warehouses, bridges, factories etc.

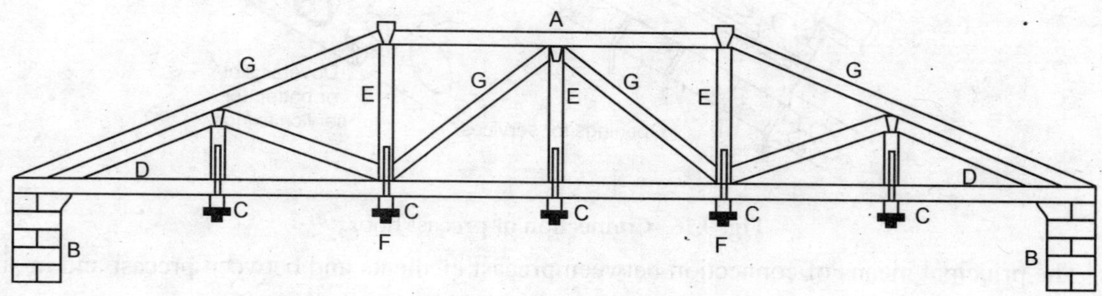

Simple Truss Bridge

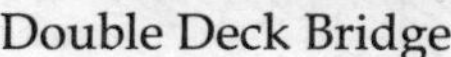

Double Deck Bridge

Suspension Bridge

The structural steel used in architectural structures as hot-rolled. They are produced in different grades which represent the yield strength of Steel. The major chemical in the gradients of Carbon manganese steels are iron, carbon and manganese. These are normally called mild structural steels or carbon steels. The strength and ductility are high, and being economical is therefore widely used. An I-beam, also known as H-Beam (for Universal column), W-beam (for wide flange), rolled Steel joist (RSJ), or double-T is a beam with I or H-shaped cross section. Mild Steel is also known as plain carbon steel. It is another type of steel used in steel building Constructions. It is very strong and durable and does ensure sturdy establishment. Carbon steel provides high strength which makes it more suitable for buildings.

The properties that need to be considered by designers when specifying Steel construction products are : strength, toughness, ductility, weldability and durability. The yield strength (YS) is the stress at which a material begins to deform plastically and will not return to its original shape when the applied stress is removed. The ultimate tensile strength (UTS) is the maximum stress that a material can withstand while being pulled before breaking.

The characteristics of Steel are judged from its properties : tensile strength, ductility, malleability, conductivity, luster, stiffness and toughness.

Types of Steel Sections

In general, a framework of suitable sections is erected to carry all the loads coming over the structure and the filler walls are constructed of other materials. Different types of steel sections and shapes are employed. Most of them are made by rolling. The commonly used sections in steel structures are illustrated in Figs. 9.17 and are explained as follows:

1. **Plates.** Plates may be rolled into any size, and thickness but generally the thickness varies between 5 to 30 mm. The material which is thinner than 5 mm is known as a sheet. The optimum area of the rolled steel plate is restricted to 30 sq. metres.
2. **Flats.** Flats are rolled as in the case of plates but have much longer lengths and very small widths. The width varies between 12 mm to 125 mm and the thickness varies between 3 to 8 mm. These are costlier than plates.
3. **Angles.** This is the most useful section. The thickness of each leg is the same irrespective of whether the angle is equal to unequal. An angle is designated by the dimensions of longer leg by shorter leg by thickness and length, *e.g.* 60 mm × 40 mm × 6 mm × 5 metres. Indian standard specifications lay down the sizes of the angle sections. Special angles with a round back, square toe, acute square roof and bulb types are also available and are used in special circumstances.

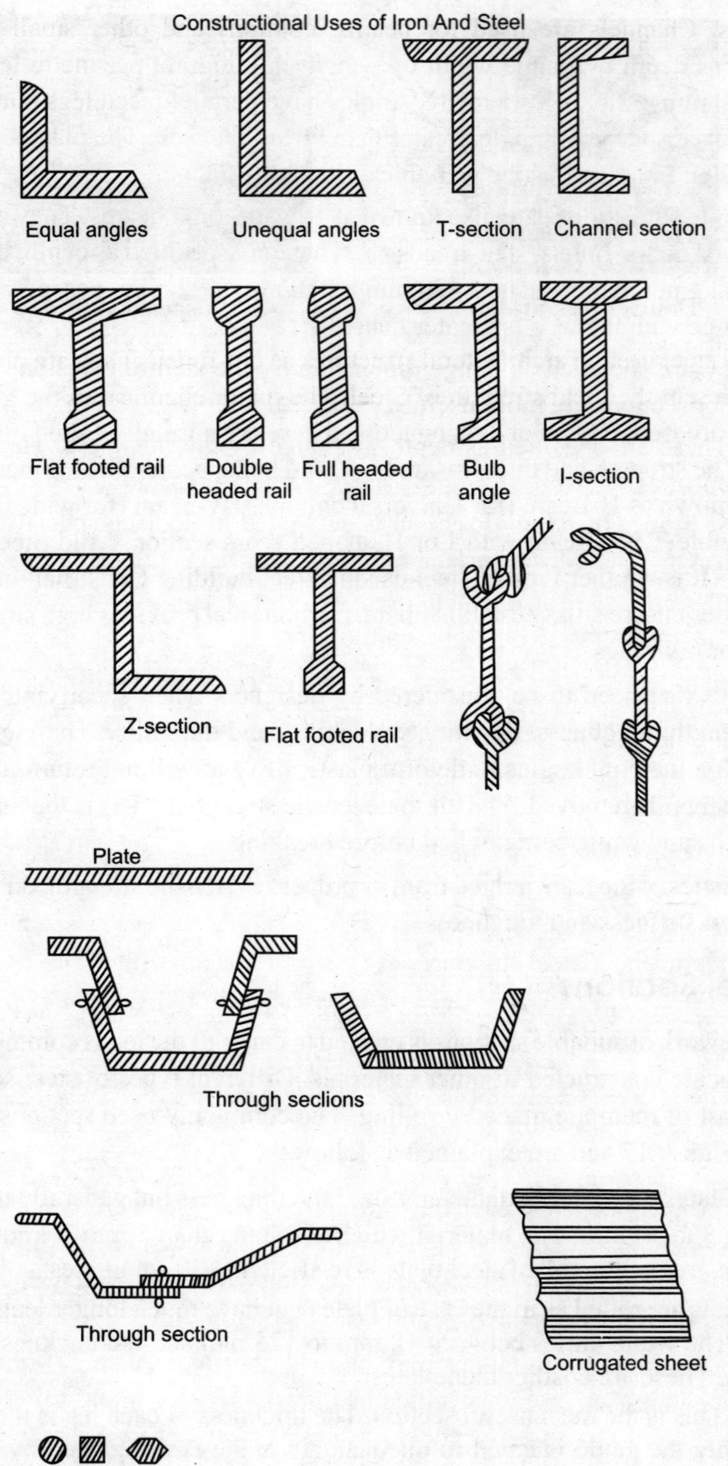
Constructional Uses of Iron And Steel
Equal angles
Unequal angles
T-section
Channel section
Flat footed rail
Double headed rail
Full headed rail
Bulb angle
I-section
Z-section
Flat footed rail
Plate
Through seclions
Through section
Corrugated sheet
Circular, Square Hexagonal Bar Section

Fig. 9.17.

4. **Channels.** Channels are used for beams, columns and other small works. They are specified as depth by flange width by weight in kilogram per metre length such as 350 mm × 100 mm × 38.8 kg per meire. Angles have parallel faced legs but the channel have the inner face splayed at angle greater than 90° to the web. The play is greater for bigger size channel. In no case is the web thicker than the flange.
5. **Joists.** Rolled steel joists are also known as I-beams or H beams. They are widely used to carry floors, doors, lintels, stair-treads etc. These are specified as depth, by flange width by weight in kg per metre, such as 450 mm × 250 mm × 92.5 kg, per metre. The inclinasion of the flange with the web is greater than 90°.
6. **Tees.** They are specified as flange dimension by the stem dimensions by the thickness. The smaller sections are more useful.
7. **Z-Bars.** These are specified as depth by flange by the thickness. These are commonly used for purlines.
8. **Trough Section.** This type of section is employed in flooring specially for bridges. The floor is brought upto a level surface by filling in with concrete and then ballast is placed.
9. **Rails.** Rail sections are also widely used. The rails may be of the following types :

   (*i*) Bridge rail. (*ii*) Tram rail.
   (*iii*) Flat bottom rail (*iv*) Bull headed rail.
   (*vi*) Heavy crane rail.
10. **Sheet Piles.** Steel plates of small widths are made U-shape to receive the other bulk head ends of steel piles.
11. **Corrugated sheets.** are used for roof covering. They may be galvanised iron or asbestos cement sheets.

Connections in Steel works. A structure is an assembly of various components which are fastened together. Members of steel structures are fastened together by means of rivets, bolts or by welding.

Various processes involved in steel works are

Straightening. All materials should be straight and if necessary before being worked should be straighted or flattened by pressure unless required to be of curvilinear form and should be free from twist.

Clearances. The erection clearance for cleaned ends of members connecting steel should preferably be not greater than 2.00 mm at each end. The erection clearance at ends of beams without web cleats should be more than 3 mm at each end, but where, for practicable reasons. greater clearance is necessary, suitably designed, seating should be provided.

Where black bolts are used, the holes may be made not more than 2.0 mm greater than the diameter of the bolts, unless otherwise specified by the engineer.

Cutting. Cutting may be done shearing, cropping or sawing Gas cutting by mechanically controlled torch may be allowed for mild steel only. Gas cutting of high tensile steel may also be permitted provided special care is taken to leave sufficient metal to be removed by machining so that all metal that has been hardened by flame is removed. Hand flame cutting may be allowed subject to the approvals of the Inspector.

Except where the material is subsequently joined by welding the loads should be transmitted into metal through a gas cut surface

Shearing, cropping and gas cutting should be clean, reasonable square, and free from any distortion and should the supervisor find necessary, the edge should be ground afterwards.

Holing: Holes through more than one thickness of material for members such as compound stanchion and girder flanges should where possible should be drilled after the members are assembled and tightly clamped or bolted together. Punching may be permitted before assembly, provided the holes are punched 3 mm less in diameter than required size and reamed after assembly to the full diameter. The thickness of material punched should be not greater than 16 mm When holes are drilled in one operation through two or more separable parts these parts, when so sopecified by the engineer, should separated after drilling and the burrs removed. Holes in connecting angles and plants other than splices, also in roof members an light framing may be punched full size through material not over 13 mm thick, except where required for close tolerance or barrel bolts. Holes for rivets or bolts should not be formed by a gas cutting process.

Assembly. The component parts should be assembled in such a manner that they are neither twisted nor otherwise damaged and should be prepared that the specified camber, if any, are maintained.

Erection

Plant and Equipment. The suitability and capacity of all plant and equipment used for erection should be to the satisfaction of the engineer.

Storing and Handling. All structural steel should be so stored and handled at the site that the members are not subjected to excessive stresses and damage.

Setting Out. The positioning and level of all steelwork, the plumbing of stanchions and the placing of every part of structure with accuracy should be in accordance with the approved drawings and to satisfaction of the engineer.

Security during Erection. During erection the steelwork should be securely bolted or otherwise fastened, and when necessary temporarily braced to provide for all load to be carried by the structure during erection including those due to erection equipment and its operation.

No riveting, permanent bolting or welding should be done until proper alignment has been obtained.

Painting After Erection

Before painting of steel, which is delivered unpainted, is commanced, all surfaces to be painted should be dry and thoroughly cleaned from all loose scale and rust.

The specified protective treatment should be completed after erection. All rivet and bolt heads and the site welds after de-slagging should be cleaned. Damaged or deterirated paint surfaces should first be made good with the same type of paint as the shop coat. Where specified surfaces which will be in contact after site assembly should receive a coat of paint (in addition to any shop priming) and should be brought together while the paint is still wet.

Where the steel has received a metal coating in the shop, this coating should be completed on site so as to be continuous over any welds and site rivets or bolts, but subject to the approval of the engineer, protection may be completed by painting on site. Bolts which have been galvanised or similarly treated are exempted from this requirement.

Surfaces which will be inaccessible after site assembly should receive the full specified treatment before assembly. Site painting should not be done in wet or frost weather It should not be done in humid weather.

9.17 ANTI-CORROSIVE STEEL AND SHEETS.

It has been found that by blending aluminium with zinc to give an alloy coating corrosion protection gets enhanced tremendiously. Hence galvenised iron is now being replaced by anti corrosive steel. USA and Australia were the pioneer countries to use anti corrosive steel, however, it has become popular now in many parts of the world including India.

This type of steel protects steel base in two ways. First, it provides tough physical barrier between extreme atmospheric conditions and the inner core of steel due to corrosion resistance of coating itself and secondly, it protects steel at cut edges and scratches. This is achieved by protection provided by zinc in coating. Galvanic action causes the zing compound to automatically buildup at cut edges and around scratches through an electrolytic reaction when water or moisture is present- This effect is sometime called as self healing property of coatings containing zinc. About 150 gm of alloy is required to cover an area of about 1 sqm.

This coating can be given on G.I. sheets and called prepainted galvahime roofing sheets and can also be used for siding panels. Unlimited spectrum of colours and long term durability enhance the appearance life and value of building. These sheets combine well with most of the other building materials and treatment. Its versatility allows to be used almost anywhere in building exteriors assuring compatibility and simplicity in the overall design approach.

9.18 WELDED WIRE FABRIC (WWF)

It is a prefabricated reinforcement consisting of parallel series of high strength cold drawn steel wires welded together in square or ractangular grids. Each wire section is electronically resistance welded. It has been used as a reinforcing material in the United States since 1915 and is included in ACI code (AO - 318 Building Code) as an acceptable building material and has various provisions regarding its design features. It is used to control temperature shrinkage stresses and to add reserve strength in slabs.

Its use in India is also becoming popular now. The use of WWF in construction has been adequately explained in IS : 1566 of 1982 and this item also appears in schedule of rates in various PWD'S. Its use in ground slabs, canal lining, rigid pavement, industrial floors has become very common due to its inherent advantages over the conventional H YSD Bars. High tensile strength, better bond, better crack width control, saving in time of building, elegance in use, quality and reliability are more important factor making it popular. Moreover it is economical also ; however it depends upon many factors. Its use for pavement slab is more or less becoming very popular, rather unavoidable. It has many advantages if used in general and pavement slab in particular.

Advantages of WWF

1. WWF provides proper and uniform distribution of steel reinforcement and stresses also get distributed uniformly. Closer spacing of thinner wires ensures significant reduction in crack development due to temperature and shrinkage effect.
2. It ensures more homogeneous two-way stress distribution throughout the concrete volume thereby restricting cracks propogation as well as crack width.
3. The mechanical anchorage provided by the WWF holds the cracked concrete of the pavement slab in position.
4. Due to increased stiffness of pavement slab, deflection is reduced in case of WWF.
5. Easy handling and placing no cutting of each wire is required. Binding of individual bars is also not required. This accelerates progress of work fast.
6. It has been proved that it has adequate fatigue and bond strength for rigid pavement slabs in particular.

9.19 PRE ENGINEERING BUILDINGS WITH BIM TECHNOLOGY

The term pre Engineered building (PEB) is traditionally associated with industrial buildings. However. the same concept and approach has been used in steel structures for industrial sheds and steel bridges. The concept of making pre-engineered or precast or pre-fabricated structure at factory and then transporting them to site, assemble using machineries and put for use has become common in last few years. The demand of PEB came into existence due to early demand for larger and heavier spans in shortest possible time This along with Building Information modelling (BIM) has lead value addition for owner and user of these structure.

The PEB concept come into existence due to huge demand of industrial buildings. Engineers and contractors developed their own standardized design for such buildings The primary frames were 2 D and used shape, angle and channel members to build trusses. Mostly bolted and welded connections were used. Then came 2 hinged, 3 hinged or fixed portals having castellated beams etc. Advances in computer aided design technology materials and manufacturing capabilities led to a growth of PEB. A three dimensional analysis by computer led in search of operation. Development of Z and C shapted member made use of these section to secondary structural elements to fasten and support external cladding. Roll formed profile steel sheets tensioned fabric, pre cast concrete masonry block and various other materials were used for external cladding. The works become very fast.

A pre-fabricated sted building consists of a variety of components, built in factory and then delivered to site. It follows a standard practice to make it easier to ship assemble and put together regardless of size. This makes building to complete it and use in a very short time. Almost 95 per cent components are built in factory premise which may be quite away from work site. It leads to saving in cost and increase in quality. However it needs specialized transportation system, erecting equipments and skilled labour. PEB are extremely strong to resist all type of forces like wind show, thermal etc. Components of pre engineered industrial buildings are shown in Fig. 9.18

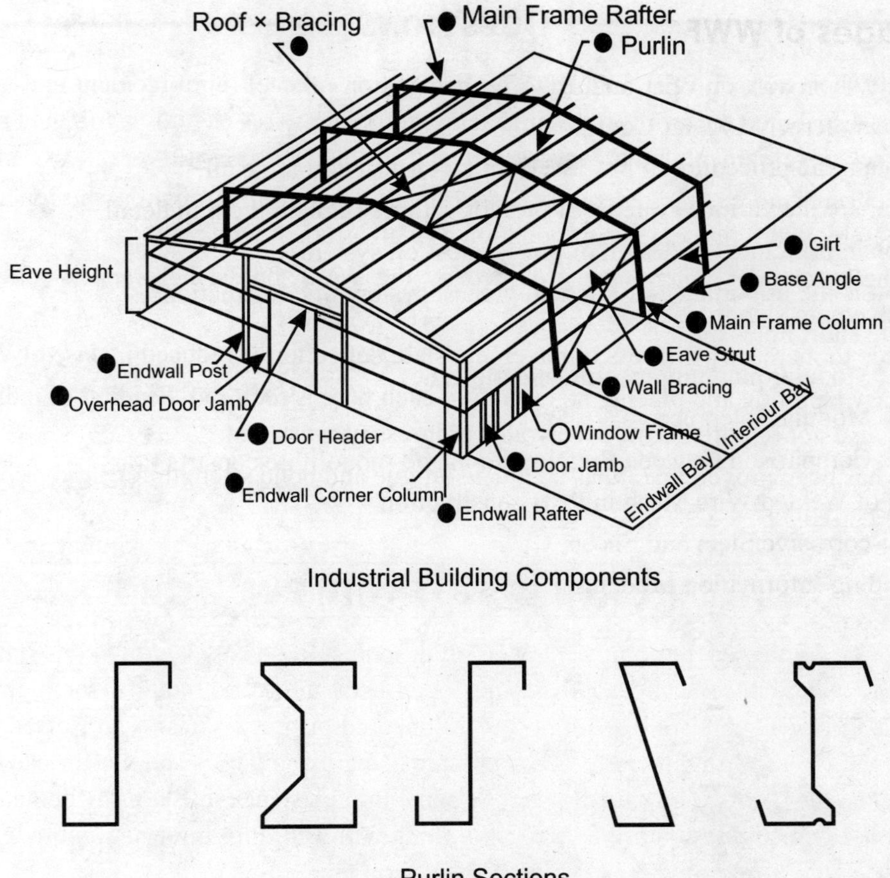

Industrial Building Components

Purlin Sections

Fig. 9.18. Components of pre-engineered industrial building

The term PEB though have been used for industrial buildings however its use in Railway bridges have been used for more then 150 years in India. Most of railway bridges are steel structures and constructed very tast.

Building Information modelling (BIM) is a recent topic in construction industry. BIM is a process for creating and managing information for a construction project life cycle. The model draws on the colla borative informations assembled and updated key stages of the project creating a digital building information model allows those who work with or interact with the building to optimise action BIM is modelling information management. The important concept of BIM is to minimize or eliminate information loss through the process of design, construction and operation of a building. The data from BIM is used to illustrate an entire building life cycle from its intial concept and design to demolition and possible materials those can be recycled, system products can be shown in relative scale related to entire project. It prevents errors by signalling conflict detection. preventing errors from popping up at various points in construction stages. Popular software Auto CAD is used in Building information modelling. New softwares have also been developed and used by engineer and fabricators as per their needs. The coming future is only of PEB with BIM technology.

QUESTIONS

1. Write short note on PEB with BIM Technology
2. Explain term 'Modular Coordination'
3. Explain the procedure of pre-fabrication construction in detail.
4. What are the various systems of prefabrication? Explain them in detail.
5. Explain principles of design of pre-fabrication system.
6. What is the importance of joints in precast system of construction?
7. Write short notes on:
   (a) Modular planning recommendations
   (b) Modular tolerance
   (c) Comparison between Prefabrication and monolithic construction.
8. Use of Welded Wire Mesh in PEB construction
9. Anti-corrosive Steel and Sheets
10. Building Information Modelling

Chapter 10 MANAGEMENT OF DEMOLITION WORKS

10.1 INTRODUCTION

Demolition of buildings and structures are required for various reasons. Demolition is the tearing down of buildings and other man-made structures. Demolition contrasts with deconstruction, which involves taking a building apart while carefully preserving valuable elements for re-use purposes. The word demolition means destruction, breaking down or removal. Demolition of building is the process of dismantling or destroying of a structure after its life of serviceability by pre-planned and controlled methods. corrosion

For small buildings, such as houses, that are only two or three stories high, demolition is a rather simple process. The building is pulled down either manually or mechanically using large hydraulic equipment: elevated work platforms, cranes, excavators or bulldozers. Larger buildings may require the use of a wrecking ball, a heavy weight on a cable that is swung by a crane into the side of the buildings. Wrecking balls are especially effective against masonry, but are less easily controlled and often less efficient than other methods. Newer methods may use rotational hydraulic shears and silenced rock-breakers attached to excavators to cut or break through wood, steel, and concrete. The use of shears is especially common when flame cutting would be dangerous.

As we know that every design of a building or a structure has a lifespan known as design life. The building is designed considering a span of life, say 80 –100 years. When this design life of the building is over, the structure is not safe for living and for neighboring buildings. There can be more reasons for demolition of a building, old structures are to be replaced by new ones. The structure lost its stability or having any structural damage. Small structures are demolished to build big structures etc. The tallest planned demolition of a building was the 47-story Singer Building in New York City, which was built in 1908 and torn down in 1967-1968 to be replaced by One Liberty Plaza.

Implosion: When explosives are used in the demolition of a building, it is termed as Implosion.

10.2 DEMOLITION OF BUILDINGS AND STRUCTURES

Many a times it becomes essential to demolish building due to one reason or the other. To demolish a stone wall inside of a building is easy; but when it comes to a structure, it is not that easy. When a building is required to be demolish; it is necessary to give a notice to Building Control Section related authorities under building act unless the same is exempt. The buildings which could be demolished without seeking permission and adopting laid down procedures include:

- An internal part of an occupied building.
- A building having a total cubic content externally less than 50 m^3
- Part of an existing large building that consists of an attached conservatory, shed garage or other similar part.
- A building which is solely used for agricultural purposes.

The notice should be served before demolition of any building started. The notice should be in writing and give details of the building to be demolished. The notice should be duly signed and should include following details :

1. Address and location of building.
2. Person or firm details undertaking demolition.
3. Demolition data.
4. Owner of building his details and reason for.
5. Type of structure to be demolished like small or big, flat or sloping, single storey or multistorey, load bearing or framed structure etc.
6. Site plan mentioning site outlined in detail.

Giving all these details; the copy of notice should also be given to :

(*a*) Occupiers of adjacent buildings for safety precautions.

(*b*) Relevant; electricity and gas suppliers.

(*c*) Police department if required to divert, the traffic

After receiving notice by council or other appropriate authorities who can grant the permission for demolition, the authorities can serve a counter notice, which sets down various conditions in respect of the manner in which the demolition is to lie carried out and other matters to secure the general health and safety of adjacent owners and general public. These conditions may relate to disconnection of services such as gas. water, electricity, drainage etc. protection of abutting buildings and the effective sealing of the site while the work is undertaken.

Demolition work should not be started until

- the municipal corporation or other appropriate authorities have served a notice on the person intending to carry out the demolition or
- six weeks (as per rule depending upon locality and power of jurisdiction) have expired since written notification of the intended demolition has been given to the authorities. However, in this case also precautions of disconnection of water, electricity, gas etc. must be taken and police department should have been informed.

Safety requirement in connection with demolition of building are laid down in the construction regulations and must be followed prior of starting of demolition work. It is also essential to discuss and plan the programme with council in case of if the building is listed or if it is in a conservation area or if it include dwellings.

Even to demolish a building which has suffered fire or storm damage, it is not possible to demolish it without seeking permission from authorities. In such cases also, after getting permission, a planning is required to demolish it. Demolition of such buildings is more dangerous as they have already gone through some demolition and have become dangerous.

10.3 BUILDING DEMOLITION PROCESS

Different steps are involved in the process of demolition of building structures are: 1. Surveying 2. Removal of hazardous materials 3. Preparation of plan of demolition of building 4. Safety measures during demotion of building.

1. Surveying of Buildings for Demolition

Surveying means study of different parameters of the structure and its surroundings. There are two types of surveying are mainly conducted. They are

Building surveying and Structural surveying

Building Surveying

In survey of buildings for demolition, following data are collected:

- Types of construction material used
- Usage of building prior and present during demolition.
- The presence of wastewater, hazardous materials, matters arising from toxic chemicals, flammable or explosive and radioactive materials, etc.
- Drainage conditions and possible problems on water pollution, flooding and erosion.
- Shared facilities with adjoining building, including common staircases, partition walls.
- Adjoining pedestrian and vehicular traffic conditions
- The sensitivity of neighborhood with respect to noise, dust, vibration and traffic impact.

Structural Surveying

In structural survey, following investigations are made before a demolition starts:

- The method of construction
- The structural system and structural conditions of basements, underground tanks or Underground vaults.
- The original structural system employed in the design
- The condition of the building.

2. Removal of Hazardous Materials

If hazardous materials like asbestos minerals, petroleum contamination and radioactive metals are found in the investigation of site for demolition, specialized personals are called for the removal of the hazardous materials from the site prior to the demolition of structure.

3. Preparation of Demolition Plan for Structures

A detailed demolition plan is made which illustrates the different process involved and they are:

- The location of the building to be demolished.
- The distances from the building to be demolished to its adjacent buildings, streets, structures an significant street furniture.
- The structural support systems of the building.
- A plan showing the procedure for the demolition of the building; detailed sequence of demolishing structural members; and the method of demolition to be adopted.
- A plan showing all precautionary measures for the protection of the public including hoardings, covered walkways, catch platforms, catch fans, scaffolding, protective screens and safety nets.
- Method of handling demolished building debris.
- Time required for the complete demolition process etc.

4. Safety Measures during Demolition of Building Structures

All the workers, site supervisors and engineers including plant and equipment operators are briefed with the potential hazards and process of demolitions.

All goods that are flammable are removed from the site unless it is used in the work involved. All the flammable materials like wood, timber, fuels etc. are stored in proper storage facilities. Firefighting appliances are stationed in the demolition site till the process is completed. Due to the demolition of structure, many problems are faced by the workers, such as. exposure to dust, chemical exposure, heat stress and ventilation, noise exposure, medical and first aid facilities, sanitation and occupational diseases. To overcome these problems suitable measures are undertaken.

10.4 DEMOLITION METHODS FOR BUILDINGS AND OTHER STRUCTURES

There are two types of demolition methods used for buildings and structures

1. Non-explosive demolition
2. Explosive demolition.

1. Non-Explosive Demolition Method

It means the demolition of a structure done with some equipment without the use of any explosive. Different equipments used for the demolition activity are

(*a*) Sledge hammer

It is a small handheld hammer used for the demolition of small wall or single column.

(*b*) Excavators and Bulldozers

These are big machines used to demolish building of small sizes. They are used for excavation of soil or transferring of debris to trucks etc. To demolish bridges, hoe rams are typically used to remove the concrete road deck and piers.

(*c*) Wrecking Balls

The building with the greater height up to (6-7 story) cannot be demolished with the help of excavators or bulldozers. In such cases crane with wrecking balls are used to perform the demolition activity. The wrecking ball crack is attached with a huge steel ball hanging from a steel rope.

The steel ball is pulled and released towards the building. The steel ball with force strikes the building and the part of the building is demolished. This method is not recommended as the trajectory of the steel ball cannot be controlled after it strikes the structure and the safety implications associated

(*d*) High Reach Excavators

High Reach Excavator machines are used in the demolition of tall building where demolition by explosion is not possible. The building of height up to 90 M can be demolished by this type of machine. High reach excavators can be used for different use by doing some attachments such as:

- Excavators with shear attachments.
- Hydraulic hammers and remove steel reinforcement.

2 Explosive Method:

It is also called as Implosion method and used in many cases now a day's particularly when whole building or its main part in total is required to be demolished. Implosion is the process of demolition of a building using explosives. If the supports of the building are removed, the structure collapses. Using implosion technique, the main supports of the buildings such as column's, beams and slab are fixed with explosives. When these explosives are detonated, the column collapse and so is the structure.

One can demolish a stone wall or a small building using excavators and wrecking balls. But when a multistorey building is required to be demolished, explosion demolition is the preferred method for safety and efficiently demolition in a shorter period. If a high rise building is surrounded by other buildings, it is always necessary to implode the building, that is make it to collapse into footprint.

Strictly speaking, an implosion is an event where something collapses inward, because the external atmospheric pressure is greater than the internal pressure. A building implosion is not truly an implosion, atmospheric pressure does not. pull or push the structure inward; gravity makes it collapse. However, implosion is a common word used for this sort of demolition.

The basic idea of explosive demolition is quite simple. If one removes the support structure of a building at a certain point, the sections of the building above that point will fall down on the part of the building below that point. If this upper section is heavy enough, it will collide with lower part with sufficient impact force to cause significant damage The explosives are just the trigger for the demolition. It is the gravity that brings the building down. When everything is planned and executed properly, the total damage of the explosives and falling building material is sufficient to collapse the structure entirely, so cleanup crews are left only a pile of rubble.

In order to demolish a building safely, blasters must map out each element of the implosion ahead of time. The first step is to examine architectural blue prints of the building, if they can be located, to determine how the building is put together. Next, the blaster crew tours the building (several times), jotting down notes about the support structure on each floor. Once they have gathered all the raw data they need, the blasters hammer out a plan of attack. Drawing from past experiences with similar buildings, they decide what explosives to use, where to position them in the building and how to time their detonations. In some cases, the blasters may develop 3-D computer models of the structure so they can test out their plan ahead of time in a virtual world.

The main challenge in bringing a building down controlling which way it falls. Ideally, a blasting crew will be able to tumble the building over on one side, into a parking lot or other open area. This sort of blast is the easiest to execute, and it is generally the safest way to go. Tipping a building over is something like felling a tree. To topple the building to the north, the blasters detonate explosives on the north side of the building first, in the same way you would chop into a tree from the north side if you wanted it to fall in that direction Blasters may also secure steel cables to support columns in the building, so that they are pulled a certain way as they crumble.

Sometimes, though, a building is surrounded by structures that must be preserved. In this case, the blasters proceed with a true implosion, demolishing the building so that it collapses straight down into its own footprint (the total area at the base of the building). This feat requires such skill that only a handful of demolition companies in the world will attempt it.

Blasters approach each project a little differently, but the basic idea is to think of the building as a collection of separate towers. The blasters set the explosives so that each "tower" falls toward the center of the building, in roughly the same way that they would set the explosives to topple a single structure to the. side. When the explosives are detonated in the right order, the toppling towers crash against each other, and all of the rubble collects at the center of the building. Another option is to detonate the columns at the center of the building before the other columns so that the building's sides fall inward.

Fig. The Hayes homes, in Newyork, N.J. : The 10-storey housing project was demolished in three separate phase, over the course of three years. Even though all the buildings had exactly the same design, blasters handled the implosions differently for each phase. These towers were blasted by Engineered Demolition, Inc. in the summer of 1999.

Preparation for Implosion

It takes several weeks or months to prepare a building for implosion. All items of value, such as copper wiring, are stripped from a building. Some materials must be removed, such as glass that can form deadly projectiles, and insulation that can scatter over a wide area. Non-load bearing partitions ana drywall are removed. Selected columns on floors where explosives will be set are drilled and high explosives such as nitroglycerin, TNT, RDX. or C4 are placed in the holes. Smaller columns and walls are wrapped in detonating cord. The goal is to use as little explosive as possible so that the structure will fail in a progressive collapse therefore only a few floors are rigged with explosives, so that it is safer (fewer explosives) and costs less. The areas with explosives are covered in thick geotextile fabric and fencing to absorb flying debris. Far more time-consuming than the demolition itself is the clean-up of the site, as the debris is loaded into trucks and hauled away.

Depending how the structure falls, there are two types of implosion:

(*a*) **Falling like a Tree:** In this type of implosion, the building is made to fall like a tree to the sideward. This is the commonly used type of implosion. When free space is available besides the building, this type of demolition is prescribed. If the free space is available on the left side of the building, the explosives are set on the lower level of the building on the left side columns. As the explosives are detonated, the columns bursts, the building tends to falls towards the left side. Steel cables are tied to the building to control the falling direction of the building.

(*b*) **Falling into its own footprint:** When the free spaces are not available around the building and the structure around the building are to be protected. This type of demolition is used. In this type of demolition, explosives are set in the floor below the middle part of the building.

These explosives are to be heavy as the explosion must demolish the building at once. If one part blast and followed by another. Then the building falls towards the first blasted part. As the explosions are detonated, the upper part of building destroys and falls upon the lower building. Due to the heavy load and force the lower part of the building also collapses and falls on its own footprint.

10.5 MANAGEMENT OF DEMOLITION BY IMPLOSION

Hydraulic excavators may be used to topple one- or two-story buildings by an undermining process. The strategy is to undermine the building while controlling the manner and direction in which it falls. The demolition project manager/supervisor will determine where undermining is necessary so that a building is pulled in the desired manner and direction. The walls are typically undermined at a building's base, but this is not always the case if the building design dictates otherwise. Safety and cleanup considerations are also required to be taken into consideration in determining how the building is undermined and ultimately demolished.

Excavators with shear attachments are typically used to dismantle steel structural elements. Hydraulic hammers are often used for concrete structures and concrete processing attachments are used to crush concrete to a manageable size, and to remove reinforcing steel. For tall concrete buildings, where neither explosive or high reach demolition with an excavator is safe or practical, the "inside-out" method is used, whereby remotely operated mini-excavators demolish the building from the inside, whilst maintaining the outer walls of the building as a scaffolding, as each floor is demolished.

To control dust, fire hoses are used to maintain a wet demolition. Hoses may be held by workers, secured in fixed location, or attached to lifts to gain elevation.

Large buildings, tall chimneys, smokestacks, bridges, and increasingly some smaller structures may be destroyed by building implosion using explosives. Imploding a structure is very fast–the collapse itself only takes seconds–and an expert can ensure that the structure falls into its own footprint, so as not to damage neighboring structures. This is essential for tall structures in dense urban areas. Another dangerous scenario is the partial failure of an attempted implosion. When a building fails to collapse completely the structure may be unstable, tilting at a dangerous angle, and filled with un-detonated but still primed explosives, making it difficult for workers to approach safely. A third danger comes from air overpressure that occurs during the implosion. If the sky is clear, the shock wave, a wave of energy and sound, travels upwards and disperses, but if cloud coverage is low, the shock wave can travel outwards, breaking windows or causing other damage to surrounding buildings. Controlled implosion, being spectacular, is the method that the general public often thinks of when discussing demolition; however, it can be dangerous and is only used as a last resort when other methods are impractical or too costly.

According to Brent Blanchard, an implosion expert, every building in the world is unique in its form. And for any given building, there are many number of ways, a blasting crew may bring it down. Blanchard notes the demolition of Hayes Homes a 10 building housing project in Newark, NJ, USA, which was demolished in three separate phases over the course of three years. Different blasting firms performed each phase. Though all the buildings were identical, though each blaster adopted different way to demolish it. They used slightly different type of explosives and leaded in varying numbers of support columns. They even brought the building down in different sequences with varying amount of time factored in between each buildings collapse.

In general, blasters explode major support columns on the lower floor first and then a few at upper floors. To quote an example, in a 20 Storied building, a blaster may blow columns of ground floor, first floor, 12th and 15th floor. However in most cases, blowing of support structures on lower floors are sufficient for collapsing whole building, but loading a few columns at upper floors helps in breaking the building materials into smaller pieces as they fall down. It makes easy to remove derbies after blasting. Once the blasters have figured out how to set up an implosion, it is up to him to prepare for blasting.

10.6 DECONSTRUCTION

A new approach to demolition is the deconstruction of a building with the goal of minimizing the amount of materials going to landfills. This "green" approach is applied by removing the materials by type material and segregating them for reuse or recycling. With proper planning this approach has resulted in landfill diversion rates that exceed 90% of an entire building and its contents in some cases. It also vastly reduces the CO_2 emissions of the removing of a building in comparison to demolition.

Safety is paramount; a site safety officer is usually assigned to each project to enforce all safety rales and regulations.

Listed buildings and buildings in conservation areas

One do not need to make a planning application to demolish a listed building or to demolish a building in a conservation area. However, one may need listed building or conservation area consent.

Elsewhere.

You will not need to apply for planning permission :

- to demolish a building such as a garage or shed of less than 50 cubic metres; or
- if the demolition is urgently necessary for health and safety reasons; or

- if the demolition is required under other legislation: or
- where the demolition is on land which has been given planning permission for redevelopment

In all other cases, such as demolishing a house or block of flats, the council may wish to agree the details of how you intend to carry out the demolition and how you propose to restore the site afterwards. You will need to apply for a formal decision on whether the council wishes to approve these details. This is called a "prior approval application".

QUESTIONS

1. Explain various precautions required to be taken while dismantling any building.
2. Explain the process of dismantling of any structure by blasting.
3. Explain various methods of dismantling of any building.
4. How management of demolition of building is carried out.
5. Explain process of demolition of building using various construction machinery

Chapter

CONSTRUCTION EQUIPMENT BASICS

11.1 INTRODUCTION TO CONSTRUCTION EQUIPMENT

Construction activity is an integral part of a country's infrastructure and industrial development. It includes buildings such as hospitals, educational institutions, townships, offices, commercial complexes, etc.; urban infrastructure such as water supply, sewerage, drainage etc.; highways, flyovers, tunnels, ports, railways, airports; power systems; irrigation and agriculture systems; telecommunications, industrial plants.

Construction has become the basic input for socio-economic development. Besides, the construction industry generates substantial employment and provides a growth impetus to other sectors through backward and forward linkages.

Construction equipment plays a significant role in the execution of modern high cost time-bound construction projects. These produce output at accelerated speed, and enables completion of tasks in a limited time. The equipment saves manpower which is becoming scarce, costly and more demanding day-by-day. In addition, equipment improves productivity, quality and safety. Thus, the construction equipment have become indispensable item of resources and their role continues to grow steadily.

The project planner/engineer on the job is required to match the right machine or combination of machines to the job at hand. For this purpose, he is required to identify the construction equipment for executing the project tasks, assessing equipment performance capability, forecasting datewise requirement of numbers and types of equipment, and then selecting the equipment to be acquired.

The production tasks which can be performed with equipment include excavating, hauling transporting, compacting, grading, hoisting, concreting, precasting, trenching, pipe-laying etc. In addition, the support equipment are also required, at the project site, which include generators, compressors, pumping sets and other utility services equipment.

Proper emphasis should be on reduction of downtime achieving optimum utilization, increase in production at minimum cost, for which consciousness has to be developed for right selection and management of these equipment. Equipment engineer should coordinate with various wings of the project in discharging his job of equipment planning, balancing, selection of equipment, its deployments, personnel selection and training, preventive maintenance and general suppression maintenance and general supervision.

Here, in this section, we shall discuss the construction equipment in brief, for detail, readers may refer the book "Construction equipment and management" by S.C. Sharma, published by m/s Khanna Book Publishing House. correction

11.2 CLASSIFICATION OF CONSTRUCTION EQUIPMENT

Construction equipment can be broadly classified on the basis of their functions as under:

1. Earthwork Equipment

(*i*) *Earth cuting and moving equipment* : Bulldozers, Scrapers, Front-end loaders, Motor graders.

(*ii*) *Excavation and lifting equipment* : Backhoes, Power Shovels, Draglines, Clamshells.

(*iii*) *Loading equipment* : Loaders, Shovels, Excavators.

(*iv*) *Transportation equipment* : Tippers, Dump Trucks, Scrapers, Conveyors.

(*v*) *Compacting equipment* : Tamping Foot Rollers, Smooth Wheel Rollers, Pneumatic Rollers, Vibratory Rollers, Plate Compactors.

2. Concreting Plant and Equipments

(*i*) *Production equipment* : Batching Plants, Concrete Mixers.

(*ii*) *Transportation equipment* : Truck mixers, Concrete dumpers.

(*iii*) *Placing equipment* : Concrete pumps, Conveyors, Hoists, Grouting equipment.

(*iv*) *Concrete vibrating equipment* : Needle vibrators, Plate compactors.

3. Material Hoisting Equipment

(*i*) *Hoists* : Fixed, Mobile, Fork-lifts.

(*ii*) *Mobile Cranes* : Crawler-mounted, self-propelled rubber-tyred, truck- mounted.

(*iii*) *Tower Cranes* : Stationary, Travelling type.

4. Special Purpose Heavy Construction Plant and Equipment

(*i*) *Aggregate production equipment* : Crushing plants, Rock blasting equipment, Screening plants.

(*ii*) *Concrete paving equipment* : Concrete paver finishers.

(*iii*) *Pile driving equipment* : Pile driving hammers.

(*iv*) *Asphalt concrete mix production and Placement equipment* : Asphalt plants, asphalt pavers.

(*v*) *Tunneling equipment* : Drill jumbos, bolters, Tunnel boring machines.

5. Support and Utility Services Equipment

(*i*) Pumping and Dewatering equipment.

(*ii*) Pipe laying equipment.

(*iii*) Generators.

(*iv*) Welding equipment.

11.3 EQUIPMENT PLANNING

Modern construction projects are complex in nature and success of a project depends greatly on proper and scientific planning. Before starting every project, its planning is done with great care, as the efficiency of the whole project largely depends upon its planning. While planning each and every detail should be worked out in anticipation and should be considered carefully.

The task of the project planner/engineer on the job is to match the right machine or combination of machines. For planning the equipment requirement, it is necessary to first determine machine productivity. After studying the plants and specifications, and visiting the project site, the planner must visualize how to best employ specific pieces of equipment to accomplish the work. Is it less expensive to top-load trucks with a wheel loader or a shovel? Several such questions must be analysed and then answered by the planner. To perform such analysis, the planner must consider both machine capacity and methods of employment.

Since no two projects are exactly alike; therefore, it is important that the planner begins each new project with a completely open mind and reviews all possible options. He should also consider new improved equipment available in the market.

Planning of a construction project involves deciding about the extent of mechanisation, equipment planning, execution planning etc. While planning a project we must very carefully decide about the extent of use of construction equipments. On major construction projects mechanisation is indispensable, while for middle and minor construction projects a compromise between manual and mechanical means has to be made. In a country like India, where man-power is cheaply and easily available, we can take advantage of this facility specially for the activities which can be easily carried out by the manpower. However, a combination of manual labour and machines is preferred. Now the question arises before the planner that : to what extent mechanisation is desirable?

11.4 EXTENT OF MECHANIZATION

Extent to which mechanisation in a project should be planned, depends upon the following main factors :

1. **Manpower :** Type of manpower required by the project for its execution, is first decided and whether this type of manual labour is cheaply and sufficiently available round the year. In case the manual labour is costly or available only for limited duration or required type of labour is not available in the area, then the solution is only mechanisation. To decide whether the labour is costly or cheap, planner must consider the output given by the labour and overhead expenditure for the labour like accommodation, transportation to be provided to them etc. on one hand and output of the equipment and expenditure on the equipment on the other hand.
2. **Completion Period :** Every project is required to be completed in a particular duration, since delay in completion of project means increase in the construction cost due to rise in prices, payment of overhead for more period, imposition of penalty for delay

etc. The overhead expenses include depreciation of equipments, supervisory expenses, accommodation expenses, light, water, power and insurance expenses etc. This completion period plays a deciding role about the extent of mechanisation, since less the completion period more the mechanisation. By employing more number of construction equipment project can be completed in lesser duration.

3. **Nature of Work :** Where the nature of work is such, which is beyond the scope of human labour to perform, use of construction equipments is unavoidable. Nature of work includes adverse weather, climatic conditions, topography, toughness, handling of large quantities of construction material for long distance.
4. **Availability of Money :** Since cost of equipment is too high and to be spent initially, it needs heavy investment in the beginning. Therefore, availability of money is main factor in deciding the extent of mechanisation.
5. **Availability of Indigenous Equipment :** The import of costly equipment and of spare parts proves a drain on foreign exchange resources of the country. After-sales-service, repair facility are also considered where equipment manufactured in our own country are not available, because time required for importation, shipment, installation and commissioning of the equipment and thereafter lead time for the purchase of repair spares will hamper the progress of the project adversely.
6. **Social Objects of the Project :** Some times projects are sanctioned with the object, of providing employment opportunities, to improve the economic condition of the masses in addition to the development in the area. In such cases manual labour is preferred over the use of machines unless otherwise any particular activity is beyond the capacity of manual labour.
7. **Labour Relations :** In certain areas, past experience indicates labour unrests, strikes, absenteeism, uncertainty, unavailability of labour or seasonal fluctuations. In such circumstances machines are preferred being more reliable to complete the project in time.
8. **Overseas Projects :** In cases where large number of projects are in hand in other countries, it is difficult and expensive to send large number of man-power required in adopting the manual means in these projects, therefore only, alternative left is to execute the project using construction equipment only.
9. **Quality and Performance :** The work carried out by means of construction equipment is of high standard of quality and performance. The quality is consistent, execution is fast, work is more durable, maintenance and repair will be very limited if projects are executed by using the construction equipment.
10. **Development of Industries :** In countries like India where industries are in developing stage, it is the need of time to use construction equipment for the development of manufacturing industry and faster industrialisation of the country.

11.5 USE OF MACHINES V/S MANUAL LABOUR

The advantages and disadvantages of using machines and manual labour (*i.e.*, Mechanized methods v/s conventional methods) in construction projects are summarized as under :

	Points in favour of machines	*Points favour of manual labour*
1.	It can handle tough works which can not be performed otherwise.	Initial heavy investments are not required to be made.
2.	It is cheaper when quantum of work is more and material is required to be transported for long distances.	Valuable foreign exchange is saved by employing manual labour instead of using machines which are to be imported.
3.	Machines when used in the project, it delivers better quality of work	Small and short duration projects are sometimes found to be economical with manual labour.
4.	It is possible to adhere to schedules.	Direct advantage to considerably large number of people by way of employment which help in social and economical betterment.
5.	It is easy to predict the behaviour of a machine as compared to that of man, hence planning is easy.	No problem due to delay in procurement of spares for the repair of machines. The delay (lead time) is substantially high in case of imported equipment.
6.	Being lesser number of persons, (operators only), it is easy to supervise and control.	Project can be started immediately, which is not possible in case machines are used. Since importation, shipment, transportation, installation and commissioning of equip¬ments require a substantial time.
7.	It helps in development of technical know-how, skill and industrial development of the country.	Disposal of equipment and its balance spare parts poses a problem after the project is completed.
8.	Availability of Manpower and seasonal fluctuations will not affect the project's progress.	
9.	Effect of climate and topographic conditions is less when machines are used as compared to that when manual labour is used.	
10.	No problem for housing and transportation of persons and also for the arrangement for supply for food, water light and medicines etc.	

11.6 FORECASTING EQUIPMENT REQUIREMENT

As discussed earlier extent of mechanisation depends upon the nature and quantity of work to be handled, the specifications to be adopted, time period of completion, availability of skilled and unskilled labour. There are certain types of works where man cannot work and the only alternative is to use construction.equipments, e.g., earthwork compaction, construction of bridges, working under-water, concreting, dewatering, construction of bituminous/rigid pavements, working in steep slopes etc.

Considering above mentioned factors, planners are required to decide the extent of mechanisation, which means out of large number of activities involved in the completion of the projects which are the activities which should be completed by using the equipments. With the identification of the activities to be carried out with the help of equipments, work of equipment planning starts.

Following are the main points planner has to decide in equipment planning :

1. **Working Shifts :** By increasing the number of shifts we can reduce the number of equipment required, as the same equipment can be used in second and third shifts. This helps in reduction in initial investment of capital. But the number of shifts in which work can be executed depends upon the nature of work, the area in which the work is spread over etc. While planning the number of shifts it must be kept in mind that in double shifts, only 60 to 70% (nearly) extra work can be achieved and not really double. Similarly, in triple shift only 100 to 125% of extra work, as compared to single shift can be achieved. This is because the efficiency in the evening and night shifts is less as compared to that in day shift, besides maintenance also becomes difficult.

   Since repair facility for the break downs occurred in the night shift is not fully available, provisions for standby equipment is desired. Experience has shown that this provision should be as below :

   (*i*) For equipment used in single shift 10% standby

   (*ii*) For equipment used in double shifts 20% standby

   (*iii*) For equipment used in triple shifts 30% standby

2. **Number and Size of Equipment :** For deciding the number and size of the equipment, following factors should be considered :

   (*i*) Quantum of work.

   (*ii*) Working days available for completion of projects.

   (*iii*) Number of shifts planned.

   When quantum of work is too much then bigger size equipment are selected. The average rated output of these equipment will help in deciding number of the equipment required keeping in view the number of working days available and number of shifts planned per day.

3. **Matching Units :** There are certain equipments, which are used together in a team with other equipments, like scraper with pusher dozer, concrete batching plant with truck mounted concrete mixers, paver finisher with road roller, asphalt mixing plant with tipper trucks, wheel loaders with dumpers/ dump trucks etc. These numbers and size of the

matching equipments should be planned for each activity, keeping in view that complete matching set must be utilised at the maximum and combination must be optimum.

4. **Procurement Scheduling :** Procurement scheduling is done considered procurement time required and time when equipment is required.
5. **Workshop and Stores Planning :** While planning for the purchase of equipments, it is necessary to plan for the workshop and stores for carrying out the repair and maintenance.
6. **Man-Power Planning :** Planning for manpower in order to operate the equipment and to carry out the repair and maintenance is also necessarily be done sufficiently in advance. In case skilled and trained manpower for the operation, maintenance and repair is not available than efforts should be made to train them before the arrival of equipments in the project.
7. **Site Conditions :** Site conditions play an important role in deciding the type of equipment. The type of equipment selected mainly depends upon soil available, and valley conditions, surface and strata where work is to be carried out. This helps in deciding whether to use wheeled equipments or crawler type equipments, to use dragline excavator or power shovel, to use crane or cable ways, to use rear-dump trucks or side-dump trucks or bottom trucks etc.

Output and Capacity of Equipment

Next phase of planning is to decide the equipment on the basis of its output and capability. The equipment must be capable to handle the work efficiently on which it is to be engaged. The output of the equipment so decided should be determined on the basis of past experienced on similar works. As we have discussed earlier that the same equipment gives different output in different conditions of work, therefore, it is very essential to use the past output data of similar works handled by the equipment in similar conditions.

Output of the equipment varies with various conditions. Following are some of them :

(*i*) Different type of soil needs different amount of effort and hence equipment engaged on them gives different output.

(*ii*) Reduction of horse power of engine due to altitude.

(*iii*) Time spent on servicing and inspection varies with the working conditions.

(*iv*) Output of the equipment also depends upon the operator's efficiency.

(*v*) Output of dumper etc. depends on the condition of haul road.

(*vi*) Time spent on the unproductive works.

(*vii*) Changing weather conditions are also responsible for variation in output.

(*viii*) Capacities of the matching machineries working in the same group.

11.7 SELECTION OF EQUIPMENT

Proper selection of equipment for a construction project is of vital importance for its speedy and economical completion. Problem of equipment selection has become more complicated, because large variety of equipments are being manufactured now-a-days. For proper selection of equipment, a considerable experience in the operation and maintenance in the field is essential. Records kept for operation, maintenance and actual output obtained under comparable conditions of previous projects will greatly help in taking decision for equipment selection.

Following are the main points which should be considered in the process of equipment selection :

1. **Suitability of Job Conditions :** The equipment must meet the requirement of work, climate and working conditions.
2. **Size of the Equipment :** Size of equipment should be such that it must be able to be used with other matching units. If the equipment selected is of larger size, that will remain idle for most of the time or shall work on part loads, which means production cost will be more. On other side, if the equipment is of smaller size than desired, the equipment will not be able to work with the matching equipments and hence other equipments will have to remain idle or to be allowed to work on part loads, which shall again be uneconomical.
3. **Standardisation :** It is better to have same type and size of equipments in the project. It means lesser spare parts reserve, more interchangeability of parts if required, easy for the operators to understand it, mechanics will be able to maintain and repair better as they become expert by handling similar type of equipment.
4. **Availability of Equipment :** The equipment which is easily available in the market should be purchased. It should also be ensured that the equipment is of repute and is likely to be continued to be manufactured in future also. This is necessary for future standardisation and ensuring spare parts supply. It is easy to dispose off such equipment after completion of projects.
5. **Availability of Spare Parts :** While selecting a particular type or make of equipment, it should be ensured that the spare parts will be available at reasonable price throughout the working life of the equipment. It should also be ensured that the downtime of the equipment for want of spare parts may not be more. This is all the more necessary in case of imported equipments.
6. **Multipurpose Equipments (Versatility) :** There are certain types of equipments which ae not utilised fully. Therefore, if possible, they must be capable of performing more than one function for example, excavator with wheel loader bucket arrangement or with rock breaker attachment.
7. **Availability of know-how :** The equipment selected should be satisfactorily handled by available operators and mechanics. A sophisticated equipment may give excellent performance but it may be difficult to handle and maintain it through available know-how.
8. **Use in Future Projects :** When equipment completes only a part of their useful life in a project, it should be kept in view that the equipment can be used in future projects and may not become obsolete.
9. **The Economical Aspects :** While selecting the equipment, it should be considered that the cost of unit production should be minimum.
10. **Reliability of the Equipment :** Equipment selected for the project must be reliable one.
11. **Service Support :** Service support should be available in the area of project where the equipment shall be used. Service after sales is a major criteria for selection of equipment.

12. **Operating Requirements :** The equipment selected should be easy to operate and maintain, acceptable to the operator and should have lesser fuel consumption.
13. **Past Performance :** If the equipment being purchased is of new make and model, it is desirable to enquire about its performance from other users, who are using this make and model.
14. The size and numbers should be such that full life is utilised in the project with very little residuals.
15. Reputation of the manufacturer.
16. Warranty or guarantee offered by the manufacturer.
17. Use of standard components in the equipment.
18. Adequacy of drive mechanism or power of the primemover.
19. Versatility of the equipment should be given due priority. This means a machine which can be used for many jobs. The versatility promises extra profit from two directions : (*i*) allows one machine to do the job of several machines and thus cutting into ownership and operating costs associated with additional plant and labour, and (*ii*) it increases utilization, which means a machine earns money when it might otherwise be idle. Now-a-days attachments can be fitted or changed quickly with the help of couplers.

A balance between reliability, investment cost, and operating cost should be found, since a policy of selecting the lowest priced equipment can often lead to overall higher costs.

11.8 EQUIPMENT PRODUCTIVITY

Productivity is the reduction in wastage of resources. The resources may be men, machines, materials, power, spaces, times, building, etc.

It may also defined as human efforts to produce more and more with less and less inputs of resources as a result of which the benefits of production may be distributed more equally among maximum numbers of people.

Productivity is a ratio between the quantity of goods and services produced (output) and the quantity of resources used to produce these goods and services (Inputs).

Productivity can be represented by,

$$\text{Productivity, } P = \frac{\text{Unit of output}}{\text{Efforts input}}$$

Productivity is the key to prosperity, Rise in productivity generates extra production, which has direct bearing on standards of living, as it reduces cost per unit and enables reduction in cost and increased profit for the organisation.

Reasons for Low-productivity.

1. Frequent machine breakdown.
2. Insufficient managerial control.
3. Idle men

4. Poor quality
5. Shortage of raw materials/work available
6. Large quantities of rework.
7. Poor planning
8. Unclear and untimely instructions
9. Lack of motivation
10. Poor working environment

Productivity Improvement

Following are the main contributors to the productivity improvement:

1. Incentives
2. Human relations
3. Cost control
4. Training about productivity
5. Application of new technology
6. Proper planning and scheduling,
7. Management skill

How of Increase Productivity

It increases productivity following major points need fullest control and should be minimized:

(*i*) Wastage of materials
(*ii*) Machine breakdown
(*iii*) Waiting on the part of men and work availability
(*iv*) Excessive handling
(*v*) Poor management
(*vi*) Poor working conditions
(*vii*) Legal effects

11.9 IMPORTANT PRODUCTIVITY ASPECTS

11.9.1. Equipment Operation Planning

Since the success of a mechanised construction shall be judged by its low cost, timely and safe completion, therefore, it is essential that operation should also be well planned. Following are the main points which should be considered while planning for equipment operation.

(*a*) Equipments must work in coordination, especially when they are required to work in a team, for example, the working of big size crusher where stones are fed by the dumpers, which are loaded at the quarry by the wheel loaders. In this case wheel loader, dumpers and crusher are required to work in a team and therefore their operation would be planned in such a way that there is no waiting at any place. The capacities of all the connected equipments should be properly balanced.

(*b*) Proper working and the maintenance of auxiliary equipments should also be ensured. Since the auxiliary equipments, may be small in size, may affect the whole activity and thereby idling all the equipments used in the team,

(*c*) Poor maintenance of service road may result in more cycle time of truck, dumpers etc, besides causing more downtime, Properly graded, sprinkled with water and then rolled service roads add the to the efficiency of the equipment and of the operator.

(*d*) In order to ensure operating of equipment and to eliminate delays, it is necessary to ensure scheduled maintenance in addition to watch the working by the operator.

(*e*) Safety on the work is a must, In order to ensure proper safety, equipment must be kept in good condition, operate them properly, working area should be well-lighted and observe proper traffic control.

(*f*) Good workmen should be given proper incentives. Incentives may be provided be way of bonus, additional increment in salary, cash awards, promotion, appreciation certificates etc.
This will help in increasing the production.

(*g*) proper records should be maintained for production equipment operations, maintenance, breakdown, consumption of stores and spare parts, repair etc. It will help future estimation, reviewing the plan in addition to the cost accounting of the equipment and evaluating the progress of work.

11.9.2. Spare Parts Management

Machines worth crores of Rupees are idle for want of spares, on other hand huge stock of spares are laying perhaps may not be used. This emphasizes the need of paying more attention on the management of spares. Improvement in the capacity utilisation and cost reduction can be achieved be better spare parts management. The objective of spare parts management is to provide right parts in right quantity, in right place, at right time and it right cost. The parts must be properly codified and classified. Various cost reduction techniques for spares help the manager to control large number of spare parts selectively and effectively, which help to utilise his energy to problem areas resulting in optimal use of his efforts. These methods demand more concentration in areas of more importance, while areas of least importance can be left to supervisory level.

11.9.3. Maintenance Management

Every machine is thoroughly tested and inspected by the manufacturers before selling. When used, it is subjected to wear and tear, hence proper attention should be given to protect the machine and its component from undue wear and thus protect them from failure. A proper attention means lubrication, cleaning timely inspection and systematic maintenance. Maintenance means efforts directed towards the up keep and the repair of that machine. Repair must be done a time when it may have least disruptions *i.e.*, machine may be repaired when it is not being used or its use may be postponed without affecting the production much. Thus, maintenance is responsible for the smooth and efficient working of equipment and helps in improving the productivity. It also helps in keeping the machine in a state of maximum efficiency and economy.

Following are the types of maintenances :

(*a*) Repair maintenance can be followed for non-critical equipment, and where failures are unpredicted.

(*b*) Corrective maintenance should be adopted for frequently failing components.

(*c*) Preventive maintenance (planned or scheduled or systematic maintenance). Periodic inspection or checking at the pre-decided frequency helps to find out the reasons leading to breakdown and ratify them when they are in minor (or initial) stages. Thus enable to carry out the repair when one wants to do it *i.e.* when it has least effect on the production schedule. Further this repair requires lesser time as compared to that of breakdown repairs and thus down time is reduced by doing preventive maintenance.

Functions of Preventive Maintenance

Following are some of the important functions of the maintenance programme:

(*i*) Inspections or check ups at carefully decided frequency.

(*ii*) Lubrication. Mechanical components like gears bearing, bushes and other friction surface give good performance for long periods, when they are systematically lubricated *i.e.* application of right type of lubrication at the right time, at the right placed and in right quantity.

(*iii*) Every preventive maintenance work should be preplanned.

(*iv*) Good record keeping is essential for good preventive maintenance.

(*v*) Training of maintenance personnel.

(*vi*) Storage of maintenance spares.

Advantages of Preventive Maintenance

(*i*) Less overtime work required as it is pre-planned.

(*ii*) Less repair costs.

(*iii*) Less production downtime.

(*iv*) Less stand-up equipments needed.

(*v*) Increased life expectancy.

(*vi*) Better spare parts control leading to reduced inventory.

(*vii*) More reliability.

Tire Maintenance

It will be needless to assert that tires are an expensive item in the cost budget and consumes a very high proportion of operation costs, and rate only next in expenditure to fuel. But these are mostly ignored from the time of initial fitment until the time ultimate failure and as a result millions of potential tire kilometres are wasted every year. Correct tire selection and proper tire maintenance are therefore very essential for ensuring economic operation. For proper maintenance of tires and tubes, following factors should be considered.

(*a*) Proper inflation.

(*b*) Proper loading.

(*c*) Proper speed.

(*d*) keeping the tire up to desired temperature.

(*e*) Proper the up to desired temperature.

(*f*) Avoiding vehicle's mechanical irregularities.

(*g*) Storing of tires properly.

(*h*) Tire rotation.

(*i*) Re-treading at right time.

11.9.4. Downtime and Availability

Let us say, an equipment failed 5 times in a year and total required to put it in operation is about 30 days, considering 300 working days, the equipment was thus not available for 10% and is known as downtime. The downtime is the duration from the time when equipment goes under breakdown to the time when it is restored to service and includes time for reporting of failure, time required for reaching to the site and initial inspection of the equipment, actual repair time, time required for final inspection and trial.

Downtime can be reduced by following ways :

(1) **Speedy Fault Detection.** Fault should be detected before it becomes serious enough to affect performance, as we known "a stitch in time saves nine".

(2) **Speedy Fault Diagnosis.** The time spent in fault diagnosis can be reduced by following a logical method of fault diagnosis.

(3) **Speed Repairs.** Time for repairs include; time required for fault detection and diagnosis, removal of assembly, stripping, adjustment, assembly and then final trial test. For speedy repairs it is essential to use latest tools,and immediate decisions should be taken for whether recondition or replace a particular part.

(4) **Reduction in Waiting Time :** Waiting time for repair is considerably higher than actual time taken for repairs due to the irregular (random) arrival of repair jobs. Queues are formed mainly because of the mismatching of arrivals and servicing rate. Waiting time can be reduced by increasing the efficiency of repair crew (though training, incentive etc.), and by adopting policy of replacing first and repairing afterwards for jobs involving heavy repairs. For this purpose some assemblies or sub-assemblies are kept in reserve. Waiting time may also be reduced by increasing the manpower. Due to the trend towards more sophisticated equipment, the cost of construction equipment is continuously increasing. Hence the importance of maintaining availability also increases, as there is a greater loss whenever the equipment is under breakdown.

Operation and Utilisation

Since this is the task directly responsible for production, maximum managerial care should be given to the 'operation'. Following are the main factors which effect the operation:

(*i*) **Availability of Trained Personal.** Since there is general shortage of trained personnel inp our country for construction equipment operators, there is need to impart proper training by organising training programmes in order to achieve optimum utilisation of construction equipment.

(*ii*) Adequate repair and maintenance facilities in the project.

(*iii*) Availability of spares required for repair and maintenance.

(*iv*) **Working Conditions.** Efficient operation depends on the capacity of the machine, operating skill, spares availability and working conditions etc. It also effects the economic operation of the equipment. Therefore before executing the project, date pertaining to the geology, quarries, geographical locations are studied in detail. The production during the operation of the equipment is effected by these factors.

(*v*) Haul roads should be well maintained for better operation. utilisation and better output of the equipment.

(*vi*) **Information Monitoring System.** Since information is an evaluated data, therefore various data related to operation should be recorded, and evaluated for information/ monitoring the operation task. The information transmitted form the site should be given immediate attention. The information which should be maintained for the purpose of monitoring are: daily production report, maintenance history reports, frequency analysis, manpower utilisation report (actual manpower utilization V/s available. category wise) list of equipments under repair and likely data of completion of their repair etc.

(*vii*) Feed back should be extended to the manufacturers in respect of performance levels, repetitive breakdowns and operation difficulties.

(*viii*) Since the tire is a costliest item and due to high cost proper care is necessary as described earlier.

(*ix*) Norms of output for each type of equipment in each project is determined considering the operational factors, The output of equipment should be compared with that of pre-determined norms to enable to improve utilisation (output) over existing levels. A care should be taken that to achieve target, it may not lead to mal-utilisation of equipment.

(*x*) Encourage operators and maintenance personal for giving long life of the equipment.

QUESTIONS

Q.1. Discuss the importance and necessity of construction equipment for country's infrastructure and industrial development.

Q.2. Classify the construction equipment on the basis of their functions.

Q.3. Discuss various factors considered for extent of mechanization in a project.

Q.4. Explain advantages and disadvantages of using machines and manual labour.

Q.5. Explain construction equipment selection criteria. Discuss the criteria for recommending a construction equipment.

Q.6. How equipment planning is done for a project of large size?

Q.7. How would you forecast the equipment requirement for a construction project?

Q.8. Describe various criteria for deciding two shift to three shift working in a construction project.

Q.9. Discuss the factors which affects the output of construction equipment.

Q.10. Discuss important points which should be considered while planing for equipment operation.

Q.11. Discuss various factors which should be considered in the process of equipment selection for a construction project.

Q.12. What is productivity? Discuss the reasons for low productivity. Explain the main contributor for productively improvement of equipment.

Q.13. Discuss various aspects related to equipment productivity.

Q.14. Explain following in brief with reference to equipment productivity:

(*a*) Equipment operation planing.

(*b*) Spare parts management.

(*c*) Maintenance management.

Q.15. Discuss about downtime and availability of equipment form the point of view of equipment productivity.

Q.16. Explain various factors related to equipment operation and their utilisation from productivity point of view.

Chapter

EARTH-MOVING EQUIPMENT

12.1 INTRODUCTION

Earthmoving is the process of moving soil or rock from one location to another and processing it so that it meets construction requirements such as location, elevation, density, moisture content etc. Earthmoving operations involve processes such as excavating, loading, hauling, dumping, spreading, compacting, grading, finishing, crushing etc.

An earthmoving operation consists of the preparation of material, the loader/truck loading cycle, haulage of trucks to the disposal place, the deposition of the material and the trucks return trip to the loading station to start another load-haul-unload and return cycle.

Depending on the scope and working conditions of each project different operation methods and machine types should be selected to maximize the overall performances of the operation. Earthmoving operations require technical knowledge and skill to operate and maintain a variety of heavy equipment such as dozers, scrapers motor graders, shovels, hoists etc.

Modern mechanical earthmoving really started when the crawler tractor was adopted from its agricultural antecedent. To meet the growing demand of minerals particularly coal, the earthmoving equipment was first introduced in 1940's in open-cast coalmines in our country. Shovels of about 2.2 cum capacity with dumpers of 12 to 15 tonnes capacity were introduced in these mines. Bhakra Nangal Dam started in 1950's used higher capacity and sophisticated earthmoving equipment like dozers, scrapers, shovels, draglines and dumpers. Almost same time in 1950's shovels of 1.5 to 2.5 m , front-end loaders, dozers with rippers, and 15 to 25 tonnes dumpers were introduced in iron ore mines, replacing manual mining and locomotive traction. With the introduction of heavy earthmoving equipment, bigger size operations were taken up in construction of Hydel dams, irrigation schemes, coal and metal mining. Kudremukh Iron Ore Project in late 1970's used 15 cu m shovels, 120 tonnes diesel engine dumpers, 15 cu m front-end loaders and dozers of over 600 H.P.

The equipment of various sizes used for various projects were imported from various countries. The indigenous manufacturers in the country are now planning to manufacture 10-12.5 m shovels, 85-120 tonnes dumpers,about 450 H.P. class dozers, 30 m scrapers, 30 m /90 m draglines etc. With indigenous manufacture, it will save high cost of importation and also help in standardisation which will solve the problems of spare parts stocking, maintenance and downtime.

With the application of heavy earthmoving equipment (H.E.M.), production of coal has increased from 25 million tonnes, similarly hydel power has also increased manifolds in last 30 years. All these activities involve movement of rock, earth and minerals in such a huge quantities that these works of irrigation, hydel, and mining (coal and mineral) projects can not be undertaken manually within the time schedule set for the projects.

Types of Earth Moving Equipments

Earth moving Equipments can be classified into following two types.

(*i*) **Production Equipment :** Equipment used for digging, moving the material to the site of construction (may be dam, irrigation project, road project, mining project or building construction project) or to the crushing plant.

(*ii*) **Service Equipment :** These are for the help of main production equipment to achieve optimum capacity. For small works dozers can be used as productive equipment. Motor Grader is used as production equipment in road construction projects, while it is used as service equipment for keeping proper gradient of service road for dams, mines and quarries.

12.2 TRACTORS

Tractors are the machines which change energy of the engine into tractive energy. The tractors are primarily used to pull or push loads, but they are also used for mounting many types of accessories like, bulldozer blades, rippers, front end shovels, hoes, draglines, clamshells, trenchers, winches, side booms, frontend bucket loaders, pipe Layers etc.

As mentioned above,the tractor is a basic equipment and available in two types; namely (*i*) crawler or track type, and (*ii*) rubber-tired or wheel-type. The tractor is a multi-purpose machine, light models of which are used for agricultural or short haulage works, whereas heavy models are employed for earth moving works, cranes,shovels or special rigs. Wheeled units are employed for light but speedy jobs, whereas crawler units are rugged machines and are used for heavy duty works, where more tractive power is required.

Tractor Construction

Main constituents of a tractor are : Engine, clutch, transmission system, ground drive and controls.

A diesel engine is generally used as prime mover and mounted on the tractor rame and delivers the power through a clutch to the main power shaft. The power is transmitted from the engine to the final drive through engine clutch, torque convertor, change speed gear box transmission to rear differential and then finally to the rear drive axle. The drive is also transmitted through front differential to the front axle.

Engine clutch is provided to disconnect the engine from the rest of the transmission system so that gears can be changed to change the speed of the tractor. Care should be taken while changing the gears that the engagement of the clutch should be gradual to avoid undue strain on the surfaces of the clutch plates. The clutch control m small and medium sized tractors is through pedal like that of a truck, but in large tractors it is through lever operated by hand.

Transmission system provides speed reduction and multiplication of torque of the engine shaft, since final drive needs a high torque at low speed. The reduction in speed is done by means of a change speed gear box. Now-a-days torque converters are also used for the purpose. The power is then given to the drive axle through the differential, ahich changes the direction of the drive to 90° by using bevel gears and also reduces the speed upto some extent. Differential also facilitate in steering. As tractor changes its direction, one of the rear wheels must travel faster than the other, and this is achieved by the differential. In the wheeled tractor the steering is done through the front wheel. The outer ends of the two half shafts carry pinions which mesh with larger ;ears of the final drive, resulting variation in speed when the tractor is taking a turn. The change in the speeds of the track is made directly by stopping power supply to one, while the other continues to rotate.

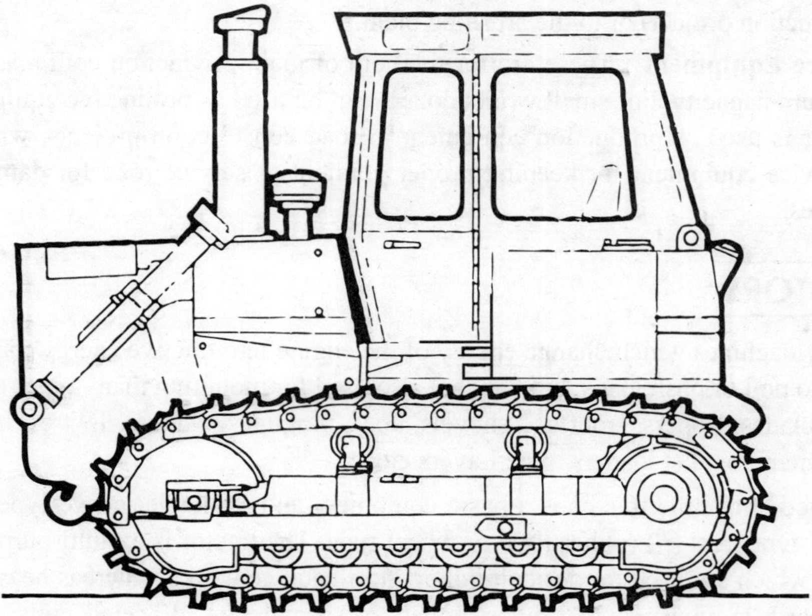

Fig. 12.1. Crawler tractor (Courtesy : Liebherr).

The power from drive axle is transmitted through wheels or crawler chains in wheeled tractors and crawler tractors respectively. Wheeled tractors generally have two large wheels in the rear and two smaller wheels in the front. For better traction under adverse load and ground conditions two wheels are fitted on each of the rear hubs. To make rear heavy for better stability of the tractor rear tyres are filled with water. The tread of the tyre should be selected carefully as it greatly affects its tractive efforts on a particular type of soil where it operates.

In Crawler tractors, live axles rotate large toothed wheels, known as drive sprockets, located at the rear of the track frame. In the front of each track frame, an idler is provided. An endless track chain is mounted on the sprockets and idlers. The chain is made of flat shoes pinned together end-to-end.

Crawler v/s Tyred Tractors

Each type of tractor has certain advantages in certain conditions. A comparative study of both the types of tractors is given hereunder to help in deciding that which of the two will suit more in a particular set of conditions.

(*a*) Advantages by using Crawler-mounted tractors :

(*i*) More tractive efforts,hence can also operate on loose or muddy soil.

(*ii*) In absence of tyres, can easily operate in rocky conditions, as there is no danger for the damage of the tyres.

(*iii*) Where maintenance of haul roads is difficult, it can easily travel, specially in rough terrains,

(*iv*) Crawler tractors are more compact and powerful and hence can handle difficult jobs as well.

(*v*) Greater floatation because of the lower pressure under the tracks,

(*b*) Advantages by using Wheel-mounted tractors :

(*i*) Can travel at higher speeds during the operation and also from one job to another.

(*ii*) Ease in operation. Operator feels less fatigue.

(*iii*) Can travel on paved roads without damaging them.

(*iv*) Can travel long distances at its own power, whereas Crawler mounted needs trailers.

(*v*) When work is spread over long area, these are found to be producing more output.

(*vi*) Operation, maintenance and repair costs less in wheeled tractors as compared to Crawler tractors. As large number of track parts are subjected to more wear.

(*vii*) Initial cost of the Crawler tractor is higher than the wheeled tractors, due to expensive track system.

(*viii*) Easy to manoeuver.

12.3 BULL-DOZERS

Dozers are machines designed primarily for cutting and pushing the material over relatively short distances. They consist of a tractor with a front mounted blade, controlled by hydraulic cylinders to very the depth of cut and rate of levelling depending on the material and application. Rear mounted hydraulic sacrifiers and rippers can be fitted to loosen hard material prior to dozing. A dozer is a frame mounted unit which a blade, curved in its section, extending in front of the tractor.

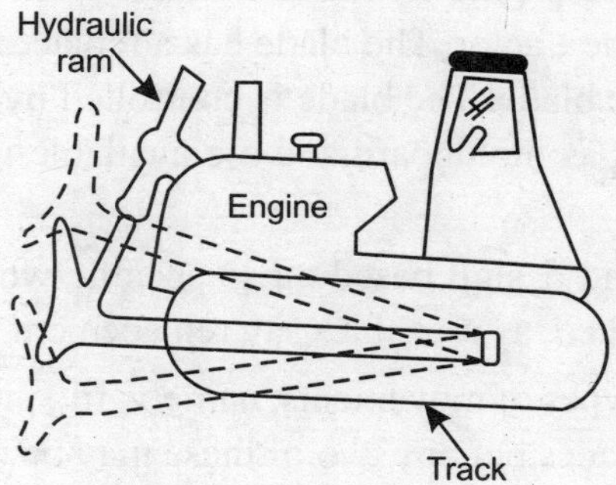

Fig. 12.2. Bulldozer with blade.

Bulldozer is a most versatile and most important equipment on construction projects. Basically it is a pushing unit, but is widely used as a multipurpose equipment and can perform large number of operations with minor changes under different names like; angle dozer, tilt dozer, tree dozer, and push dozer. Bulldozer scraps soft and granular material, pushes it in front or stock-piles. These are used for short hauls upto 100 metres only.

Angle dozer is used to push its load at an angle of nearly 25° to the direction of travel of the dozer. This is helpful where the material is pushed down the slope on hill work or where the material to be collected in a long windraw on one side of the line of travel of the dozer. The tilt dozer is used where blade is required to be tilted by raising one end upto 25 cm above the other, so as to start excavating a ditch or a trench or for excavation in hard ground. The push dozer is used to push the scraper unit during digging or loading operation by means of a rectangular plate, called the pusher plate. The tree dozer is used to remove the trees.

Dozer Specifications

At the time of purchasing of dozens following points should be clearly specified after examining the exact job requirements and the manufacturer's limitations.

Engine horsepower at the flywheel, total operating weight, speed range in various gear positions; blade type and size and also maximum tilt position or angle position; number, type and size of rippers; turning radius, fuel tank capacities. In case of wheel dozer also mention, number and sizes of the tyres, wheel base, ground clearance etc.

Following main factors should be considered while selecting and tractor dozer :

(*i*) Size of the dozer for a given job.

(*ii*) The type of work expected from the tractor dozer, e.g., bulldozing, ripping, land clearing, pulling a scraper etc.

(*iii*) The type and condition of haul road.

(*iv*) Gradient of the haul road.

(*v*) Distance to be moved.

(*vi*) Type of work expected to be taken from the equipment after the present job is completed.

Dozer Blades

A heavy blade of slightly concave profile is attached in the front of the tractor. It is connected through two arms and a yoks to the tractor. The blade has a replaceable cutting edge, which wears out with the use of the blade. The blade. The blade is controlled by means of hydraulic cylinders. The dozer blades are also known as moldboard and are available in size from 2 m to 7.5 metres wide and 0.8 m to 1.5 in height.

Dozer blades vary in sizes and design based on so specific work applications. The hardened cutting edges and side bits are bolted, as it enables say replacement when worn out.

Dozer blades can have three types of movements, namely; tilt, pitch and angling. All these three features are not available to all blades,but any two of these may be incorporated in single mount.

1. **Tilt :** In this case dozer can raise or lower the either end of blade in the vertical plane. Tilting permits concentration's length.

2. **Pitch :** In this case the top of the blade can be pitched forward or backward varying the angle of attack of the cutting edge. This is a pivotal movement about the point of connection between the dozer the blade. When the top of the blade is pitched forward, the bottom edge moves back which increases the angle of cutting edge attack.
3. **Angling :** In this case blades mounted on a C-frame can be turned from the direction of travel. In angling the blade is turned so that it is not perpendicular to the direction of travel of the dozer. This results in pushing the material to roll of the trailing end of the blade.

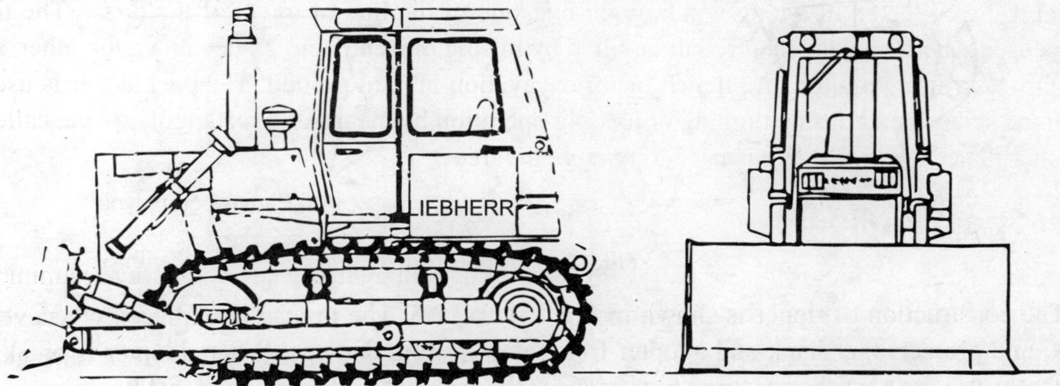

Fig. 12.3. Crawler dozer with blade (Courtesy : Liebherr).

Rippers

Crawler-tractor, mounted with ripper is finding increasing use in construction, mining and quarrying. Cost of ripping vary from material to material to be ripped and some times very hard rock formations cannot be ripped. Cost of ripping should be compare: with the cost of drilling blasting' to consider economical way of excavation. While considering ripping cost, life of ripping type as well as shortening of life c undercarriage should be considered.

Characteristics of the material to be ripped influence the selection or ripper type, number of shanks required, ripping speed and amount of ripper penetration. For hard material single shank is preferred, but where material is easily penetrated and fractured into small pieces, a two or three shank ripper is used.

Ripper Types

Various types of rippers are used with the dozers. These may be towed units, integrated units, back rippers. Towed units are generally cable operated and are hard to manoeuver, whereas integrated ripper units are hydraulically operated. With crawler mounted tractors these integrated units are easy to manoeuver and provides a more compact and efficient units,while damage to tyres is expected in rocky conditions. Integral rippers are of two types–hinge or parallelogram. Back rippers are mounted on dozer blades, so that as the dozer proceeds forward, the back rippers pivot in the mounting brackets and drag over the ground. Those are used in light duty works.

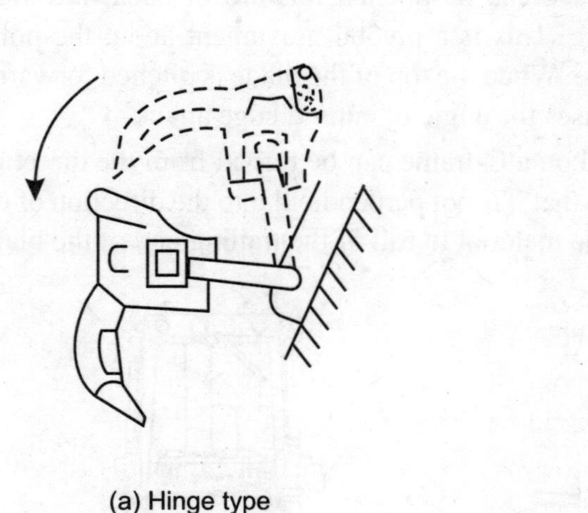
(a) Hinge type

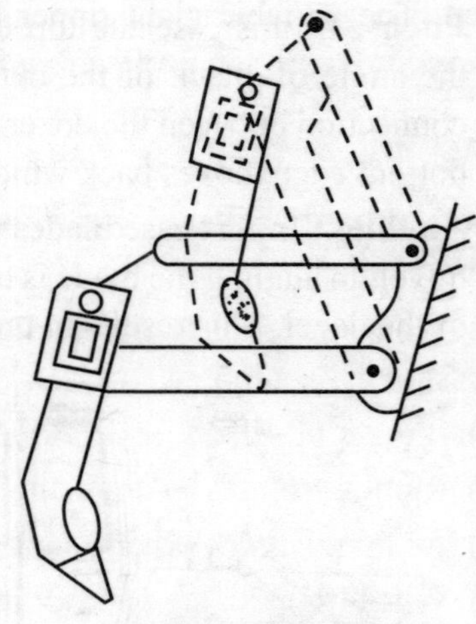
(b) Parallelogram type

Fig. 12.4. Rippers.

The construction of ripper is shown in Fig. 2.4 and 2.5. The tip can be replaced whenever it break, and protects the shank and adopter from adopter from damage. When adopter is breaking frequently, use the next shorter tip so as to reduce the load on the adopter. The protector is provided to support the tip as well as to provide wear provided to support the tip as well as to provide wear protection for the adopter. This protector is replaced when it is adopter. This protector is replaced when it is worn out to 6 mm, so that any likely protector failure (Which may cause adopter damage) may not occur.

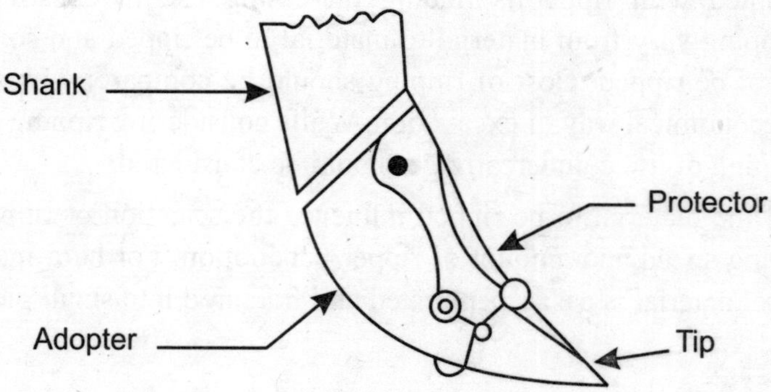

Fig. 12.5. Ripper Shank with tip.

Rippers can also be classified as follows :

(*i*) Fixed Multiple-shank Ripper.
(*ii*) Variable Multiple-shank Ripper.
(*iii*) Fixed Giant Ripper.
(*iv*) Variable Giant Ripper.

The fixed multiple shank ripper is used for comparatively simple ripping operations. The shank moves only up and down during digging operation. Digging depth is adjustable by changing shank pin position or shank. Fixed giant ripper has only single powerful shank, but is not common as compared to variable giant rippers.

The variable multiple-shank ripper can change the angle of its shanks by means of ripper tilt cylinders. These are used where there are many boulders, or where the quality of rock is not consistent. The variable giant ripper has a single power shank. The angle of this ripper can be changed according to the ground requirements.

Ripping Efficiency

In order to achieve high ripping efficiency, following suggestions should be considered during ripping operations :

(*i*) When material is not broken, loosen the surface in one direction, then cross-ripping should be adopted.

(*ii*) If material is soft, use more than one shank ripper.

(*iii*) Ripping should be done downhill.

(*iv*) Avoid reversing when shank is in the ground.

(*v*) When both dozing and ripping operations are required to be done, it should rip going out and doze on the way in.

(*vi*) Shank tip should be sharp.

(*vii*) Use low gear range so as to get as low speed as possible.

(*viii*) When tracks start slipping, ripper and shank should be raised.

(*ix*) For hard material, use shorter tips, and for ordinary work use longest and sharpest point.

(*x*) Do not uncover the unripped portion of the rock during dozing of the material loosened by ripping. This will reduce the wear of under-carriage parts.

Applications of a Dozer

Following are the main dozer application :

(*i*) Land Clearing : On very project, bulldozer is the first equipment and its first task is to clear the ground for camp-site, and site or installing various plants.

Clearing of the land involves several operations depending upon the condition of soil and topography, the amount of clearing is done. The operations may include, removal of soil to remove unevenness of the ground, removal of tress and stump including roots, removal of vegetation etc. For this purpose various types of equipment can be used. These may be tractor mounted bulldozers, tractor mounted special blades in which the stringer penetrates the ground to cut the main horizontal roots of the tree or a blade which slides along the surface of the tree or a blade which slides along the surface of the ground for cutting vegetation, tractor mounted rakes to be used to grab and pile trees, bounders etc. without transporting excess amount of soil, tractor mounted clamp rakes to pick up felled trees and transport them to placed of dumping.

(*ii*) Striping : It is the removal of top soil that is not usable. Stripping is planned in such a way so that haulage is for minimum distance. Upto the distance of 100 meters dozer itself is used, but for more distances scrapers should be employed.

(*iii*) Sidehill cuts : Sidehill cuts should always be started from the top and then worked downward.

(iv) **Backfilling :** In backfilling material is pushed ahead of the dozer over embankments into ditches or against a structure. Angle dozers are found more suitable for this job as they can drift material into the trench while maintaining forward motion.

(*v*) **Ditching :** The rough ditches can be prepared by the dozers. These are constructed with a straight blade by working at right angles to the length of the ditch.

(*vi*) **Spreading :** These are also used for spreading the material hauled by trucks, scrapers etc. Depth of spread is adjusted considering the thickness required.

(*vii*) **Dozing rocks and frozen ground :** Rocks are generally removed by using one corner of the blade, as the full power is thus applied on this section of the blade. Large sections of rocks are removed by using the blade to lift the rock, simultaneously applying power to the tracks.

(*viii*) Maintaining haul roads.

(*ix*) Clearing of the floors of borrow and quarry pits.

(*x*) **Digging :** A dozer while moving forward with the blade inserted in the soil digs it. If the blade is lowered further, more work can be done, thereby resistances will increase, as a thick slice requires more digging power than a thin one. The blade, during operation should be lowered or raised gradually to avoid bumps in the path.

(*xi*) **Breaking Pile :** A pile may be knocked down by walking into it with the blade at the desired grade, after which it may be spread or piled elsewhere. If the heap is too large, it may be cut away in parts, either directly, or if it is not possible, the side cut should be repeated from different angles.

12.4 SCRAPERS

Scrapers are the devices to scrap the ground and load it simultaneously, transport it over the required distance, dump and then spread the dumped material over the required area in required thickness level, and return to the pit for the next cycle. Thus, the scrapers are designed to dig, load, haul, dump and spread and some times called as carry all.

The scrapers are of three types : (*a*) towed type, (*b*) self propelled or motorised or conventional type, (c) self loading or elevating scraper.

Scrapers can also be classified as : 1. Crawler tractor scrapers or Wheel tired tractor scrapers, 2. Pusher loaded or self loaded scrapers.

The towed scrapers are provided with either cable or hydraulic control. Althoug these are becoming obsolete but even then some contractors use them because when coupled to a suitably powered crawler tractor, they can operate in extremely adverse conditions.

Self-propelled or motorised or conventional scrapers are most popular now-a-days. These are generally manufactured in ranges from 10 to 25 cubic metres. A motorised scraper needs push loading by a crawler mounted or a wheeled tractor. Wheeled scrapers have more fauling speed and hence are suitable for long distance hauling while crawler-towed scrapers travel at slower speed and can be used for short hauls only.

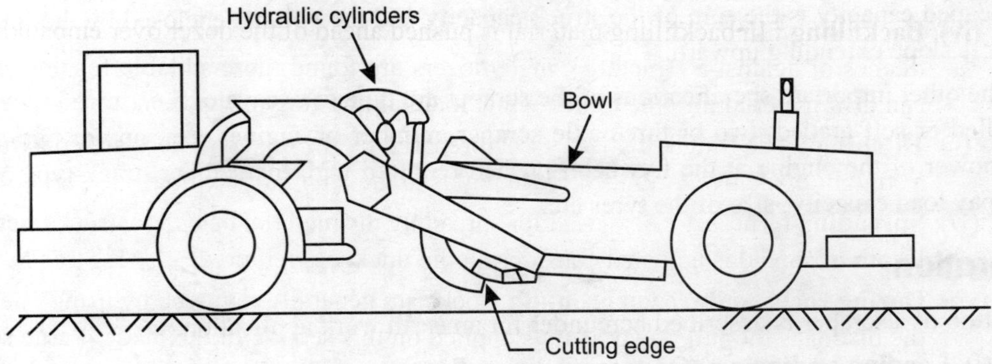

Fig. 12.6. Self Propelled Scraper.

The problem of loading by a pusher is overcome by the third type of scraper known as self-loading or elevating scrapers. These are twin engine scrapers and can work completely independently of all other plants. It does not need a pusher. To make them more effective and competitive with dump trucks, some of the manufacturers are producing three axle-scrapers of 40 cubic meter pay loads. However, these have certain limitations, firstly, the elevator cannot load particularly free flowing sands or handle rock infested soils. Secondly pay loads are restricted because of the additional weight of the loading elevator and its drive system. Thirdly, increase in loading time, the inability to obtain a full load, indicates that for powerful units a pusher is still required.

Construction

A self-propelled or motorised type scraper has following main parts :

(*i*) **Bowl :** The bowl is a pan to hold the scraped dirt. It is hinged at the gear corners to the rear axle inside the wheels, and is capable of tilting down for digging or ejecting. The bowl size is specified to indicate the size of the scraper.

(*ii*) **Cutting edge :** The bowl has a cutting edge attached at the bottom. The cutting edge is lowered into the dirt to make a shallow cut.

(*iii*) **Apron :** This is a wall in front of the bowl, which opens and closes to regulate the flow of the earth in and out of the bowl. This can open or closes in carrying position as well.

(iv) **Tail gate or ejector :** These are the rear of the pan which is capable of forward and backward movement inside the bowl. During loading it remains at its rear wall, while moves forward to help in ejecting the load during dumping.

All these items are controlled by the hydraulic system. Movements of these parts are controlled hydraulically. In hydraulic system, oil pressure works in the double acting rams provided separately for bowl, apron, and tail gate. The flow of oil in these rams is controlled by valves.

Specification of a Scraper

Usually the capacity of a scraper is given in terms of cubic metres struck and cubic metres heaped. Struck capacity of a scraper is the actual volume enclosed by the bowl and apron, struck off by a straight line passed along the top edge of the side plates. In other words it is the volume of the material levelled with the top edges of the bowl.

Heaped capacity is the sum of the struck capacity and the volume enclosed by the four plates at a 1 : 1 slope extending upward.

The other important specifications of the scraper are type of excavator, i.e., towed type or self-propelled or self-loaded; two or three axle scraper, number of engines, i.e., one or two engines; horsepower of the engine at the flywheel; pushed required with the scraper; track-type or wheel type; pay load capacity; size of the tyres etc.

Operation

Operation of a scraper is described hereunder for an earth work application :

(*i*) **Loading or digging :** Operator moves to the cut with the ejector at the rear and the apron raised approximately to 40 cm. The bowl is then lowered to the desired depth of cut, increase engine speed, move forward in first gear keeping optimum depth of out. When the bowl is full, the apron is closed and the bowl is then raised.

(*ii*) **Transporting :** The bowl is transported in high gear in raised position to provide sufficient clearance, during transporting, apron should be fully closed to prevent loss of the material, and the ejector should remain in the rear position.

(*iii*) **Unloading :** Unloading in a scraper is also termed as 'dumping and spreading'. The bowl should be positioned to spread the material to the desired depth during this operation. A partial opening of the apron during the initial unloading will help in even spreading. For wet and sticky material, the apron should be raised and lowered repeatedly until the material behind it is loosened and drops out of the bowl. Then the ejector is moved forward to push the remaining material out of the bowl at a uniform rate.

When the dump is complete, the tail gate is fully retracted, the apron dropped and the 'bowl' raised to transporting position. The finishing touch during spreading is done by grading it with the cutting edge, and the unit then hurries back and starts next cycle.

In hydraulic system the operations as described above are affected through oil pressure working in double-acting rams, usually two for apron and one for tail gate.

For better efficiency in the scraping operations following points should be taken care of:

(*i*) Depth of cut should commensurate with the power available for the cut.

(*ii*) Keep the haul roads maintained.

(*iii*) Pushers should utilise their waiting time in dressing the cut.

(*iv*) Loading operation should be carried out downhill wherever possible.

(*v*) Excessive turning of a pusher should be avoided, and scraper pusher balance should be maintained.

Economics of Scraper Operation

To examine the economics of scraper operation two main factors need detailed consideration, they are : material and haulage distance.

(*i*) **Material :** Scrapers can conveniently be used in all types of soils, like sand, gravel, silt, loam clay and their mixtures.

(*ii*) **Haulage distance :** It is generally seen that for very short distances say 50 to 100 metres, a bulldozer is most economical earth moving equipment, whereas a scraper is most economical when the haulage distance is less than, say, 1500 metres and more than 100 metres.

Number of Scrapers per Pushdozer

For attaining the maximum hauling capacities,wheel type tractor-pulled scrapers need the assistance of pushdozers during loading to reduce the loading time. Similarly crawler-type tractor pulled scrapers, which are claimed to be self loading can also be loaded by means of push dozers. If the use of push dozer is found to be economical, *i.e*.. the increase in production is sufficient enough to pay back the cost of the tractor dozer, it is better to use it.

When pusher dozer is used, it is desirable to match the number of pushers with the number of scrapers. Waiting of any one, reduces its efficiency and increases the production cost. The cycle time for a pusher includes the time required to load one scraper plus the time needed to move into position to load another scraper. It depends on the conditions in the loading pit, sizes of the pusher and scrapers and loading method.

Following formula should be used to determine the number of scrapers that a push dozer can serve :

$$\text{Number of scraps served} = \frac{\text{Cycle time for scraper}}{\text{Cycle time for pushdozer}}$$

12.4.1. Elevating Scrapers

Scrapers described earlier depends upon a pusher tractor for loading, the elevating scraper is self loading one in addition to self propelled capabilities. It is similar to a conventional scraper except that its apron has been replaced by an 'elevator', also known as 'ladder'. This ladder is made up of chains with buckets bolted to it. The chain is rotated by power independent of the travel of the scraper. It has a variable speed and also reversing of the chain and buckets is possible. Loading of this scraper is done by forcing incoming earth over the cutting edge and subsequently ladder will carry it up into the bowl. They are capable of loading most materials except rock and boulders which are too large to pass between the cutting edge and the elevator buckets, and cohesive material with high moisture content.

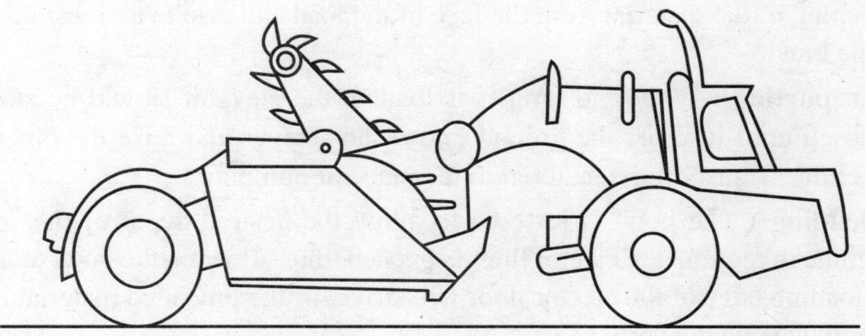

Fig. 12.7. Elevating scraper.

By using elevating scraper, cost of a pusher and its operator is eliminated. Elevating scrapers also have very good finishing ability, and hence very useful in preparing final layer of earth work in road or airfield construction work.

The operating cost and maintenance cost of the elevating scraper is higher by nearly 30% as compared to that of conventional scrapers. Due to dead weight of the loading mechanism, its transportation efficiency is reduced.

Fig. 12.8. Elevating scraper (Courtesy : John deere).

Operation

Operation of an elevation scraper is described hereunder. The working cycle is similar to that of conventional scraper.

(*i*) **Loading :** For making a cut, the bowl is first lowered to a depth which can permit the elevator and tractor to operate at high and constant engine speed.

The cut will force material towards the bowl and elevator buckets will sweep the material into the bowl. The elevator has 4 forward and one reverse speed. Soft surface like, sand, silt and topsoil are dug and loaded at high speeds. Low speeds are used for loading tough materials, like, hard clay etc. When the bowl is fully loaded, Stop the elevator to avoid sweeping of the material from the face of the load and also to avoid spilling over to sides of the bowl.

(*ii*) **Transporting :** When the scraper is loaded, the elevator should be stopped and bowl is raised until it clears the ground, move the scraper and raise the bowl further during travelling. Transport the material to the place of dumping.

(*iii*) **Unloading :** The bowl is lowered to allow the desired depth of the spread. With the machine in motion, the ejector floor is opened, thus allowing the bowl to unload itself and the loading edge of the ejector floor will strike off the unloaded material in the bowl, in a smooth and even layer.

When unloading is complete, the bowl can be raised, ejector floor is closed and the scraper is returned for the next cycle.

Optimising the Scraper Production

Following are the some of the suggestions which will increase the scraper production :

(*i*) Construct and maintain smooth haul roads for faster and safer travel with less driver fatigue.

(*ii*) Depth of cut should be according to the type of soil being handled.

(*iii*) Use ripper teeth in hard or abrasive material for easy loading. While ripping a rock, the depth ripped should be more so as leave a loose layer of material under the tracks and tyres to provide good traction and reduced wear.

(*iv*) Where possible, loading be done in down grade.

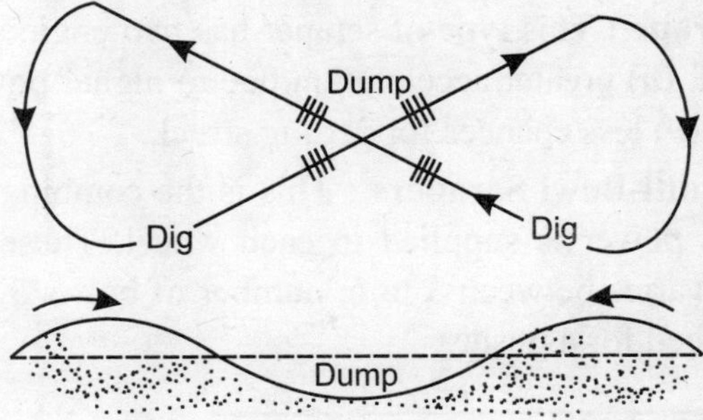

Fig. 12.9. Method of loading and hauling.

(*v*) To increase stability of the scraper during travel, carry the bowl as close to the ground as possible.

(*vi*) If necessary, pre-wetting of the soil is done so that soil is reasonably moist, since some soils will load easily when they are moist.

(*vii*) Whenever possible, plan the work to eliminate all avoidable turns. For example, 'Figure 5.8' loading, hauling and unloading can be used to great advantage. By using this method, only two turns are required while two full loads are delivered, whereas normally two turns are required for each load.

(*viii*) Dumping should be done in an orderly pattern in the lifts of desired thickness so that compaction equipment work in effective manner.

(*ix*) A spotter should always be employed to eliminate confusion and traffic congestion and coordinate the movement of scrapers for spreading the soil in the fill and that of compacting equipment.

Advantages of Elevating Scrapers

Following are the main advantages of elevating scrapers :

(*a*) Better loading ability in loose free flowing materials.

(*b*) Good finishing ability.

(*c*) Can be operated independently.

(*d*) Smooth and complete unloading of bowl by reversing the elevator rotation.

(*e*) Pusher tractor dozer is not required.

(*f*) Addition of ripper teeth to the cutting edge permits self loading in hard compacted material.

(*g*) Pulverising and mixing action by the elevator places material in uniform and homogeneous state for compaction.

Special Types of Scrapers

Other then the tractor towed, motorised, and elevating scrapers explained earlier, following are the other types of scrapers :

(*i*) **Tandem Scrapers :** In this type of scraper two bowls are attached to a single prime mover unit. It gives higher capacity per cycle.

(*ii*) **Twin Engine Scrape :** This type of scraper has two engines with single bowl. This has the advantages of, (*a*) greater acceleration due to higher power, (*b*) higher speeds under sever conditions, (*c*) less chances for getting struck.

(*iii*) **Multi-Engine Multi-Bowl Scrapers :** This is the combination of the above two ideas. In these scrapers power is supplied to each wheel. These are generally self loading; number of axles varies between 3 to 6, number of bolws 3, capaity upto 80 cu m. This eliminatescthce need for a pusher.

12.5 LOADERS (FRONT-END LOADERS)

A bucket is attached to the arms and capable of being raised, lowered and dumped through mechanical or hydraulic controls. The loaders having bucked in the front, known as 'Front-end loaders' are very common. The loaders are versatile, self propelled equipment mounted either mounted either on crawler or wheel type running gear. These loaders are fitted with front mounted general-purpose bucket operated through hydraulic rams, with which they dig, scoop, lift, transport, carry, dump or load into the hauling units, hoppers, conveyors and stockpiles. With the help of additional frontad rear mounted attachment these can doze, scrap, grub, fork-lift, trench, grade ditch rip, clamp, and winch. Application of a loader very from handing coal, sugar sand, salt, stone-etc, to earth moving and digging work in quarries.

Types of Loaders

Loaders are of the following types :

(*a*) Crawler loaders (*b*) Wheel loaders

(*a*) Crawler Loaders

Crawler track types are generally preferred for digging and loading jobs where ground conditions are poor and low pressure characteristics are required, and are preferable for applications involving rock and sharp stony ground as there is no possible of tyre damage. These are best employed for short moves between loading and dumping points. These should be transported from one site to another after loading on the trailers.

(*b*) Wheeled Loaders

As a result of the development of more capacble power trains (axles and tires), there was steady and towards wheel loaders at the expenses of crawlers. Protective chains for tires, that considerably increases tire operating life, even in quarry applications, additionally favoured the wheel loader against the rather more difficult to manoeuver crawler.

Wheeled loaders are generally four wheel drive. However, for handling light jobs on good conditions two wheel drive variants are also used. Four wheel drive loaders are generally used for construction job, whereas, two wheel drive loaders are used for bulk handling of coal, cement, fertilizer etc. into hoppers and trucks.

From manoeuverability poits of view, wheel loaders are following types :

(*i*) Articulated type (Pivot steer).

(*ii*) Rigid frame-two wheel steer or all wheel steer.

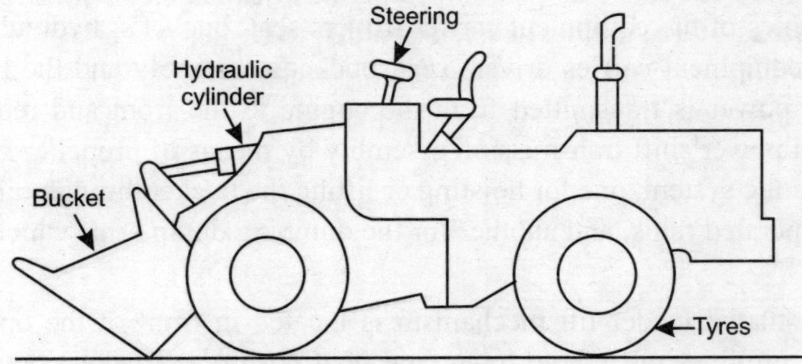

Fig. 12.10. Line diagram of wheel loader.

1. **Articulated type of loaders :** This types of loaders is hinged in the middle of the front and rear axles, as shown in Fig. 5.14(*a*). Due to their operational characteristics and capabilities of working in limited spaced and short turning radius, resulting higher speed of work, these have become very popular. Because of the advantages of quick manoeuvring and thereby easier spotting of loads, less rolling resistance on turns, better mobility on soft surface, these loaders give better performance and high productivity. The articulation permits the loader to pivot 30-34 degree either side to the centre.

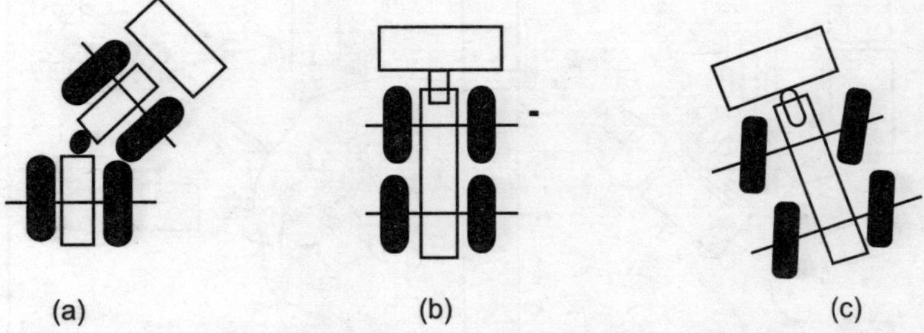

Fig. 12.11. Types of loaders : (*a*) Articulated, (*b*) Rigid frame-two wheel steer, (*c*) Rigid frame 4 wheel steer.

2. **Rigid frame type-loaders :** These are comparatively cheaper than articular frame type of loaders. With this type of loader sometimes operators misjudge hopper or truck body width and therefore he has a manoeuver back and forth before dumping. Rigid frame loaders are of two types, namely two wheel steer and all wheel steer. By designing all

wheel steer loader, manoeuvrability is gained along with the advantages of rigid frame construction. In all wheel steer loaders, operator can which frame 4 wheel construction. In all wheel steer loaders, operator can switch from 4 wheel to 2 wheel steer or vice-versa by using a lever. On stockpile operations, all wheel steering allows the operator to align the bucket in the pile quickly to ensure a heaped load in the minimum possible time. While moving on the roads operator can set his loader on front steer only and then handles it like a normal road vehicle.

Construction

As we have discussed earlier, articulated frame type of wheel loaders are very common in common in construction engineering, hence these are discussed here. The loader is consisting of two frames, the front frames of the equipment carries drivers seat, bucket's hydraulic rams etc., The front frames of the equipment carries drivers seat, bucket's assembly and the hydraulic and the hydraulic tank. The power is transmitted from the engine to the front and rear axle through a torque convertor and power shift transmission assembly by means of propeller shaft. The loaders have two stage hydraulic system, one for hoisting or lifting the bucket through raising the lift arms with hydraulically-operated rams, and another for the dumping dump arms which are operated by another set of arms.

The hydraulic actuated loader lift mechanism is located in front of the operated for safety reasons. Lift control provides raise, hold lower and float positions. Bucket control has roll-back, hold and dump positions. Additional hydraulic controls are also provided for optional attachment.

Applications

Loaders with its general purpose bucket and with additional attachment can be used for.

(*i*) Loading with its general purpose buckets and with additional attachment can be used for:

(*ii*) Loading the material in the hopper of crushers, conveyors etc.

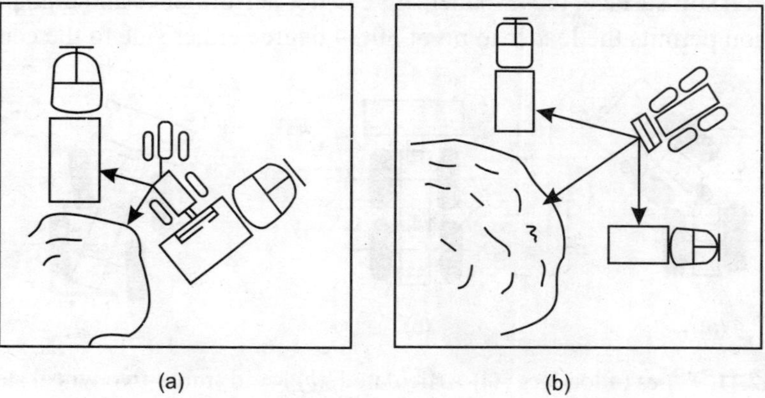

Fig. 12.12. Loading arrangement with (*a*) Articular, (*b*) Rigid frame loaders.

(*iii*) Loading the waste material, clearing debris.

(*iv*) To move bulky objects.

(*v*) Removing snow.

(*vi*) Clearing the areas.

(*vii*) For forklifting.

(viii) For dozing, trenching, ripping etc. with the help of suitable attachments. Thus the loader is very useful equipment and is used in mining, quarrying, industrial work in addition to the constriction engineering works.

Operations

Loaders are used carry out following main operations:

1. **Loading :** Loading operations in the main operations performed by the loaders. Loading consists of scooping, lifting, turning and dumping material such as sand, gravel and crushed and crushed material from stockpiles, bank or constructions site into the hauling units. For better output the bucket should be filled full and the number of hauling of units deployed for the work should be such that there is no waiting time for the loader and there should not be more than one hauling unit in the queue.

Fig. 12.13. Loaders : (*a*) Crawler (*b*) Wheeled, types (Courtesy : Liebherr)

2. **Hauling :** Wheel loaders are excellent for moving loose materials over short distances to dump into hauling units, hoppers, conveyors, bins or any other place of work in the construction sites.

A cycle time for a loader consists of the following :

(*a*) Time for scooping, lifting and dumping the bucket along with time required for changing the directions and taking the turns.

(*b*) *Hauling time* : It is the travel time for load carrying and empty returning. Generally the high reverse speeds reduce the returning time enabling the reduction in cycle time and thus more work in one hour.

3. **Excavating :** Crawler loaders and heavy duty wheel-loaders are excellent for many excavation jobs, like basement digging etc. With excavator attachment it can also handle several other types of excavation works. These loaders can excavate as well as lift the excavated material and dump it into trucks or on the stockpiles. Incase surface is hard for excavating and scooping, the earth ripper or sacrifier attachment is mounted on the year of the loader to loosen such surface. The operator can then easily dig and load the material into bucket for removal.
4. **Clearing :** Loaders can scoop up and load the debris of demolished buildings into hauling units. They also knock down small and temporary buildings; loaders are also used for clean up the areas after snow-storms, tornadoes, floods etc. A grubber blade attachment is used in place of bucket for large-scale clearing work. Thus the loaders are the first equipment to prepare the site for building and construction operations; and also the last equipment in order to backfill, spread, level and top up with selected good soil for grass and landscaping.

12.6 EXCAVATORS

Excavators also called diggers, mechanical shovel or 360-degree excavators, back-hoe. These are mainly used for following operations :

1. Digging of trenches, foundations.
2. Material handling.
3. Forestry work.
4. Demolition.
5. River dredging.
6. Snow removal with snow plough.

For all construction projects, excavation of earth is most important operation.

Such projects may be roads, railways, dams, irrigation and other projects. The machines used for earth excavation are called excavators. Excavators include different category of excavating machines, which are broadly categorized as 'hydraulic excavators' and 'shovels'. The shovel family include dipper or power shovel, dragline, clamshell, hoes, dredgers. To convert one type of shovel into another, only the front end attachment is changed while the basic structure of the machine remains the same.

Hydraulic excavators are now becoming more and more popular and these enjoy various advantages. This can be fitted with various multipurpose attachments for various earthmoving jobs, like, laying pipes, removing trees, excavating drains, general earth moving jobs of cleaning areas, loading etc. The hydraulic system is very productive and efficient due to its mobility and lightness and lower maintenance costs, and hence contributing towards popularity. A most important advantage is the hydraulic wrist action of the excavator which directs the bucket teeth at their most efficient angle during digging operation. The hydraulic system also contributes for the smooth, shock-free operation during digging and traction. The hydraulic controls are easy to operate and respond quickly.

Mechanical or cable operated excavators are of either 'Hoe' type or 'Shovel' type. In 'Hoe' equipments the bucket opening faces towards the machine and the bucket is pulled towards the machine and are used for excavation below the standing level of the machine like, excavation of house and building foundations; trenches for irrigation, sewerage, cables, gas and oil pipe lines; maintenance and cleaning of rivers, canals and irrigation channels. Whereas in shovel the bucket opening faces away from the machine and the bucket is pushed away from the machine and is used for excavating or loading operations above or on the standing level of the machine, i.e., for excavating of rock or earth; loading of material like earth, coal, aggregate etc. 'Hoe' and 'Shovel' are described separately in this book in the chapter on 'Shovels'.

Types of Excavators

Excavators are of following four types based on the type of carriers on which they are mounted:

(*a*) Crawler mounted excavator.

(*b*) Truck mounted excavator.

(*c*) Self propelled excavator.

(*d*) Excavators mounted on barge or rail.

(*a*) Crawler Mounted Excavator

These excavators are mounted on the crawler system and are very suitable fc: carrying out large works in rough terrain. Crawler excavators have following macharacteristics:

(*i*) Can be used for work on soft or wet grounds.

(*ii*) Can be used on sharp rocks or other adverse conditions.

(*iii*) These can climb steep grades (even upto 40 per cent).

(*iv*) Requires very less turning space.

(*v*) Have very less speed for travelling.

(*vi*) Can be shifted from one site to another only on trailers.

(*b*) Truck Mounted Excavators

These excavators are mounted on truck chasis and has following main characteristics :

(*i*) High road speed and mobility it the main advantage.

(*ii*) Lower stability over sides and hence require counter-weight or outriggers.

(*iii*) Requires more operation space.

(*iv*) Tyres or outriggers give high ground pressure and hence it requires firm and smooth operating locations.

(*v*) Have two engines and two cabs, separately for truck chasis and excavator.

(c) Self Propelled Excavators

These excavators are self propelled and has rubber tyres. The main characteristics are :

(*i*) Medium travel speed generally between 10 to 30 km per hour.

(*ii*) Has one engine and one can for control by one operator.

(d) Excavators on Barge or Rail

These excavators are mounted on barge or rail to carry out work of excavation in water or near railway line respectively.

Difference between an Excavator and a Back-hoe

Backhoes and excavators have a common purpose–digging–but the two machines have different designs that reflect specific intended uses. Generally, attached to a front loader or tractor, a backhoe has a two limbed arm with a digging bucket on its end. The limbs are called the boom and the dipper, the dipper holds the bucket. An excavator also have a boom and dipper with a bucket on the end, but is attached to a cab that site on a vehicle with wheels or tracks.

An excavator can be a regular rubber tired back hoe or a track hoe, which is also an excavator having tracks like a dozer.

12.7 HYDRAULIC EXCAVATORS

Hydraulic excavators are basically diggins machines, having following three main components:

(*a*) An undercarriage to give mobility. This may be crawler track mounted or wheel mounted.

(*b*) A superstructure with operator's cabin mounted one either a slew ring to traverse through 360° or on a rigid frame.

(*c*) Hydraulic articulated boom and dipper arms with bucket.

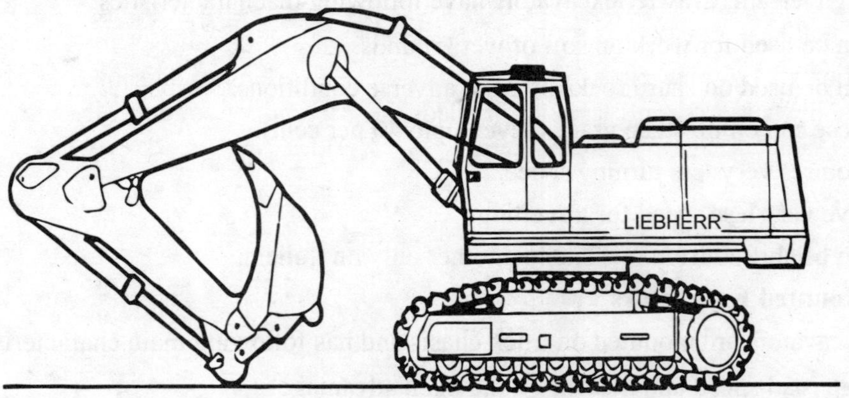

Fig. 12.14. Hydraulic excavator (Courtesy : Liebherr)

Hydraulic actuated excavator have rotating superstructures which are somewhat different from those of the cable-actuated machine like back hoe, power-shovel, dragline and clamshell (explained in next chapter). However, attachments of backhoes, straight line telescopic hoes, and telescopic boom lift cranes are available for mounting on hydraulic actuated excavator. This increases the suitability, applicability and popularity of hydraulic excavators.

Fig. 12.15. Hydraulic excavator in loading operation (Courtesy : Liebherr)

The various digging actions are powered by hydraulic cylinders and motors and are controlled by hydraulic valves or variable-stroke pumps. The hydraulic excavator has the advantages that its digging forces do not tend to lift bucket out of a deep trench, as does the cable pull on a cable-powered hoe. But hydraulic excavators need slightly larger engine as compared to that of cable operated, because hydraulic pumps are slightly less efficient than gear and chain drive, and also due to pressure deep in line and valve.

Hydraulic excavators are troubled by gradual errosion of pumps and valves from line dirt or by sudden failures due to large metal particles in the flow. Oil filteration is therefore very important. Care should also be taken that, foam or air must not be pulled into the pumps because air even in minute quantities can soon destroy a pump.

Now-a-days hydraulic excavators of various sizes upto 30 m^3 bucket capacity and 1730 kW engine output are available. These large sized, heavy hydraulic excavators are mostly used in exploitation of raw materials, open cast mining, for loosening and loading of large material quantities, and where transport vehicles of adequate size can be economically employed.

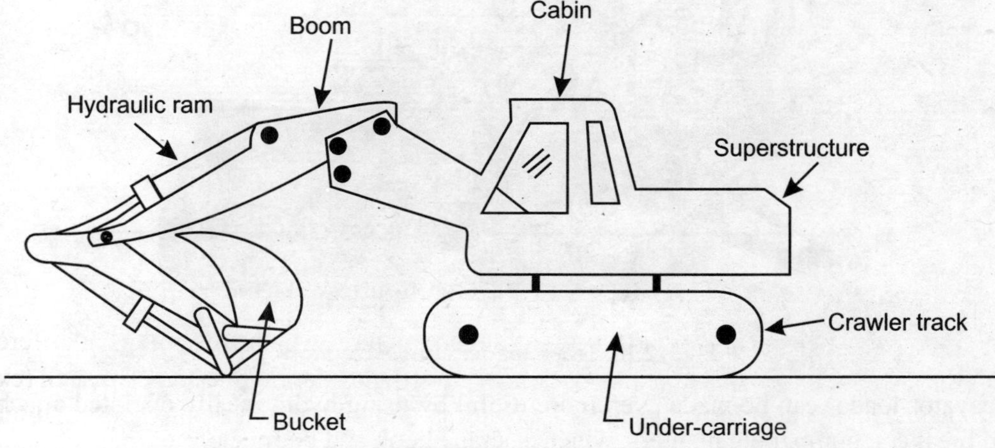

Fig. 12.16. Hydraulic excavator.

12.8 EXCAVATOR-LOADERS

Excavator-loaders something also known as backhoe-loaders, are basically digging and loading machines, and comprise following main components :

(*a*) A tractor with two-or four-wheel drive.

(*b*) A rear mounted digging arm and hydraulic operated excavator bucket, able to slew through a 180° arc.

(*c*) A front mounted loader arm assembly with loader bucket.

Excavator end is used for side tipping; excavating of house and building foundations, trenches or irrigation; sewerages, gas and oil pipe line work; and also for irrigation channel cleaning and maintenance. Whereas the loader end is used for loading and rehandling of materials such as excavated soil, aggregates, coal etc.

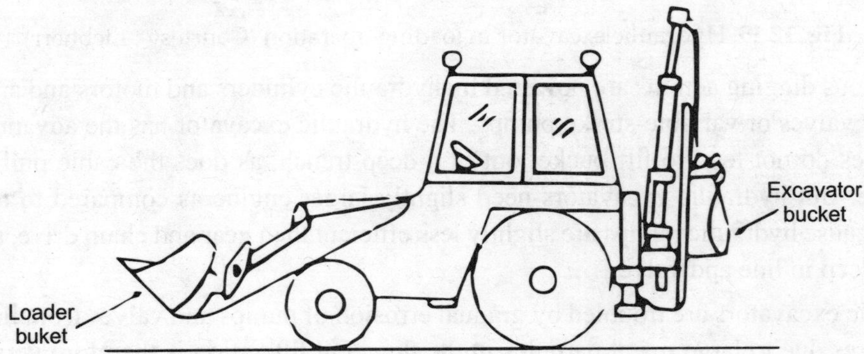

Fig. 12.17. Excavator-loader.

Fig. 12.18. Track loader cum excavator.

Excavator-loader can be made even more useful by using hydraulically operated attachments for road breakers, pump, pole planters, winches, crane hook and boom etc.

Excavator-Loader for Civic Utility Works

Excavator-loader is a versatile equipment, since in addition to the standard excavator and loader buckets, a vast range of special purpose attachment like the forklift, the rock breaker, the ripper, the jib crane, the ditch clearing bucket, can be provided. With these attachments the excavator loader can be used for handling numerous tasks of our cities, including laying of water pipelines, trenching, backfilling can done, ground can be levelled, an collect and load garbage, can be used in road maintenance works, can be used on hot mix and crushing plants for loading of gravel and stone.

Fig. 12.19. Versatility of excavator-loader (Courtesy : JCB)

QUESTIONS

Q.1. Justify the statement, "Tractor is a basic equipment for earth moving machinery'. Describe its construction in brief.

Q.2. List out application of a bulldozer.

Q.3. Describe 'dozing' and 'ripping' action by bulldozers.

Q.4. Describe construction and working of scrapers.

Q.5. Describe construction and working of elevating scraper. Differentiate it with conventional scrapers.

Q.6. Illustrate various applications of loaders. Mention merits and demerits of wheel loaders and crawler loaders.

Q.7. Explain the relative advantages and disadvantages of a 'crawler tractor' and a 'wheel tractor'

Q.8. What are the operations that can be performed by bulldozers? Describe briefly the procedure for clearing of an area with bulldozers.

Q.9. Discuss the factors which influence the output of earth moving machinery.

Q.10. What is a tractor? For what purpose it is used?

Q.11. What factors influence the selection of a tractor? Explain in brief.

Q.12. How the power is supplied to the dozers?

Q.13. Write a note on the blades of a dozer.

Q.14. What is meant by rock ripping? How the reliability of rock is determined?

Q.15. What is a scraper? Discuss dozer pulled scrapers.

Q.16. Give the classification of scrappers and discuss them in brief.

Q.17. Explain different types of ripper attachments with sketch.

Q.18. Suggest methods for increasing the scraper production.

Q.19. Write short notes on the following :

(*a*) Advantages of elevating scrapers.

(*b*) Cycle time for a scraper.

(*c*) Articulated type of loaders.

(*d*) Loader applications.

Q.20. What is an excavator? Describe working of a hydraulic excavator.

Q.21. Mention different types of excavators.

Q.22. What is an excavator-loader? Explain various applications of excavator-loader.

Q.23. Explain construction, working and application of hydraulic excavator.

Q.24. Differentiate between excavator and back-hoe.

Q.25. Discuss various applications (uses) of excavating equipmets.

Q.26. Discuss main components of hydraulic excavator.

Chapter

HAULING (TRANSPORTATION) EQUIPMENT

13.1 INTRODUCTION

Hauling uits transport the earth, aggregate, rock, ore, coal and other materials. Hauling units may be road vehicles or rail-road locomotives. Haulage mainly consists of trailers pulled by tractors or trucks. Trucks have been mobility and can haul various types of materials. But their performance and output is greatly hampered on rough roads and in bad weather. For better performance, greater road maintenance with lesser grade is required. Whereas rubber tyred tractors with separate wagons have better mobility and haulage speeds on less maintained rough roads and can carry heavier loads than that of trucks.

Crawler tractors with wagons are employed when roads are very rough and maintenance of roads is very costly. There are used for short hauls due to low speeds. Locomotives are suitable for long distance haulage on a fixed route, which is likely to be continued for a long period. These can be used in any weather and rail-road equipment require less maintenance.

The use of trucks provides a high degree of flexibility, as their number in service can usually be increased or decreased easily to permit modification in the total haulic capacity of a fleet.

Classification

Haulage trucks can be classified in many ways, including :

1. On-highway or off-highway.
2. The method of dumping the load—rear dump, bottom-dump, side-dump.
3. The type of frame—rigid frame or articulated frame.
4. The kind of drive—two-wheel, four-wheel or six-wheel.
5. The class of material hauled—earth, rock, coal or ore.
6. The capacity—gravimetric (tons) or volumetric (cubic meters).

On-highway vehicles are designed to be used on public highways, whereas off-highway vehicles are designed to be used on construction sites and are designed and manufactured to preclude their use on public roads and thus they may or may not comply with legal on highway limitations.

13.2 DUMP TRUCKS

Dump trucks are open vehicles which are capable of carrying and dumping earth, aggregate or other loose material to construction sites on various projects such as dams, highways, ports etc. Generally heavy-duty machines of more than 10 tonnes payload are intended for on-site haulage of construction or mine materials. Dump trucks in many sizes with pay loads ranging upto more than 300 tonnes are available. These may have two axle or three axle, and may be 4 × 2 (2 wheels driven out of 4); 4 × 4 (all 4 wheels driven out of 4); 6 × 4 (4 wheels driven out of 6).

Special features of dump trucks usually include : heavy, all-welded chassis to sustain rough ground conditions, extra large wheel to support load and give good ground clearance,ultra heavy duty tipping bodies constructed by high tensile steel to withstand damage from high impact and highly abrasive materials.

As mentioned earlier the dump trucks are of two types, namely on-highway and off-highway.

13.2.1. Highway Dump Trucks

For a construction project dump truck is a very common and essential equipment and is available in various sizes and types. These on-highway dump trucks are generally classified by the volumetric or tonnage payload capacity. Generally dump trucks of capacities ranging between 3 to 25 cu m are commonly used with a speed upto 80 km per hour. However dump trucks of more than 200 cu m (300 tonnes) capacity have also been manufactured.

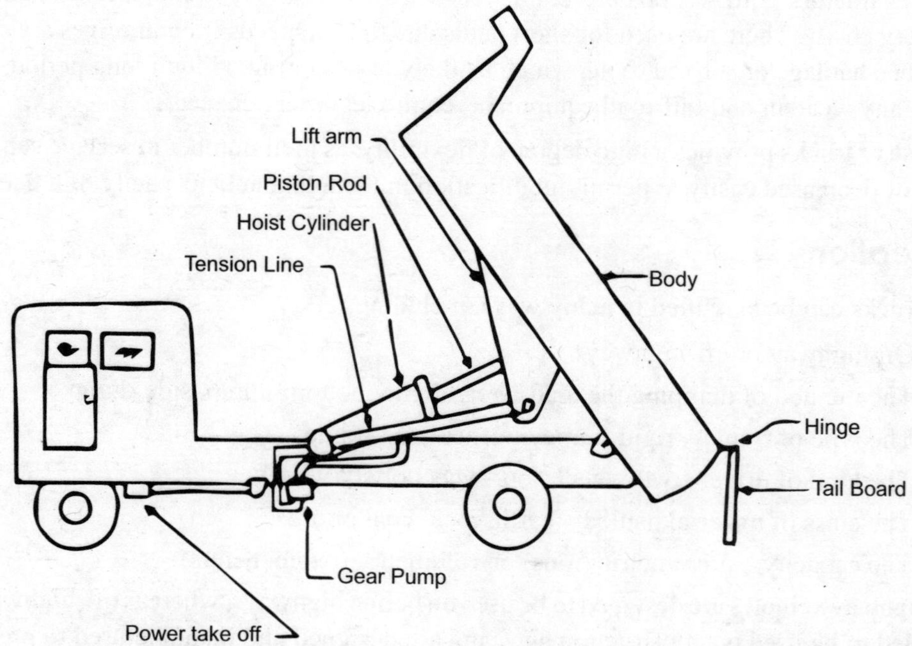

Fig. 13.1. On-highway dump truck.

On-highway dump trucks have chassis, prime mover and power transmission system similar to that of a truck, whereas the body of the dump truck is heavy duty type and is capable of tilting towards one side by means of hydraulic jacks and can be operated from the cabin. In on-highway dump trucks a tail gate is provided at the rear so that while moving on the highway the materials do not fell on the road. The tailgate opens when the body is doing the tipping operation for dumping the material. The speed with which on-highway dump truck travels is high and can be upto 80 km per hour or so depending upon the condition of the road.

13.2.2. Off-Highway Dump Trucks

Off-highway dump trucks do not have tail gates, therefore, the body floor slopes upward at a slight angle toward the rear. The floor shape perpendicular to the length of the body for some models is flat, while other models utilize a "V" bottom to reduce the shock of loading and to help center the load.

These are used off the highway specially in earth moving and mining applications. These are commonly known as dumpers. The body construction is specially designed to absorb high-impact loading. The floor of the body is of high-tensile steel to reduce abrasion damage. The ratio of gross vehicle weight to horsepower is generally lower than for an on-highway dump truck. Top speed of off-highway dump truck is also slow (30-40 km per hour). Since these are used on rough roads, oversize air cleaners and filters, shock proof mounting of all instruments, dust-proofing of electrical system, cushioned mounting for radiators etc. are provided.

Fig. 13.2. Off-highway dump truck.

To satisfy a particular set of job requirements, an off-highway dump truck can be selected out of various types available. Main types of off-highway dump trucks are :

(*i*) Rear dump trucks.

(*ii*) Bottom dump trucks.

(*iii*) Side dump trucks.

1. Rear Dump Trucks

These are very common and heavy duty trucks and are capable of handling even quarry rocks. The body of these trucks is hinged at the back and fitted with a hydraulic ram (multi stage jack or a telescopic hoist) on the underside to lift the front of the body and tilt it to the dumping position. In on-highway dump trucks tailgates are provided at the rear whereas in off-highway dump trucks the shape of the body at the rear is such that it facilitate dumping. Top front of the body is extended to form a cap over the cabin to protect it from hitting while loading. The body rest on the main truck chassis frame provided with a number of rubber cushions to absorb shocks while loading and travelling.

These are suitable to handle many types of materials, but are most suitable for quarry rocks. These are also used when :

(*a*) Dumping is in restricted hoppers or fill.

(*b*) Loading under a large shovel or dragline is subjected to sever loading impact.

(*c*) Maximum flexibility is required for hauling a variety of materials such as earth, sand and gravel and more bulky material such as blasted rock, ore, coal etc.

(*d*) Maximum gradeability is required.

In colder climates, where there are chances for freezing the material being transported, the dumpers are available to prevent material from freezing. In these dump trucks, exhaust gases from the engine are carried through ducts in the front slope sheat, in the bottom of the body, and on a portion of the side walls. The heated body can also help when handling wet, adhesive clay soils, hauxite and other sticky materials.

Fig. 13.3. Dump truck in unloading operation (Courtesy : Ashok Layland)

2. Bottom Dump Trucks

These are suitable for long hauls of easy flowing materials like sand, gravel, dry earth and which are to be spread to layers as on a fill of a dam. The material is discharged through bottom while the vehicle is moving at a controlled rate by means of 2 longitudinal gates. These are unsuitable for big size material or wet or sticky materials due to limited openings.

In bottom dump trucks the material can be discharged while the vehicle is in motion and laid in a long windrow. This way of dumping will not hamper the passage t£ other vehicles and also reduce the levelling effects of the dumped material.

These sometimes known as bottom dump wagons, are the semi-trailers, front end of which is supported on the rear of the hauling tractor, and the rear end is resting on their own wheels. Wide top area facilitates quick loading under shovels. Due to low long bodies they have fast speeds on suitably maintained roads, but longer turning radius is required and manoeuvering becomes more difficult in restricted spaces. With these it is difficult to negotiable steep grades.

Therefore, bottom dump trucks should be used, when :

(*a*) The material to be hauled is relatively free-flowing.
(*b*) The road is suitably maintained, thus permitting high speed travel.
(*c*) Long and steep grades are not existing to the way.
(*d*) The load is to be spread in windrows.

3. Side Dump Trucks

These dump trucks are most suitable when dumping is required to be done in a i long narrow length or on on one or both sides of a road (for dumping the material on road shoulders). These require very less time to dump its material, whereas the rear dump truck must stop, reverse and then dump is load which needs a considerably more time. Side dump trucks are available in self-propelled that is truck type models or in a fully trailer type models (which is commonly known as side dump tractor-trailer).

13.2.3. Articulated Dump Trucks

Articulated dump trucks are made on two or more separate machine elements joined together to permit mandeuverability. It permits large machines to be separated into elements for transportation and increases operating efficiency. Now-a-days large dump trucks are gaining popularity because of their advantage in manoeuverability through articulation.

The articulated dump trucks are specially designed to operate in confined working locations where a rigid-frame trucks have problems. An articulated joint and oscillating ring between tractor and the dump body enable to turn the vehicle in limited space. Articulated trucks can climb on steeper grades (upto about 35°) than right-frame rucks/dumpers.

Fig. 13.4. Articulated dump-truck.

13.3 DUMP TRUCK OPERATIONS

For transportation of material by a dump truck, following operations are involved :

(*i*) Loading the dump truck.

(*ii*) Hauling.

(*iii*) Dumping the material at desired place.

Fig. 13.5. (*a*) On-highway dump truck.

1. Loading

The dump trucks are loaded by draglines, shovels, or by wheel loaders. The dump truck or dumper driver, for best efficiency and more production, should keep it in most efficient working area of the shovel or dragline or wheel loader. Less the swing and lance moved by the shovel boom between the cut (from mine quarry) and the truck body the more amount of loading will be done by the shovels or a loader in a given amount of time.

After loading the truck, shovel operator signals the truck driver to move his vehicle from under the shovel. The driver then starts his vehicle for the haul road and reach the higher practicable gear range as soon as possible. The driver should take owing precautions for his own and vehicles safely:

(*a*) Stay in the cabin during loading of the truck.

(*b*) Drive carefully in the working area and avoid hitting sharp rocks and shovel overflow.

(*c*) Do not move the truck until the shovel operators gives a signal.

2. Hauling

Drive the vehicle as per ground condition and the condition of the road, and use highest gear adjusting the road condition. Driver should ensure precautions for the safety of his own and for long life of the vehicle.

Fig. 13.5. (*b*) Off-highway dumper.

3. Dumping

Dumping operation depends upon the type of work where dumping is required to be done. For instance, for dumping at exact point, a second person, known as 'spotter' is required to control the dumping work, who directs the rear dump to its dumping area. For dumping, the ore or limestone etc. into a hopper of crusher, marks should be made for the guidance of the driver so that he can dump the material into the hopper easily. In order to prevent any cut or abrasion in the tyres, he should prevent hitting the protecting rail at the hopper. When dumping is done to open area or in earth fill dam the load is usually dumped by the operator alone and picks his own dumping place. For earth fill dams the dumped loads are moved by the crawler tractors to the desired position. In order to save the tractor bulldozer efforts, load should be dumped as close as possible to the proceeding load.

13.4 SELECTION OF DUMP TRUCKS

Selection of type and size of a dump truck is very important task and should be carefully decided considering various factors. Some of the important factors are :

(*a*) **Nature of Work :** For earth hauling purposes rear, front or bottom dump trucks are generally used. The rear and front dump units are self-powered and are employed for loading by large shovels or draglines. The front dump units are used where space is restricted. The bottom dumps are generally semi-trailer type supported on rear wheels, the front of which is carried on the rear of the hauling tractor. For the transportation of rocks and other heavy materials under adverse road conditions, the rear dump trucks are most suitable. Bottom dump trucks are suitable only for free flowing materials and are used to spread the material on the fill.1 The side dump units are suitable for transporting and dumping the large quantities of material on the one or both sides of the road.

(*b*) **Nature of Haul:** On steep grades, the self-propelled rear dump tracks are selected as the uniform load distribution causes greater weight to be placed. Whereas, in the bottom dump units, which are semi-trailer type, while moving upgrade causes concentration of load towards the rear. Thus these bottom dump units can be limited to the level ground.

(*c*) **Matching with other Equipments :** Capacities of the hauling should be decided after considering the size of excavator or loading unit. Generally, as a thumb rule size of the hauling unit is selected between 4 to 6 times the bucket capacity of the loading unit. The number of hauling units is decided by time required in one cycle, *i.e*., loading, hauling, unloading and returning back. The care should be taken that the waiting time for a truck by a loading unit or by a track for its turn for loading is kept minimum.

(*d*) **On-Highway or Off-Highway :** As explained earlier dump tracks are of two types, on-highway and off-highway. For the selection of a dump unit it is necessary to analyse the requirements. If an on-highway type unit is applied to a ragged off-highway situations or vice-versa, it will be too expensive and progress of work will be adversely affected. Generally, on-highway units are suitable for long, high speed hauls over roads with low rolling resistance. Loading in these vehicles be done, by relatively small capacity power shovel or a loader. Whereas off-highway dump tracks also known as dumpers are used for short, low speed hauls over rough roads. They are capable of moving on steep grades, and has high power engines often fitted with turbocharger and torque converters. They can work in dusty atmosphere of quarries, crashers etc. as they are fitted with oversize air clearness and filters. These dumpers have body suitable for absorbing high impact leading and generally fitted with steel wear resistant plates to reduce abrasion damage.

The section of proper hauling unit should be decided on the basis of proper estimation, various proposals should be seen based on economical considerations. It has been seen on the basis of experience that, for short haul excavation upto 100 metres, crawler dozers are most economical. For longer haul where construction of haul roads are not practicable, crawler scrapers are economical. For still longer hauls, dump trucks on or off-highway types are economically employed.

From above discussions, it is clear that there is no straight formula to determine which equipment and which method is the best. In practice based on these considerations, an entire job economy has to be studied from various aspects for all possible alternatives. Most economical, feasible and efficient system under the required job conditions should then be adopted. Since this selected is most important decision which will effect overall cost efficiency and time requirement of the project, an utmost care and attention should be given to this decision.

Balancing of Hauling units with the Size of Excavator

The capacities of hauling units should balance with the output of the excavator; power shovels, draglines or loaders. If this is not done, either the hauling units or excavator will remain idle, and the combined cost of excavating, loading and hauling of the material shall be higher than when balanced units are used. For an excavator loading the earth or material into the truck, the size of the trucks must be carefully decided considering the following factors.

Factors in Favour of Large Size Trucks

(*i*) Lesser number of trucks required, which will result in less total investment in hauling units and lesser cost of maintenance and repairs.

(*ii*) Less number of drivers, thus reduction on operation cost.

(*iii*) More excavator output due to lesser time required in spotting the vehicle and ease in loading.

Factors in Favour of Small Size Trucks

(*i*) More flexible in manoeuvering, hence more advantageous on short hauls.

(*ii*) May have higher speeds.

(*iii*) Lesser loss of production when one truck in a fleer breaks down.

(*iv*) Easy is balance the number of trucks with the output of the excavator.

(*v*) Cost of maintenance of haul roads is less when small trucks are used, as heavier loads may cause more damages to the haul roads.

(*vi*) Easy to repair.

13.5 COMPARISON BETWEEN VARIOUS TYPES OF DUMP TRUCKS

Type		*Advantages*	*Disadvantages*
Articulated dump trucks	1.	Can be used in most climates and terrains round the your.	More expensive
	2.	Able to cope with extreme grades and difficult ground conditions.	Top speeds are generally lower.

	3.	Due to lesser heights, use of variety of loading equipment is possible.	Not usually roadable.
	4.	High manoeuverability.	
Rigid frame off-highway dump trucks	1.	Less expensive than articulated dump trucks.	High axle load require good quality haul routes.
	2.	Generally higher top speed especially on good quality haul roads.	Downtime increases in wet ground and difficult terrain.
	3.	Long economic life under moderate condition.	Not usually roadable.
On-highway dump trucks	1.	Lever purchase price.	Net suitable for continuous work in rugged off road conditions.
	2.	Roadable.	Poor traction in wet grounds.
	3.	High operating speeds.	Restricted payloads.

13.6 TRAILERS

The two types of trailer are used namely, full trailers and half (Semi-trailers). Full trailers, also known as wagons are self supported and are provided with a swivelling axle and a draw bar for connecting the prime-mover.

Semi-trailers are supported on their own wheels only at the rear, while front end is supported on the prime-mover. The semi-trailers have more manoeuverability as compared to full trailers. The front end of the semi-trailers has a goose neck resting on the rear of the prime mover (driving unit). The semi-trailers can be turned and operated to a more limited space than that needed for a truck or for a full trailers. This can turn upto 90° either to the right or to the left without moving forward. These are suitable on narrow-roads and where there are no steep grades.

Semi-trailers are usually attached to the hauling unit through an attachment which operates through an hitch. The hitch has two portions one mounted on the rear of the hauling unit, and the other under the goose neck of the trailers. On joining them together, a necessary coupling is provided. It provides a large surface area of contact between the pulling and the trailing units resulting in a good load distribution at the connecting end.

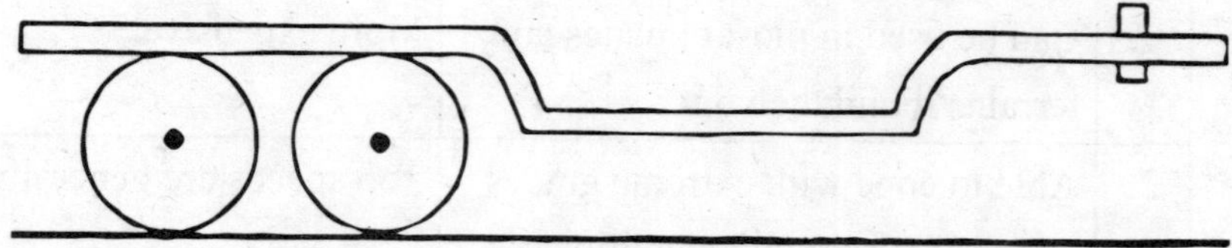

Fig. 13.6. Drop frame type semi-trailer.

The braking system of a trailer or semi-trailer may be operated by either hydraulic or by vacuum or by air pressure. Trailer brakes are synchronised with that of hauling unit brakes. A provision is also made to apply trailer brakes automatically when its connection of brakes to from hauling unit.

Trailers for haulage of heavy equipment and machinery may have drop deck or level deck.

In drop deck (or drop frame) trailer loading and unloading is done from sides. These have lesser road clearance to allow more head clearance for tail machines. Level deck or flat bed trailers have abundant space on the deck, and it is easy to load and unload them from the rear. In order to keep the vehicle stable and safe during haulage, the height of the trailer should be such that centre of gravity of the loaded trailer remains near the ground.

Some times the machine mounted on the main frame work (girders) can transported directly without the deck to reduce the height of loaded trailers.

The platforms of the trailers can also be made tilting type with hydraulic system for easy loading and unloading. The trailers are tandem axle (two axle) arrangement with 8 wheels, and 3-axle arrangement with 12 wheels for haulage of heavy loads.

QUESTIONS

Q.1. Name different types of dump trucks indicating their applications.

Q.2. What is a truck? Give the classification of hauling equipment, mentioning their applications.

Q.3. Explain rigid frame type trucks.

Q.4. What is a dump truck, where these are used?

Q.5. Explain on-highway and off-highway rear dump trucks.

Q.6. What is an articulated dump truck? Where are these used? Explain articulated rear dump trucks.

Q.7. Explain in detail about bottom dump wagons.

Q.8. Explain major activities involved in the dump truck operation.

Q.9. Explain different criteria for selecting the dump trucks.

Q.10. Write short notes on:

(*a*) Factors in favour of large trucks.

(*b*) Balancing of hauling units with the size of excavator.

(*c*) Haulage tractors.

Q.11. Explain, what do you understand by full trailor and semsssi-trailor.

Chapter

CONCRETING EQUIPMENT

14.1 INTRODUCTION

Concrete is basically cement, aggregate, and water mixed together and then deposited and permitted to solidify. Sometimes admixtures are used for various purpose such as to improve workability, accelerate setting, to produce desired colour. Operation involved in concrete production vary with the type of work, but generally involves; batching and mixing handling and transportation, placing, compacting, finishing and curing.

Various types of concrete batching and mixing plants are available to suit various requirement. These may be operated manually semi or fully automatic. These are either static. fully mobile or transportable and are available upto 300 cm m per hour capacity.

The concrete plants consists of all arrangements for :

(*a*) receiving all ingredients of concrete, *viz*.: aggregates and sand from storages bins, cements from soils and from tank;

(*b*) weighing each ingredient for each batch of the mix; and

(*c*) mixing these ingredient thoroughly to from a concrete of required specification.

For small jobs, portable concrete mixers are used which are fed with measured quantities of each ingredient either manually or through hoppers attached to the mixers. For large jobs, automatic, and mixed product is then handled by mechanical means. For carrying out the works at certain distance, dry mix batching plants are used, which supply dry batch to the truck mixers; the mixing is then done in these truck mixers while travelling to the site of works.

14.2 CONCRETE MIXERS

Concrete mixers are used to mix the ingredients of concrete. Efficiency and quality of concrete depends largely on duration of mixing, speed of the mixing drum, timely supply of water, correct setting of the machine, proper cleaning of the mixing blades and of inner surface of the drum. These mixers are available in various sizes and are specified by the volume of mixed concrete produced after mixing of each batch.

Sometimes the mixers are specified by two quantities. Total volume of unmixed ingredients is given first cu m or lit, whereas volume of mixed concrete is given afterwards in the same units, e.g,

10/7 mixed takes 10 cu ft of unmixed material and gives 7 cu ft of mixed concrete in each batch; 285/200 litres takes 285 litres unmixed material and gives 200 litres of mixed concrete in a batch.

The concrete mixers are made either stationary or portable. Portable mixers are common in construction industry and are mounted on either metal or tubber tyred wheeled trollies and has one drawbar for towing. The power to drive the concrete mixer is obtained from a diesel engine of from an electric motor. Diesel engines are generally used due to its reliability and problem of availability of electric power supply.

Following **types of concrete mixers** are generally used for construction works.

1. Hand Fed Tilting Drum Mixers

These are the smaller capacity mixers. Aggregate, cement and water are fed into the drum by hand. The revolving drum with its mixing blades gives a lifting, combing and agitation to mix the concrete.

Fig. 14.1. Hand fed tilting drum concrete mixer.

The drum with cast base and steel plate body rotates on bearings through gear mechanism. The inside of the drum is generally fitted with renewable liners. A set of blades provided in the drum gives continuous agitation of the material while mixing. It is fed by hand through buckets directly into the drum.

2. Loader Fed Tilting Drum Mixers

These mixers are of large capacity as compared to the hand fed tilting drum mixers and are fitted with a loading hopper operated by a wire rope for feeding the aggregate and cement into the drum. The hopper are provided with brake to stop and hold skip automatically. Water is supplied to the drum from a measuring tank operated by hand lever provided above to discharge water directly into the drum. The principle of mixing is same as that of hand fed mixer. The mixing drum is supported on solid cast iron yoke and can be tilted on either side. Mixers are mounted on study chassis with either 3 pneumatic or metallic wheels. A draw bar with turn table is provided on front axle.

Fig. 14.2. Loader fed tilting drum 10/7 concrete mixer.

3. Reversing Drum Mixers

This type of mixer is used on comparatively larger works. The mixing drum in in horizontal positions and is non-tilting type. It has two sets of blades, one blade mixes the mix while drum is rotated in one direction, the other set of blades dischargers the mix when the drum rotation is reversed. The drum is fed by a loading hopper. The hopper commonly mounted on four rollers moves on channel run-ways. The hoppers is lifted upward by wire-rope winding and unwinding on operated drum. Discharge gate of hopper opens through a lever when it reaches its discharge position. Safety arrangement is provided to prevent over running of hopper. Water storage tanks, measuring tanks and pumps etc. are provided to supply desired quantity of water at desired time. These mixers are driven by diesel engines or electric motors through reductions gears and reversing gears. Like other mixers these are also mounted on a chassis with 2 axle having 4 pneumatic wheels with draw bar and turn table arrangements.

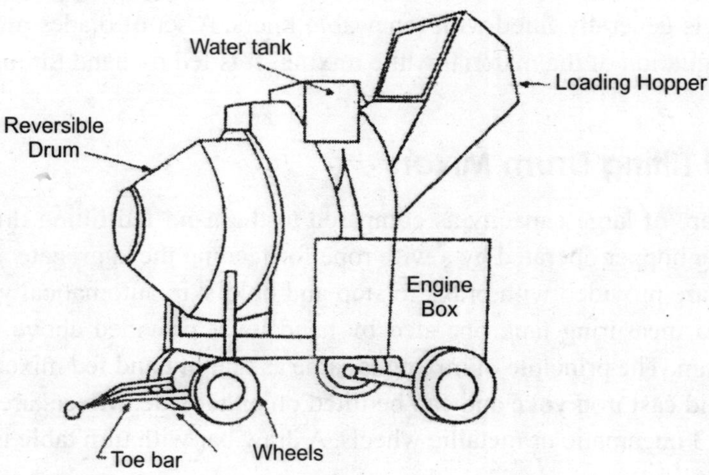

Fig. 14.3. Reversing drum mixer.

4. Roller-Pan Mixers

These mixers are primarily used fo morter. The rollers and mixing blades rotates in a pan and not only mix the material but also produce a kneading and crushing action. The mix is discharges through a gate in the floor of the pan. The drive to the rollers and mixing blades provided by either diesel engine or electric motor (generally diesel engine) through the bevel gears.

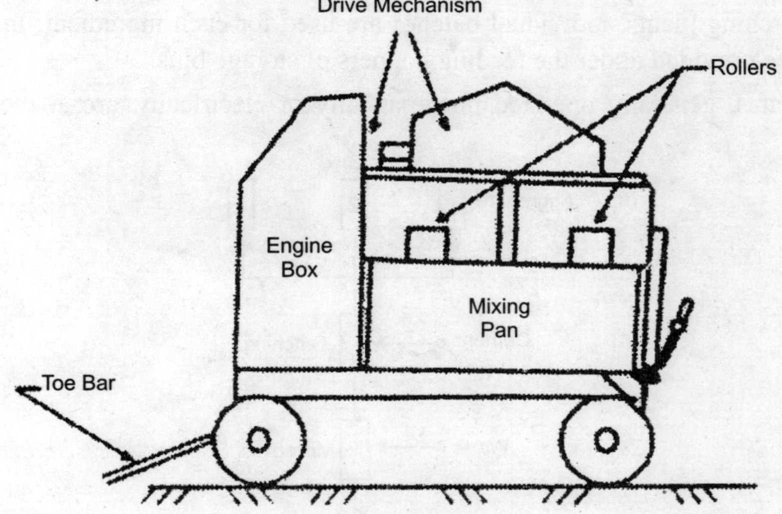

Fig. 14.4. Roller pan mixer.

14.3 CONCRETING

Concreting consist of the following activities :

(*i*) Proportioning and Batching
(*ii*) Mixing
(*iii*) Transporting
(*iv*) Placing
(*v*) Compacting
(*vi*) Curing

A flow diagram for concreting operation is shown in Fig. 4.5.

Concreting Design

Aggregate, cement, and water plus desired admixtures are mixed in proper proportion to produce as concrete having desired properties. Determination of this proportion in known as design of concrete mixture. The aggregates are joined together into a solid body by the solidification of a paste made of cement and water. The function of water is to hydraulic the cement and water. The function of water is to hydrate the cement into side mass and give sufficient plasticity to permit it to be formed into the desired shape. Excess water dilutes the paste and weakens the concrete, while insufficient water not be able to make paste to join all the particles of aggregate together to solidify.

Weigh Batchers

Ingredients of concrete *viz.*, coarse aggregates, sand, cement and water for a particular mix of concrete weighed separately for each batch in the weight batcher.

For small concrete mixers, the aggregates, sand, and cement are weight one after the other cumulatively in the same weigh batcher and dumped into the feeding hopper of the mixer.

For large batching plants, individual batches are used for each ingredient. In such pants the weigh buckets are provided under the feeding hoppers of storage bins.

Discharge gates, generally operated pneumatically or electrically, are at the bottom of the buckets.

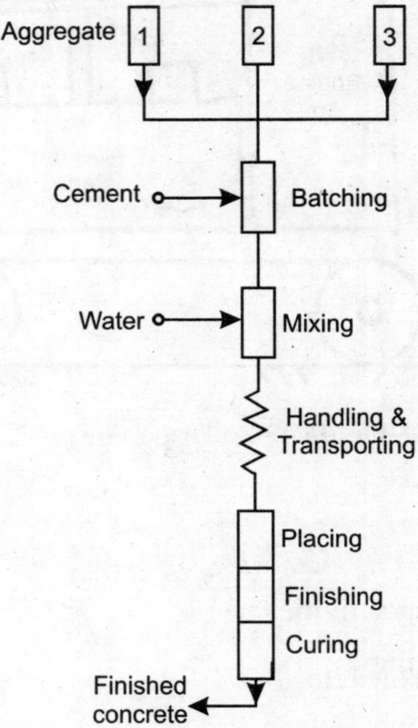

Fig. 14.5. Flow diagram for concreting.

Measurement of Water

Accurately measured water is required to be supplied to the mixing drum for each batch of concrete. A water tank with measuring arrangement, which can hold a required quantity of water for batch and arrangement to deliver it is required time is provided to the next batch. The filling of water in the tank should take place by the time the concrete is mixed.

Water measuring tanks are mainly of following two types :

(*a*) Horizontal types

(*b*) Vertical type

(*a*) Horizontal Water Measuring Tanks : These tanks deliver the water to the mixer through a syphon, which is adjustable to regulated the discharge. Water level indicator is provided to indicate level of water in the tank. After each discharge the tank is filled by operating a 3-way valve by means of a handle. Main draw back of this type of tank is that when the tank is not horizontal, the discharge quoted will not be correct.

(*b*) **Vertical Water Measuring Tanks :** These has a central syphon discharging vertical down through the pipe connected to the mixer drum.

This avoid the problem of incorrect discharge when the mixer is not standing on a horizontal in inclined water discharge when the lever is pressed.

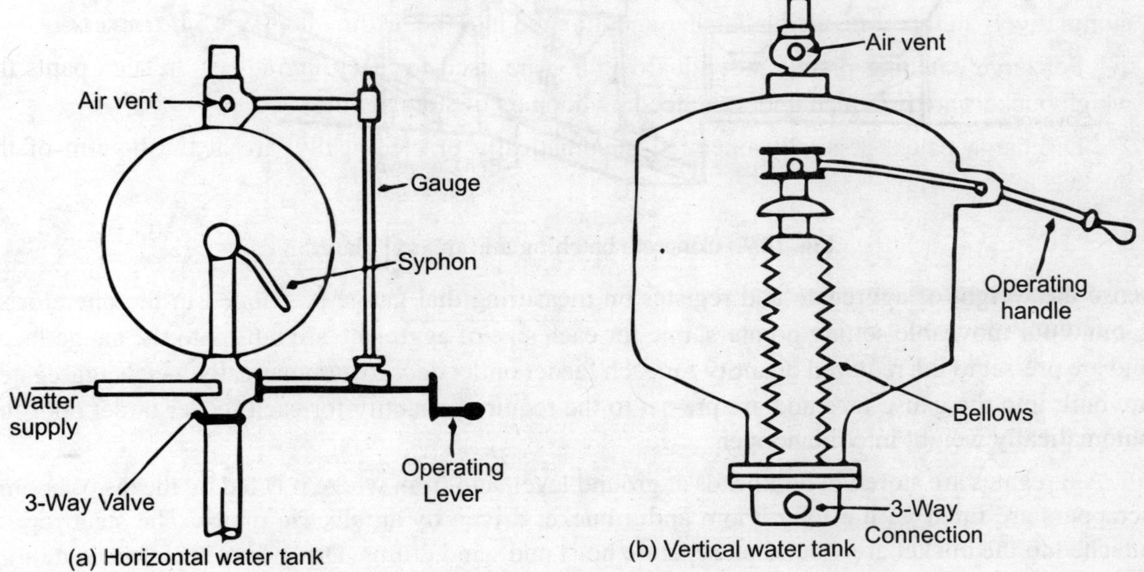

Fig. 14.6.

14.4 CONCRETE BATCHING AND MIXING PLANT

As the name suggest the plant has main function of 'Bathing' and then 'Mixing'. Batching means proportioning of ingredients of concrete, *viz.* : aggregates, sand, cement and water separately for each batch. Batching and mixing plants used in construction field generally consists of the following main constituents :

1. Aggregate Feeders

Separate aggregate feed bins are used each size of the aggregate and sand, and controls their quantity. Weighing system is used for important jobs, since control by volume is not perfect as it varies due to closeness of packing of the material.

The aggregate feed bins are loaded by any of the following methods :

(*a*) By shovels directly into the bins.

(*b*) By lorries tipping directly into the bins.

Each of these aggregate feed bins fitted with pneumatically operated outlet door. The bin for said is provided with an electric an electric vibrator to ensure free flow through the outlet door. Aggregate weighing is done by means of a hydraulic capsule located in the weigh frame which

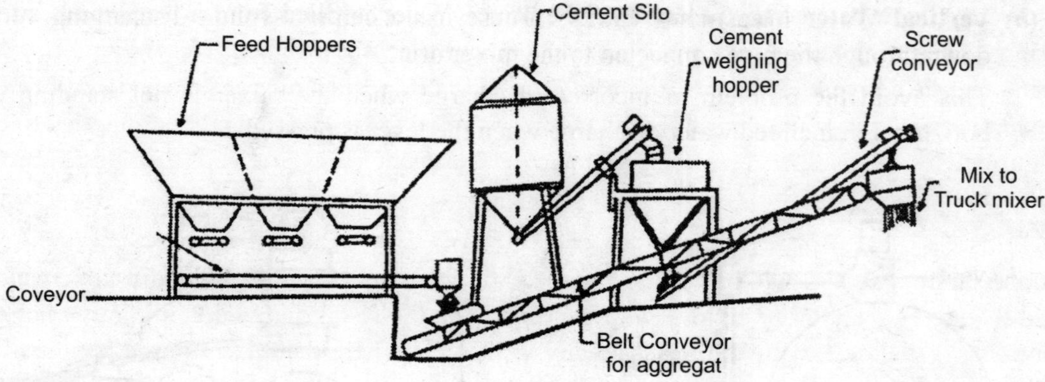

Fig. 14.7. Concrete batching and mixing plant.

sense the weight of aggregate and register on measuring dial gauge positioned in the operator's cabin. Four moveable setting pointers, one for each size of aggregate are built into the gauge face and are pre-set to the required quantity for each feeder outlet door to automatically weigh aggregate are built into the gause face and are pre-set to the required quantity for each feeder outlet door to automatically weight into the hopper.

Aggregates are stored in bulkheads at ground level, and from where it is fed by means of boom scrappers are fitted with a long boom and a bucket driven by an electric motor. The steel ropes attached to the bucket are wound on separate hoist and wind drums. Drive from the electric motor to appropriate drum is engage by separate electro-magnetic, multi-plate clutches. Foot operated brakes are provided to prevent over-run.

2. Cement Silo

Cement is stored ins silo which is filled from a cement carrier pumping the cement into the silo under pressure. Weight batching is provided for measuring the quantity of cement to be fed for each batch mix. This weighed cement is carried to the mixing unit through the conveyor belts which are totally enclosed to avoid any loss.

During filling under pressure by bulk tanker, cement laden air is vented to atmosphere through a filter. Cement/air filters are generally, 'multi sock' type having number of easy-to-renew nylon socks giving mixing filteration area. These nylon socks trap cement. Access door allows inspection and cleaning when necessary. Dust from the filters falls back into the silo when the shaker deviceis operated level control.

Cement enters the silo through a valve provided on an inlet pipe connected either to a screw conveyor or rotary blower pressurising the pipe line, carrying cement, at about 0.4 bar. These are used to inject the cement received from either of the two source from a bag splitting hopper, or from standing bulk cement carrier. The bag splitting hoper is of steel construction and is fitted with a bag splitting knife to cut the bag and a grill which allows cement to fall through into the hopper whilst trapping the bag and other destruction.

3. Water Measuring Device

Since concrete production required exact quantity of water, a reliable and accurate water measuring device is required in the plants. Measured correct quantity of water is fed during the period of charging a batch into the drum of the mixer. Water is filled from a water tank into a measuring tank during the period when mixing takes place in the mixer. This water from the measuring tank is delivered to the mixer during the next batch when aggregate, sand and cement is charged into the drum of the mixer. Thus after each charge the measuring tank is supplied to the mixer, through the adjustable spray bar so as to achieve homogeneous mix within a shortest time.

4. Mixing Unit

Mixing unit consists of two parallel shafts made of cast steel provided with adjustable paddle mounted to external supports provided with bearings. The mixer discharge the mix into a hopper for delivering it to the concrete dumpers directly or into bin where the mix is stored temporarily. The mixer is taken to the construction site. The walls of mixing units are made of wear resistant steel.

Sometimes pan-mixers are used in place of paddle mixers, which uses mixing blades for uniform mixing. Split drum mixers are also becoming common in batching plants.

5. Batching and Mixing Controls

Central console housed in operator's cabin has necessary electrical equipment to allow the following three methods of operation:

(*a*) Manual operation of the individual push-button enable production of single batch or test mixes.

(*b*) Automatic weighing, loading, mixing and discharge into the storage hopper of a single batch using only one push-button.

(*c*) Fully automatic operation production successive or any predetermined number of batches unassisted by the operator.

6. Washers

We have already discussed about washing of aggregates in a barrel in the chapter on "aggregate production" here we are discussing other methods of washing the aggregates and sand.

Washing of aggregate or sand can be done of following 3 ways :

1. Screw washer
2. Spray washers
3. Rotary washer or scrubbers.

1. **Screw Washers :** Rotation of a screw washes the fine material like sand and conveys it up inclined rectangular tub out of water. Thus these washers are used for washing, classifying and dewatering of fine aggregate or sand.
2. **Spray Washers :** In these washers, spray bars are provided over the vibrating screen. Water is sprayed at pressure over the aggregate as it passes over the screen and shakes off large portion of loose and impurities due to vibration action of screen.
3. **Rotary Washers or Scrubbers :** These are most effective when impurities are such cannot be easily removed by above two methods. Vigorous scrubbing in water is an efficient and economical method of cleaning sand and gravel. Processing takes place in a

rotating horizontal drum fitted internally with lifting flights. The deep bed of material in water becomes a churning mass inside the rotating drum, creating an intense grind which scrubs away dirt, clay and other adhesive from the aggregate. Caked deposits are broken down by impact as material cascades from lifters high in the drum. A constant level of material maintained in the drum maximises scrubbing action.

This system of washing have been discussed in detail in the chapter on "Aggregate Production".

The drum is lined with wear resisting rubber or steel plates. Rubber lining is highly resistant to abrasion, corrosion and noise. Scrubber drum is driver by an electric motor through a reduction gear box. Relatively low speed of the dump develops maximum scrubbing action. Scrubbed material passed from the drum for screening is simultaneously rinsed by high pressure water sprays, giving exceptionally clean product. Time spent in the drum in dependent on the feed rate. Very dirty material requiring longer treatment is fed at a slower rate compatible with condition. When the material is very dirty, a vibrating screen with water sprays removes a proportion of dirt from the feed prior to scrubbing.

14.5 TRANSPORTING THE CONCRETE

Concrete may be handled and transported by several methods, such as truck mixers, concrete dumpers, concrete, buckets handled by cranes, hoisting towers or cables, chutes, and belt conveyors. The method to be adopted is selected depending upon the condition of work.

Selection of the method of transportation of concrete should be done in such a way that the properties of the concrete are maintained and following objections are achieved :

Selection of the method of transportation of concrete should be done in such a way that properties of the concrete are maintained and following objections are achieved :

(*a*) Economy in transport.

(*b*) Segregation of materials of concrete is prevented.

(*c*) Concrete should be finally placed at the site before it starts setting.

Transfer of Concrete

Concrete can be transferred central batching and mixing plant to the work site of mass concreting by following ways. The method can be selected considering the size of job, height and other condition of the worksite.

1. **By Buckets :** Concrete is discharged by mixers directly into the buckets which are then either (*i*) hoisted to the place of dumping by cranes, or (*ii*) hauled over flat bed carriers to a place of work,or (*iii*) combination of these two, or (*iv*) by cableways.
2. **By Dump Trucks :** In this method, concrete is discharged directly into the dump truck body, which is then taken to the site of work. The trucks either discharge the concrete directly to the work place or into the buckets handled by cranes or cable ways.
3. **By Belt Conveyors :** In this system, belt conveyors are employed for dry mixes. Deep troughing idlers are used to prevent spillage and segregation. Concrete is first poured

into buckets from conveyors and then taken to the place of work as explained in previous paragraphs.

4. **By Monorail :** Transportation of concrete for a moderate distance in large quantities can be handled by a especially constructed single truck rail.

5. **Chutes :** These are made of metal with sound bottom. These are given a slope so as to have a uniform speed of flow of all materials of concrete in order to prevent segregation. This method is used only for limited distances.

6. **Truck-Mixers :** Concrete mixed batch is transported from the batching and mixing plant to the construction site. Fully or partly mixed concrete is transported through the truck mixers. Where the construction site is far from the bacthing plant, entire mixing operation is carried out in a transit truck mixer during the journey.

For a transporting the concrete mix following means are adopted.

(*a*) Transit truck mixer (*b*) Truck agitators

(*a*) Transit Truck Mixers : Dry mix material are fed into the truck mixers from a dry batching plant, after measuring the quantities of various constituents other than water, *i.e.*, aggregate, sand and cement. The transit mixers then transport it to the work site and mixes during the journey using its own water supply. These transit mixers are provided with a suitable water system having a sufficiently large water tank, meter and a pump for supply water for mixing and for cleaning the drum.

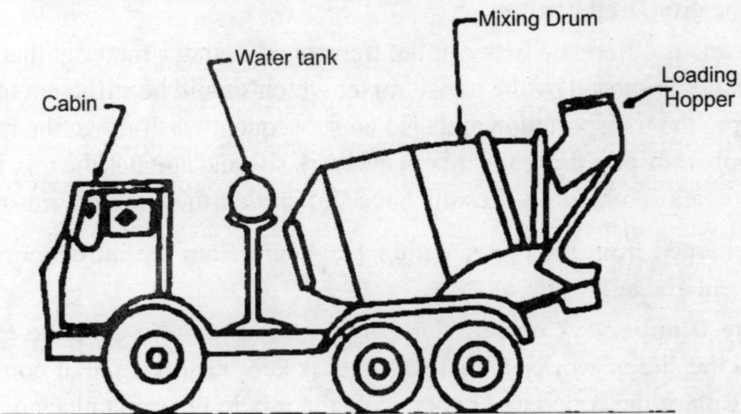

Fig. 14.8. Concrete truck mixer.

The mixing drum is supported on a truck chassis in such a way that it can revolve and also can be raised or lowered at one end. The size and capacity of the truck chassis is selected depending upon the capacity of the mixing drum, which is varying from 4.0 cu m to 12 cu m. For smaller capacity 2 axle chassis and for larger capacity 3 axle (1 front and 2 rear) chassis are preferred.

The mixers have mixing drum with mixing blades made of abrasion resistant and corrosion proof steel. Whereas feed chute and discharge chute are lined with wearing plates.

The drive for mixing blades is taken from the truck power take off (PTO). The mixing blades mix the material when it is rotating in one direction and discharge the material while rotating in other direction.

Fig. 14.9. Concrete truck mixer.

(*b*) **Truck Agitators :** Truck agitators are used solely to transport ready mixed concrete prepared by batching and mixing plants. The fully mixed concrete is kept in plastic state and saved from segregation or drying out or setting through continuous agitation of the drum of the mixer.

The main difference between the transit and agitator mixer is that the large volume of water is carried by the transit mixer which should be sufficient to mix the concrete during the transportation and also no subsequent wash-down the mixer. The agitator simply transport the ready mix to the work site and agitates the mix in order to prevent segregation or setting. Results have shown that the concrete must be delivered and discharged from the mixer within $1\frac{1}{2}$ hours from the introduction of water to the cement and aggregate.

7. **Concrete Dumpers :** Concrete dumpers are used to carry concrete mix from concrete mixer to the site of works. Concrete mixer is kept near the site at convenient place and from this place the concrete dumpers take the mix to the exact place of work. Thus these are used for short leads only. These dumpers may be two wheel or four-wheel drive propelled by a diesel engine and has low centre of gravity for stability even in adverse site conditions. It has a skip of various capacities generally between 750 kg to 2000 kg capable of unloading the mix either in the front or in the sides.

14.6 CONCRETE PUMPS

Concrete is either mixed at the site itself or transported into truck mixer for placing into required position. This placing of concrete is traditionally done the cranes, skips, dumpers, or wheel barrows. Now-a-days concrete pumps (also known as pumpcrete) are used for placing the continuous flow directly from the truck mixer on the column or section of the building, tunnels, subways, underwater works etc.

The concrete pump have a boom for placing the concrete in position. The booms of various sizes are available as per requirement. The pumps capable of handling volumes upto 45 cubic metres per hour and giving vertical lifts upto 50 m and for horizontal distance upto 300 m are generally available. Concrete pumps require a good grading or materials in the concrete mix. Pumps with a maximum output of 150 m^3/hr and with maximum pressure of 200 bar are available in the market.

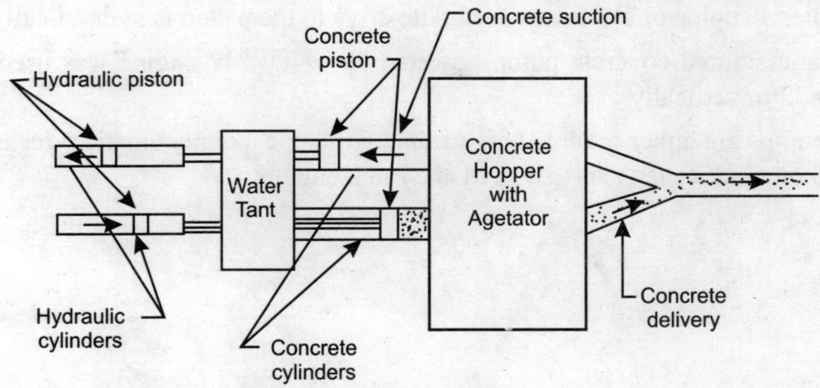

(a) Concrete pump with dual cylinders

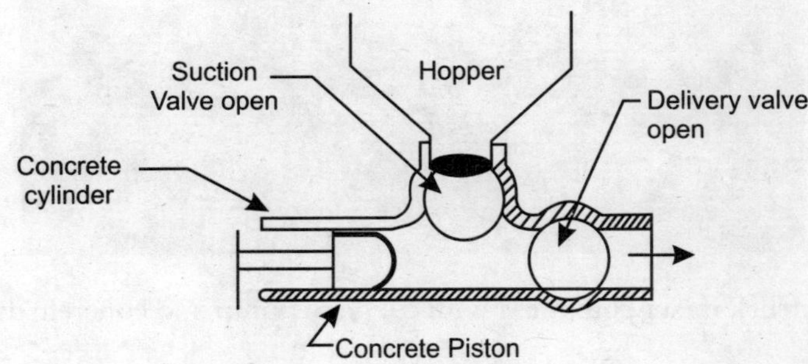

(b) Operation principle of pumpcrete

Fig. 14.10.

In the concrete pump, pumping is done by means of 2 parallel long stroke cylinder system. These pumps are basically reciprocating pumps having two cylinders so as to maintain continuity. Pumping cylinder dia go from 150 upto 230 mm and stroke from 500 upto 2000 mm. Each cylinder is hardened and connected to distribution. Each cylinder system has two valves one on the suction side located at the bottom of the hopper and the other on the discharge sides, controlling the concrete entering to the distribution pipe. The opening and closing of the valve is timed with respect to the movement of the piston, *i.e.*, in suction stroke when piston is moving away from the hopper side the concrete enters by gravity, suction valve is opened and discharge valve opens and the piston forces concrete into distribution pipe. The two cylinders are timed in such a way so that when cylinder has suction another has discharge stroke. Thus the flow of concrete from pump to work place remain continuous. An overhead hopper is provided with agitator bladed so as to enable feed of homogeneous mix of concrete into pump suction.

During pumping care should be taken that :

(*i*) all joins should be tightened.

(*ii*) concrete level in the hopper is never allowed to fall below the neck level.

(*iii*) at the time of starting nearly 0.5 cu m of grount should be fed to the pump, as cement mortar works as lubricant in the pump.

(*iv*) at the end of day's work, inside of the pump and the line should be cleaned.

Concrete pumps are either powered by the engine of the trucks on which they are mounted or by a separate diesel engine or electric motors. The drive to the piston is hydraulically controlled.

Largest manufactured concrete pump powered by a 400 kW engine was used for pumping concrete upto 432 m vertically.

Concrete pumps are either mounted on a frame with axle and pneumatic tyres and are towed, or are mounted on truck chassis and given by the truck engine.

Fig. 14.11. ELBA truck mixer combined with concrete pump and concrete distributor boom.

For selecting a pump, engineer should known that maximum output, Q_{max} and also the maximum pressure, p_{max}. Q_{max} and p_{max} can not be achieved simultaneously, asin this case required power would be just too much. At a certain power P the production is Q depending on the displacing pressure p which again depends on the resistance.

$$Q = \frac{P}{p}$$

or

$$P = Q \times p = \text{Constant.}$$

Therefore at a certain and fully utilised power, the production of a pump will decrease when the pressure increases.

As stated earlier pressure p, depends on the resistance, which ultimately is depends on following:

(*a*) Length of pipe line and vertical head.

(*b*) Stiffness and richness of mix.

14.6.1. Pumping the Concrete

By using the different type of pumps, concreting can be placed in different circumstances. The pump configurations suitable in different circumstances are discussed hereunder :

1. **Pump with a Pipe :** In this configuration pipe line is assembled and connected to the pump before the pumping operation begins. The pipeline is laid from the pump location to the concrete casting area, where the pump is located such that the ready-mixed concrete trucks have the easy access to it.

   Where the concrete is placed in more than one locations, several pipelines has a flexible rubber pipe of about 3 to 10 m length connected to it for better control of the concrete discharge location and for earlier handling.

2. **Pump with Boom :** In this configuration the pump is mounted in a truck and equipment with a evolving boom to which a fixed-length delivery line is connected. Similar to the earlier configuration, here too the free end of the line has a flexible pipe (hose) of about 3 m length is connected to it. In these, hydraulically operated and articulated booms of 25 to 40 m length are commonly used. In the truck-mounted pumps these are generally available in three to five sections.

3. **Pump with Pipeline and Tower-Mounted Boom :** This configuration combines the elements of both the earlier systems. In this system, the stationary/trailer pump is at the concrete-loading end, a separate boom at the concrete-placing end, and a pipeling running from the pump to the boom. The high rise structures which are beyond the reach of a boom, similar to that used in hydraulic articulated placing boom in the boom pump, is mounted on a tower (mast).

Concrete transportation and placement, through concrete pumping should be considered mainly in construction projects requiring heavy concreting like; bridges, retaining walls, dams, raft foundation, high-rise buildings, etc. These are also used to sites with accessibility limitation such as:

(*a*) difficult site conditions, rough terrain, or muddy ground,

(*b*) physical obstructions like ponds,

(*c*) no vehicular access is possible etc.

14.6.2. Concreting Pump Applications

As concrete pumps have high placing rates, work convenience and the ability to deliver the concrete while by passing project site obstructions. The concrete should be considered for on site concrete transportation and placing mainly in the following cases and types of projects :

1. Project where concrete is main material to be transported like bridges,retaining walls, and other concrete structures.
2. Large concrete structural elements like dam and raft foundation construction. In exceptionally large concrete volume placement, more than one pumps are used simultaneously.
3. Site having accessibility limitations e.g.,

   (*a*) difficult on-site conditions, rough terrain, muddy ground and obstruction by water bodies.

   (*b*) sites where vehicular access is not possible, and

   (*c*) any other obstruction.
4. Projects where large productivity is required which is not possible with other means.

14.7 PLACING OF CONCRETE

Concrete should be placed as near as possible to its final position, which may be any building, underground or overground or any other construction site. While placing the concrete care shouldbe taken separation of coarse aggregate from the concrete care should be taken prevent separation of coarse aggregate from the concrete. During placing operation the heavier materials tend to settle and the lighter materials tend to rise toward the top. The separation of the aggregate from the concrete generally occurs at the ends of chutes and conveyor belts, at hopper gates, and during transportations.

It is also necessary to avoid the entrapment of air in concrete. The placing work should proceed from the end towards the centre. The batching, mixing, transporting and placing must be done is such a way that work of concrete placing is continued without interruptions.

The placing can be done by using chutes, buckets, pumps or belt conveyors. It large scale concerting, placing is generally done by means of concrete distributor booms. The booms of different reaches between 10 to 40 metres and dia 125 or 250 mm are generally. The booms of large reaches are mounted onto the turntable of tower cranes by of a special frame.

Placing Boom Sections

Increasing the number of boom section, reduces the length of the vehicle (on which it is mounted) movement on the road and makes it easier to unfold the boom around site obstructions and reach into inaccesible corners. It is also capable to eliminate dead areas where the boom can bot reach, or can not safely operate.

Placing of Concrete in Cold or Hot Weather

Concrete should be placed between 5° to 28°. Lower temperature reduces the rate of setting, while higher temperature reduces the ultimate strength. For placing the concrete in cold weather, water is preheated for providing the required temperature as it will ensure an early setting and gain in strength. Freshly laid concrete shall be maintained at a temperature of not loss than 20˜C for 3 days or 10°C for 5 days after it is placed.

For placing the concrete in hot weather, the concrete is required to be precooled to keep it within prescribed range. This can be achieved by installing an ice-manufacturing plant the concrete-mixing plant. Flacked ice is added to the concrete as part of the water.

Placing the Concrete in Water

For placing the concrete in the water, pipe long enough to reach the bottom of the water should be used. The bottom of this pipe should be kept continuously below the surface of the concrete so that the fresh concrete do not come in direct contact with water.

For the construction of deep foundations, dams, bridges etc. concrete can be placed under water either by Tremie or by Buckets.

(*a*) **Placing of Concrete by Tremie :** A water tight pipe of 25 cm to 40 cm diameter is extended from the level of working platform to the lowest point where concrete is to be laid. Concrete is poured is poured through a hopper fixed at the top to this pipe. The tremie is raised gradually while concreting proceeds. The bottom end of the trimie is to be kept inside the freshly laid concrete to prevent the entry of water into the trimie pipe.

(*b*) **Placing of Concrete by Buckets :** A bucket with open top and collapsible bottom shutter is used for concreting at a considerable depth. Fresh concrete is filled up to bucket and the top of the bucket is covered with canvas to protect from water. The bucket is then lowered to the desired depth and the bottom shutter is removed to deposit the concrete. The bucket is then withdrawn. Lowering of bucket is carried out by crane.

14.8 CONSOLIDATION BY CONCRETE VIBRATORS

Concrete normally contain large quantity of voids when placed. The purpose of consolidation is to remove these entrapped air voids, as entrapped air can render the concrete totally unsatisfactory. Consolidation is normally achieved through the use of mechanical vibrators. Only in case of very small/thin wet concrete mixtures is consolidated by hand tools. Properly vibrated concrete is higher-quality concrete, mainly in terms of strength, but also in terms of reinforcement protection, resistance to aggressive agents, and overall appearance. Consolidation is something also known as compaction.

For compacting the concrete after its placement, concrete vibrators are used. With the help of these vibrators it has become possible to place large volume of concrete quickly and easily particularly in intricate positions. The vibrators help in producing concrete of higher density by reducing air voids in the concrete. The concrete thus produced has better resistance to the seepage and gives better strength.

Concrete vibrators of following two types are generally used for construction purpose :

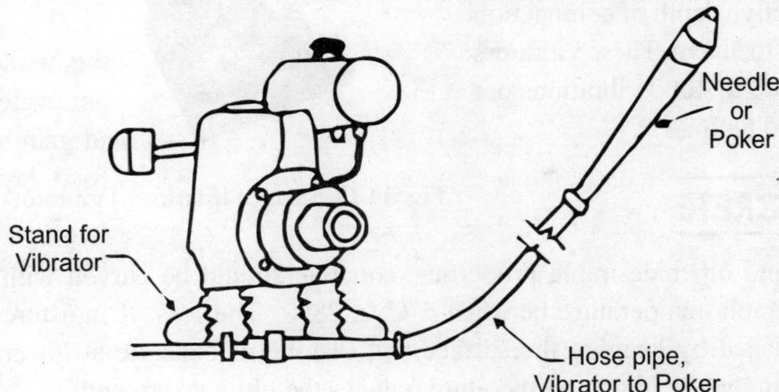

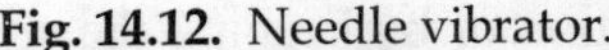

Fig. 14.12. Needle vibrator.

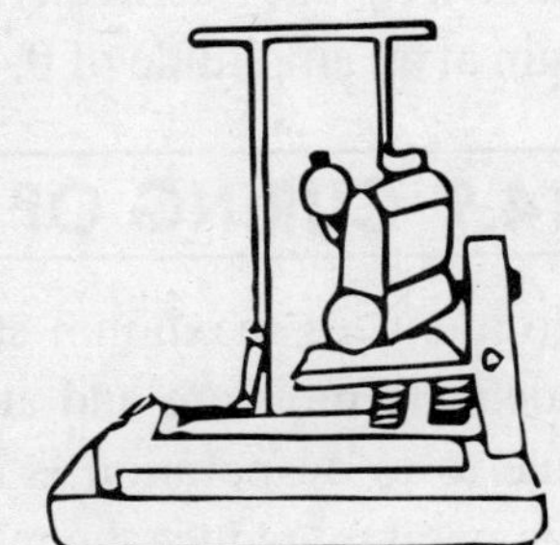

Fig. 14.13. Surface vibrator.

(*i*) Internal Vibrators

These are also known as immersion, needle or poker vibrators. These are immersed in the freshly placed placed concrete and it vibrates the concrete while in immersed position. Generally vibrators with 10,000 to 15,000 vibrations per min. at amplitudes between 0.8 mm to 1.6 mm are available for providing rapid and reliable compaction. The pokes of size 25 mm to 60 mm in dia are commonly used. These vibrators can be lifted and can be placed manually near the work place.

The vibrators are driven by diesel engine, compressed air or electric motors. Pneumatic vibrators are lighter and easy to handle, when compressed air or electricity is not available at work site diesel engine driven vibrators are used.

These vibrators have a flexible shaft, which gets its drive from the engine, and is enclosed in a hose. The shaft produced vibrations in the needle through an eccentric. Care should be taken that while using it, avoid bending of the flexible hose through small radii.

Fig. 14.14. Pocker vibrator.

(*ii*) External Vibrators

These are also known as surface or platform vibrators. External vibrators are placed on the surface of the concrete through which vibrations are conveyed to the interior. These are generally used of internal vibrator is not convenient, for example, long post, concrete products with mesh type reinforcement etc. where lack of space restricts the use of internal vibrators.

The surface vibrators consists of a rectangular pan, which is placed on the surface of concrete to be compacted. The driving unit which may be either an engine or an electric motor, is mounted above the vibrator and is carried on antivibration springs. The power is supplied to the vibrator, securely fixed to the pan. The effective depth of compaction by these vibrators vary from 20 cm to 30 cm. These vibrators have frequency generally from 3000 to 4000 vibrations per min at an amplitude of 0.4 mm to 0.5 mm.

Fig. 14.15. Surface (platform) vibrator

14.9 CURING OF CONCRETE

To attain its maximum strength and other desirable properties, concrete should be curved with adequate moisture and at a favourable temperature between 5°C to 28°C. The loss of moisture due to to evaporation is accomplished by keeping the surface wet Placing of concrete at lower temperature reduce the rate of setting while higher temperature reduces the ultimate strength.

If efficient curing is done, either by preventing evaporation or by keeping the surface of the concrete continuously wet, the strength of the concrete increases with the passage of time. This increases in strength is rapid in initial stages. Curing also increases the impermeability, durability of concrete, and resistance to abrasion. This also helps in achieving a hard dense surface and prevent cracking.

Heat is generated by concrete during its setting and curing is done to extract heat. Amount of heat generated per cu m depends on quality and quantity of cement used in the concrete. In small works since placement is at a slow rate the heat is lost mainly through convection and radiation.

Now-a-days with the latest development since high rate of placement is possible, therefore colling of concrete becomes essential specially where mass concreting is done.

In order to place the concrete within its most effective temperature range, cooling of various ingredient like cement, sand, aggregate and water becomes essential, Use of flaked ice has become popular in lowering down the concrete temperature. Some of the methods for cooling in mass concreting are :

(*a*) **Coarse Aggregate Cooling :** Coarse aggregate can be cooled either at the stockpile or on the conveyor by evaporation of controlled amount of water sprayed on it. The aggregate can be cooled by the blast of cold air injected near the bottom of bins and passed up by the blast of cold air injected near the bottom of bins and passed upward through the aggregate.

(*b*) **Cement Cooling :** Cement is cooled by chilled water by circulating it through the flies. The temperature if cement leaving the cooler is maintained at required level.

(*c*) **Sand Cooling :** Cool brine of about minus 7°C is circulated through the sand cooler in sufficient quality to cool the sand leaving the cooler before it is fed to the batching plant.

(*d*) **Water Chilling :** Water used for making concrete mix is cooled to temperature between 1°C to 5°C.

(*e*) **Ice Making and Storage :** Ice for this purpose is manufactured in a crystalline form with a consistency very similar to snow. Ice is manufactured at ground lever and then elevated to a refrigerated storage bin. This ice is fed to the weight batcher by means of a screw conveyors.

The aggregates, sand and cement are pre-cooled and chilled water is used during mixing at the mixers in the concrete batching and mixing plant supplement with ice, during hot weather. Some times cooling is done for aggregates only, while cement and sand are not cooled.

QUESTIONS

Q.1. Explain the methods usually adopted to compact concrete at site and discuss their merits.

Q.2. Discuss the method of transporting the cement concrete from the site of mixing to the site of placement.

Q.3. Describe different methods of placing the concretes.

Q.4. Discuss methods of curing the concrete.

Q.5. What is under water concreting? Explain in detail.

Q.6. What is transit mixer? Where will you recommend its used?

Q.7. Write a note on concrete batching equipment (plant).

Q.8. Explain the need of 'compaction of concrete', and describe various method commonly used for the same.

Q.9. Give a brief account of equipment used for handling and transporting concrete. State their merit and demerits.

Q.10. Compare the tilting and non-tilting concrete type concrete mixers.

Q.11. Describe construction and working of concrete pumps.

Q.12. Give the characteristics of the pumpable concrete.

Q.13. Describe different types of concrete vibrators.

Q.14. Describe construction and working of truck mixers.

Q.15. Discuss the advantages of pump transportation of concrete.

Q.16. Describe the procedure for concreting.

Q. 17. Write short notes on :

(*a*) truck agitators,

(*b*) washing of aggregate and sand,

(*c*) reversing drum type concrete mixer,

(*d*) concrete placing boom.

Q.18. Explain the procedure for placing of concrete in cold, and hot weather.

Chapter

Pumping and Dewatering Equipments

15.1 INTRODUCTION

In excavating below the ground surface, construction often encounter groundwater prior to reaching the required depth of excavation. Dewatering, temporarily lowering the level of groundwater the level of groundwater, is necessary for construction activities. After the construction operation are completed, the dewatering action can be discontinued and the groundwater will then return to its normal level. While planning for dewatering, the planner should consider that groundwater levels change from season to season.

The water from excavated areas are diverted into sumps from where it can be removed by pumping. The pumps are used mainly for following important operations :

1. Removing water from pits, tunnels and other excavations.
2. Lowering water table for enabling excavation.
3. Dewatering coffer dams.

The pumps in construction works are generally used under severs conditions, such as those resulting from (*a*) variations in the pumping head, or (*b*) from handling muddy water, or (*c*) highly corrosive water.

Information Required for Selecting a Pump

Following information should be considered for selection a pump :

1. The type off fluid to be pumped.
2. Height of lift from the existing water inlet to the point of discharge.
3. The rate at which the water is to be pumped.
4. The altitude of the project.
5. The suction head.
6. The variation in water level at suction and discharge.
7. The pressure head at discharge.
8. The size of pipe (at suction and delivery).
9. The type, size, number of fitting and valves to be used in the pipe line and length of pipe.

15.2 PUMPS

Pumps is a mechanical device to increase the pressure energy of a fluid. The pumps are used as water-handling device,in device in construction projects. The water is required to handle for human consumption, drainage and dewatering. The liquid handled by the pumps may be water, oil, sludge etc. In construction works the water may be pure or dirty.

Pumps are extensively used in construction projects and are required to perform various operations such as :

(*a*) Removing of water from pits, tunnels etc.

(*b*) Unwatering cofferdams.

(*c*) Lowering the water table for excavation.

(*d*) Supply of water for may types of utility services.

These pumps are required to perform their functions under severe conditions and handling of water which may be muddy, sandy or corrosive. The pump selected should, therefore be able handle all these types of works and dependable.

Types of Pumps

Pumps are of following types :

1. **Position displacement pumps :** These may be :
   (*a*) Reciprocating pumps.
   (*b*) Rotary Pump : These may be gear pumps, vane pumps, screw pumps etc.
2. **Rotodynamic pumps :** These may be volute pumps, circular pumps, diffuser pumps, vertical turbine pumps, centrifugal, mixed flow and propeller type of pumps.

15.3 CENTRIFUGAL PUMPS

A centrifugal pump is a hydraulic machine used to raise the liquid from a lower to higher level by creating a required pressure by means of centrifugal action. Centrifugal pumps are the machines which increases energy of a fluid. These may either lift the fluid or just boost the pressure in a pipe line.

Working of Centrifugal Pump

Fig. 15.1 explains in detail the working of a centrifugal pump. Before starting a pump, priming is done. In priming, water is filled in the suction pipe, casing and into a portion of delivery pipe upto delivery valve. This is done for the removal of air. With the presence of air, it is not possible to create pressure because the pressure generated by the impeller is run air, negligible pressure will be produced of very less density of air and hence on liquid will be lifted by the pump Thus, it is very essential to prime the pump properly and it should be noted that during priming valve is kept closed. The various method of priming will be discussed in this chapter later on.

After priming, the delivery valve is kept still closed and prime move (generally electric motor or diesel engine) is started to rotate the impeller. The rotation of.impeller in the casing full of liquid produces centrifugal head in the liquid and thus pressure in whole of the liquid is increased.

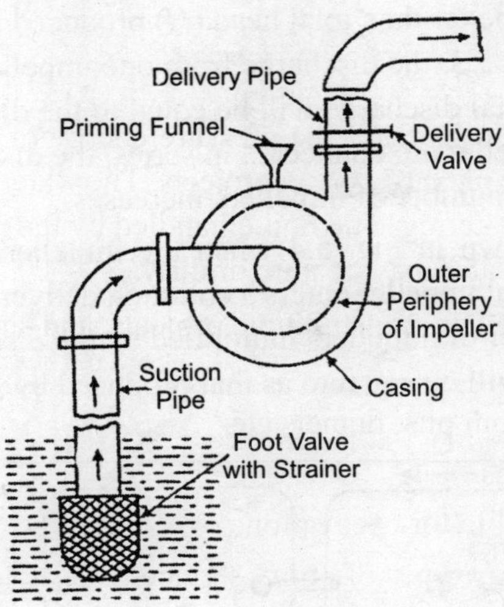

Fig. 15.1.

After a short period, when pump attains, a constant speed, delivery valve is gradually opened and thus liquid is allowed to flow in a radially outward direction through the impeller vanes and attains a higher velocity and pressure at the outer periphery. With the help of special shape of casing (Volute, Vortex or Diffuser) this high velocity energy is converted into pressure rise through the delivery pipe into a desired height. At the same time centrifugal action creates partial vacuum at the centre of impeller, known as EYE. The partial vacuum cause the water to rush through the suction pipe towards the impeller eye, to take the place of water which has left the impeller vanes, from the sump (where the pressure is atmospheric).

Thus, is this way water reaches and leaves the impeller continuously and water is raised to a certain height at constant rate.

Number of Impellers

According to the number of impellers provided, the pumps may be classified as single stage or multi-stage pumps.

In single stage pumps, only one impeller is fitted (Fig. 5.2 while in multi-stage pumps two or more than two impellers are fitted on the shaft either in series or in parallel.)

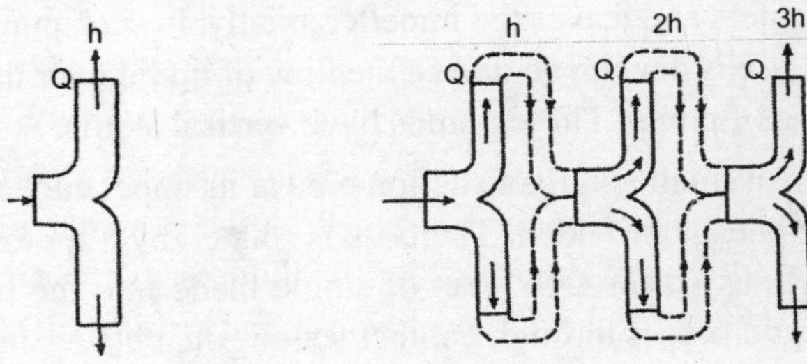

Fig. 15.2. Fig. 15.3.

(*a*) **In series :** When two or more than two impellers are in series, head production by the pumps will be the sum of the head produced in each impeller. For example if *h* is the head developed by one impeller then total head (*H*) produced by *n* impellers to series will be $n \times h$. As shown in Fig. 5.3. the discharge from one impeller is allowed to enter in the next impeller. Thus, the total discharge will be equal to the discharge through single impeller. Therefore, when impeller are connected in series, the discharge remains the same but the head increases as the number of impellers increases.

(*b*) In Parallel : As shown in Fig. 5.4 when the impellers are connected in parallel, the discharge will be equal impeller enters a common delivery pipe. Thus total discharge will be equal to the number of impellers multiplied by the discharge of one impeller. Here the total head produced will be equal to as that produced by a single impeller.

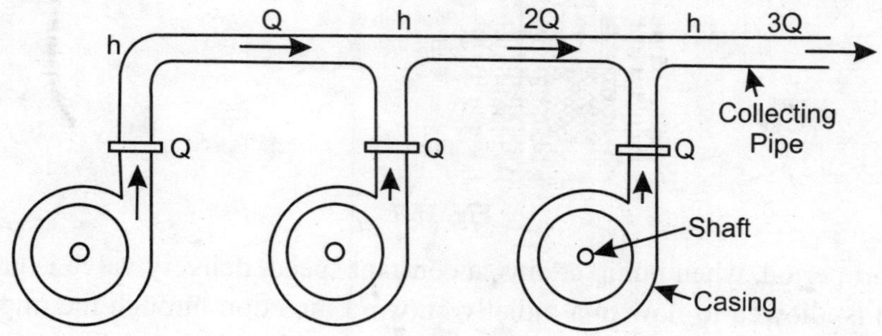

Fig. 15.4.

Generally, the term multi-stage pump is known for two or more impellers connected in series which are used used for producing high heads.

Impellers are made of following materials :

(*i*) Low carbon steels coated with stainless steel.

(*ii*) Stainless steel.(cr-12 to13% and Ni-1%).

(*iii*) Aluminium, bronze and zinc alloys.

(*iv*) Blades with cast stainless steel sheets.

(*v*) Bronze is preferred where water is chemically impure.

15.4 AXIAL FLOW OR PROPELLER PUMPS

This type of pump resembles to a reverse propeller turbine. This pump is fitted with an axial flow impeller in which water enters and leaves the impeller axially. In such pumps no centrifugal head is created in water but pressure is developed during the flow of liquid over the blades. Fig. 5.4. shows a line diagram of this type of pump. These pumps have vertical shafts.

As shown in the figure a small length of suction pipe at its upper end, ends in a casting in which a boss carrying about 4 vanes is provided. The boss is rotated by a motor connected at the top of boss shaft. Sucked liquid passed through a set of single blade into the boss vanes, where it gets pressure energy. Then liquid passes through another set of guide blades into the delivery pipe.

The first set of inlet guide blades, the incoming water axially without any velocity of whirl into the impeller. Guide blades at the outlet takes the whirl action again. (This whirl velocity is imparted by impeller blades). Thus allowing the water to flow axially in delivery pipe.

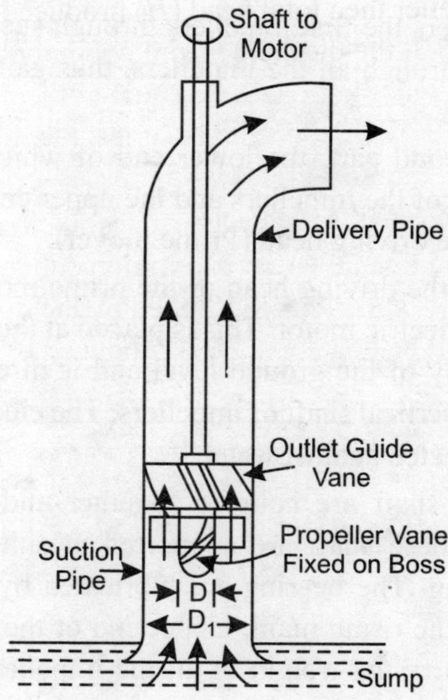

Fig. 15.5.

These pumps are suitable for very large discharge against small heads (generally 9-12m). As the speeds of these pumps are high hence danger of cavitation is more and, therefore, very low suction is kept.

These pumps are used in sewage handling, land drainage, flood control, condenser circulation in power houses and irrigation etc.

15.5 DEEPWELL OR BOREHOLE PUMPS

These pumps are used where water or any other liquid like oils etc. are required to be raised from a large depth. These are generally used for pumping from depths more than 6 to 7m. Special type of pumps can lift the oils even from a depth of 3000 m but generally ordinary deepwali pumps can lift upto 300 m. These pumps can have capacities upto lit/sec and have pipe dia upto 0.6 m nearly.

These pumps can be plunger pumps, multistage centrifugal pumps, jet pumps and air lift pumps etc. Out of these, plunger and multistage vertical type turbine pumps (centrifugal type) are most common. Plunger pumps are reciprocating pumps and are used for very high lifts.

15.5.1. Deepwell or Turbine Pump

These are multistage centrifugal pumps and used for lifting the liquid upto 300 m and have a capacity varying form 2 to 500 litres/sec. The number of stages in the pump depends upon the head required. These pumps, having 20 stages in series are in use. Fig. 5.6. shows the construction of this type of pump. These pumps consists of the following three main parts:

(*a*) It consists of a bowl which has suction case and impeller. The lower end of the suction pipe is connected with the strainer at the footvalve and this end is placed below the water level. Generally all the impellers have alteast 3 m of suction with strainer remains submerged in the water. Water enters the first impellers through suction pipe and then passes through all the impellers, thus gaining head one after the other.

(*b*) Rising main is the second part, the lower end of which is connected to the outlet of the impellers and the upper end of these is connected to be driving head (Prime mover).

(*c*) The third main part is the driving head or the prime mover, which is generally an electric motor. This is placed at the top of the line shaft usually of the ground level and is directly coupled with the long vertical shaft of impellers. The electric motor is properly protected against water.

The pumps shaft and line shaft are coupled together and are enclosed in the rising main. These shaft are supported at suitable intervals with the rubber bearing. The bearing are lubricated by the delivery water passing through the rising main. Upper end of the line shaft is fitted with bearing to carry the weight of all the impellers at the shaft.

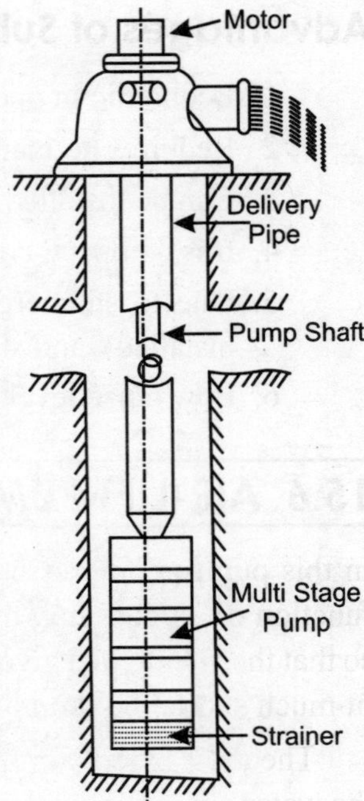

Fig. 15.6. Deepwell pump.

15.5.2. Submersible Pump

This is the modification on the deepwell pump explained above. As the name indicates, in this electric motor and pump both are submerged in the water. By submerging electric motor, large economy can be made by avoiding long shaft, large number of bearings and large sized rising main etc. The complication due to thrust (arising due to the weight of the shaft and impeller etc.) are also avoided.

Since the motor in submerged in water, the only problem is to prevent the motor winding and other electrical connection to be spoiled by water coming in coming in contact. For this purpose either of the following two solution can be made :

(*i*) To use a normal type of motor but to prevent the water from reaching it by enclosing in water-tight diving ball. This system of enclosing in known as "Dry motor" type system.

(*ii*) To allow the water to have more or less free contact with motor windings but to give them special protection by suitable type of insulation. This is known "Wet motor" type system.

Application of the Submersible Pumps

These pumps can be used for :

1. Drinking water supply installation for cities and rural areas.
2. Service water supply installations for industry and commerce.
3. Lowering the ground water level on building sites.
4. Containing the ground water level in mines.
5. For irrigation purposes.

Advantages of Submersible Pumps

1. Economic in operation, because of high overall efficiency of these pumping sets.
2. Requires no maintenance, because of its glandless construction and its water lubrication.
3. Can be installed in crooked wells.
4. It is compact, comparatively light to weight, has no longdrive shaft and no coupling.
5. Due to submerged installation, priming of the pump is not required (after once it is installed), and since foot valve is not required, hence no suction trouble.
6. Low noise level because of journal bearing.

15.6 AIR-LIFT PUMPS

In this pumps compressed air is used to lift the water up. Here the function of air is to mix the water and to reduce the mixture density so that the weight of a given column of mixture is less than the weight of much shorter column of "solid water".

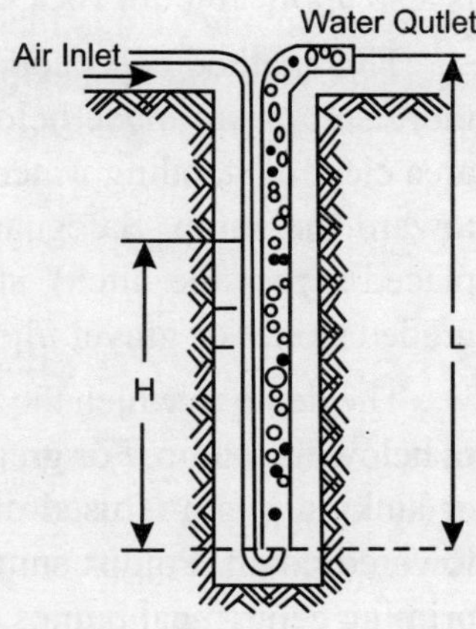

Fig. 15.7. Air lift pump.

The air is passed through one or more nozzles at the bottom into the water at the foot of the rising main fixed in a wall. As soon as the column in the rising main has become impregnated with the bubbles, the pressure due top the column of length L is less than that due to the static pressure H, thus causing the flow water and air issue from the top of the rising main so long so long as the supply of air is continued. Fig. 16.8. shown a line diagram of this type of pump.

To obtain better results the usefull lift $(L - H)$ is less than the submergence H. The ratio$\left(\frac{L}{L-H}\right)$ vary from 4 when H is about 30 m to 1 when H is about 90 m.

Advantages

1. As it has no moving below water level, therefore, can not be damaged by solid suspended in the water.
2. It can raise more water through a bore-hole of given diameter than any other from of pump.

Limitation : Its efficiency is low say 20 to 40%.

15.7 DEWATERING SYSTEMS

Seepage of water in excavation something becomes most difficult problem. Since heavy and continuous pumping in such cases in not only a costly item but may also cause settlement of adjacent structures, and may also result in erosion or collapse of the side of open excavation. Many times it so happens that after an excavation is started ground water is met in too quantities than anticipated. Such heavy inflow of water may cause collapsing with danger to adjoining structures. In these circumstances installation of bored well system or a well point system is the only solution.

Methods of Ground Water Control

Following methods of ground water control are adopted :

1. Pumping from open sumps.
2. Pumping from well points.
3. Pumping from bored wells.

15.7.1. Pumping from Open Sumps

This method is most commonly used where area is large enough for allowing excavation to be cut stable slopes and where there are no important structures close to the excavation to affect their stability by settlement resulting from erosion due to water flowing towards the sump. This method is also applicable for rock excavations.

This method costs comparatively low for installation and maintenance. In this method one or more sumps are made below the general level of the excavation. In order to keep the excavated area clear of standing water, a small grip or ditch is cut around the bottom of the excavation falling toward the sump. Safeguard against erosion can be achieved by providing check weirs (board placed across the ditch), stones or concrete paving, or laying open jointed pipes surrounded by graded stones or gravel filter material.

The depth to which the water table may be lowered by the open sump method is not more than 8 m below the pump. For greater depths of excavation the pump should be reinstalled at a lower level or sinking pump is used or submersible deepwell pump suspended by chains and progressively lowered down. Pumps suitable for operating from open sumps are: Pneumatic sump pumps, self priming centrifugal pumps, monopump, sinking pumps.

Such pumping is simple and less expensive, but has serious limitations. When fine sand or cohesionless soil lie below the water table, this type of pumping removes the fine material from the surrounding soil and result in settlement of adjacent structures. To protect it, sumps lined with gravel filter are something used.

15.7.2. Pumping from Wellpoints

This system comprises the installation of a number of filter wells (also known as well points), generally 1 m long, around the excavation. These filter wells are connected by vertical riser to large dia header main at ground level which is under vacuum from a pumping unit. The water flows to the filter well by gravity and then drawn by the vacuum upto the header and discharged through the pump. This system has the advantages that the water is filtered as it is removed from the ground and carries almost no soil with it once steady discharge conditions are attained. This system has the limitation of limited suction lift (generally upto about 5m below pump level.)

Therefore, for deeper excavations the wellpoints are installed in two or more stages.

The filter wells or well points (refer Fig. 16.9) are usually 1 m long and 60 to 75 mm dia gauge screen surrounding a central perforated riser pipe. The capacity of a single wellpoint with 50 mm riser is about 10 lit/min. Spacing between two wellpoints depends on the permeability of the soil and on the time available to effect the draw down. In fine coarse sand or sandy gravels a spacing of 0.75 to 1 m is required, while in silty sands of low permeability a 1.5 m spacing is sufficient. In permeable course gravels spacing should be as low as only 0.3 m. A normal set of wellpoint system comprises 50 to 60 points is single 150 or 200 mm pump with a separate 100 mm jetting pump.

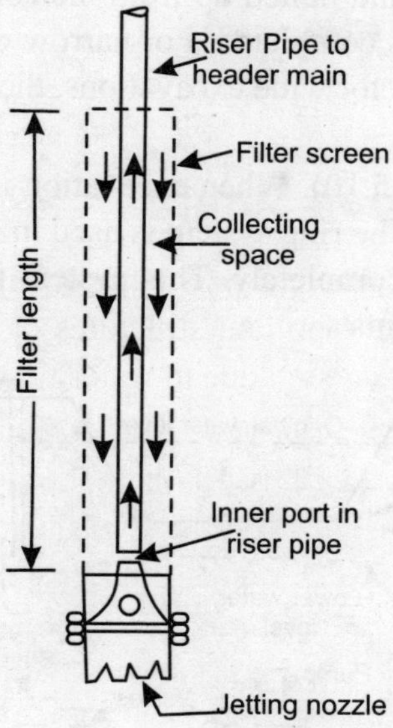

Fig. 15.8. Well point arrangement.

Types of Wellpoints System

Wellpoint system are installed in two ways :

(*a*) Line system

(*b*) Ring system.

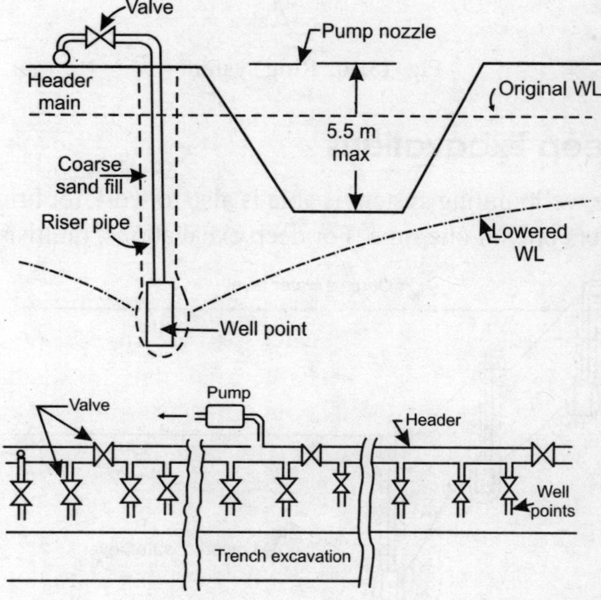

Fig. 15.9. Line system.

(*a*) **Line System :** (Refer Fig. 5.9) This system is employed when excavation area is long. The header (a larger diameter pipe) is laid out along the sides of the excavation, and the pumping is continuously is progress in one length as further points are jetted ahead of the pumped down section and pulled up from the completed and back filled lengths and repeated till entire length is completed. For narrow excavations, like trenches, header is laid only on one side, while for wide excavations, the header are required to be placed on both sides of the area.

(*b*) **Ring System :** (Refer Fig. 5.10). When excavation is done is area of appreciable width, line system is inadequate. The ring system is used in such condition, and the header main surrounds the excavation completely. This system is used for rectangular excavations such as for piers or basements.

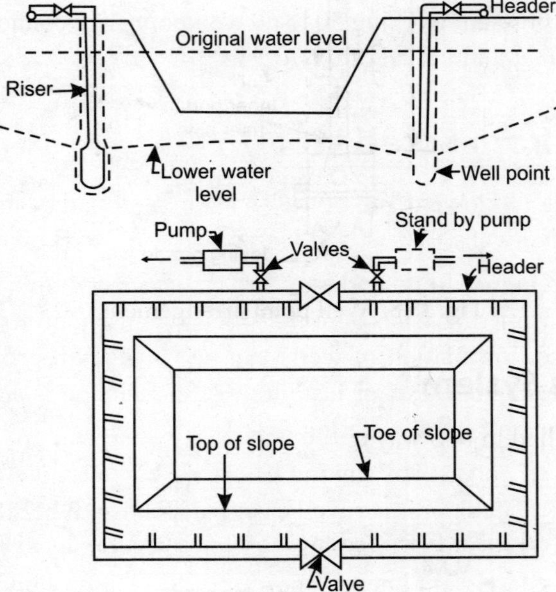

Fig. 15.10. Ring system

Wellpointing of Deep Excavations

As mentioned earlier, the wellpointing system is able is also to work for bringing water level down by about 5 to to 5.5 meters only at one time. For deep excavations, multistage wellpoints

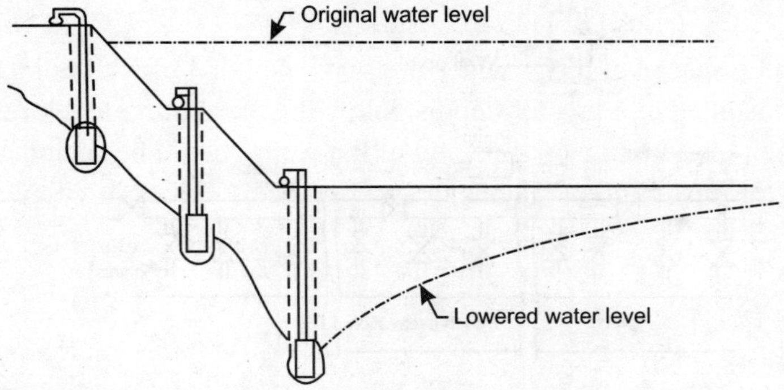

Fig. 15.11. A three stage well point installation.

with separate pumps for each stage must be provided. A three stage installation is shown in Fig. 5.11. Although deeper depths can be excavated by adopting multistage wellpoints, but the overall width of excavation at ground level become very large.

When well pointing is used in sheet piled excavations, the well points are placed closed to the sheet piles, in order to lower between the sheet pile rows.

Wellpoints Operation

For efficient working of the system, riser pipes must run full so that air is not sucked into them. This can be achieved by adjusting the stop cock, provided at the top of riser pipes. To prevent leakage in the header, air tight joints are necessary. For small leaking sealing compound is used for repair. In order to achieve effective vacuum header pipes most be kept at intake level of the pump and at the level of tops of risers. For efficient working 50 to 60 well points are connected to each pumping set. These are generally placed at an interval of 0.6 to 1 m.

15.7.3. Pumping from Bored Wells

Pumping from wells, for draw-down depth of less than 8 meters can be undertaken by surface with their pipes installed in bored wells.

When dewatering is required to be undertaken from a considerable depth, electrically driven submersible pumps are installed in deep bore holes with rising main to the surface. Since heavy boring equipment are used, installation of wells can be done in all ground conditions including boulders and rocks. Due to higher costs or installation, this method is adopted where constriction period is long and other methods of dewatering are not possible.

Installation of bore well consists of sinking of casing having a dia of about 20-30 cm larger than the inner well casing. The dia or inner well casing depends on the size of submersible pumps. This inner well casing is inserted after complete sinking of borehole screen over the length where dewatering of the soil is required and it terminates in a 3-5 length of unperforated pipe to act as a sump to collect any fine material which may be drawn through the filter mesh. Screen having slots are preferable to holes, since there is less risk to blockage from round stones.

After the installation of well casing, graded gravel filter material is placed between it and the outer borehole casing over the length to be dewatered. As the filter material is placed, the outer casing is withdrawn in stages. The space above the screen is backfilled with any available material. The submersible pump is then installed and commissioned. The completed installation if shown in Fig. 15.12.

Bored wells can be spaced at much wider intervals than that of well points, because the bored well pumps can be installed at a greater depths below the excavation level, resulting in wide area of draw down for each individual well. Spacing of the wells should be such that, due to shutting of an individual pump for repair and maintenance; water level does not rise above excavation level.

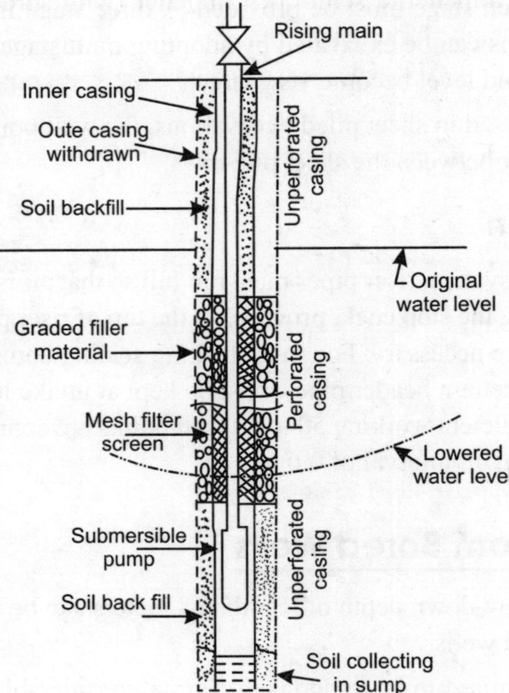

Fig. 15.12. Bored well installation.

15.8 MINING PUMPS

Mine waters are generally abrasive, acidic or alkaline, and corrosive. Mining pumps are required to remove water continuously and generally in large quantities. Therefore, low speed pump with more number of stages are employed, but this arrangement poses problem of shaft, deflection. Due to development of new abrasion resistant materials (Austenitic stainless steels, martensitic stainless steels, and their alloys) higher speeds upto 3000-3600 rpm are successfully employed. This enhances the pump reliability and strength, and pumps with lesser number of stages can be made.

15.9 NON-CLOG PUMPS

These pumps are used to handle liquids having suspended impurities even solids. These are also used for transport of coal slurries, sewerage disposal etc. These pumps take care of abrasive, clogging and chemical effect.

In designing such pumps following guiding factors are considered:

1. Arrangement for clearing the guiding factors are considered.
2. Arrangement for easy replacement of parts subjected to high wear.
3. Keeping high velocity, as at velocity less than 3 meter per second there is a tendency of desetting of suspended particles.

4. Instead of closed impeller, a semi-open impeller (with shroud on one side only) or an open impeller (without shroud) are used.
5. Pitch distance between the blades is kept such that largest possible solid matter can move along with the liquid. This causes lesser number of blades.
6. Material used for the manufacture of the impeller should be according to the purpose for which this is to be used.

QUESTIONS

Q.1. What is a "dewatering system"? Why it is necessary?

Q.2. Discuss different methods of ground water control?

Q.3. Describe line system and well system of wellpoint dewatering system.

Q.4. Discuss wellpoint operation system.

Q.5. Explain 'wellpoint' method of dewatering with neat sketch.

Q.6. What factors should be considered while selecting a water pump for dewatering purposes?

Q.7. Give classification of water pumps. Discuss their applications.

Q.8. Discuss the working and advantages of the following pumps :

(*a*) Reciprocating pump (*b*) Centrifugal pump

Q.9. Write short notes on the following :

(*a*) Multi-stage pump (*b*) Submersible pump

(*c*) Deepwell pump

Q.10. What is a pump? Discuss application of pumps in construction works.

Q.11. Discuss important components of a centrifugal pump.

Q.12. Explain the working of centrifugal pump.

Q.13. Explain in brief:

(*a*) Axial flow pump (*b*) Air lift pump

(*c*) Bored-well pump (*d*) Non-clog pumps

Chapter

Hoisting Equipment (Lifting, Lowering and Elevating Equipments, and Cranes)

16.1 INTRODUCTION

Need of the present day is for the equipment to handle heavy loads with past speed, reliability, safety and economy. In order to meet the variety of requirements, equipments of different types and sizes have been developed and are being manufactured. Material handling equipment are mechanically, electrically, hydraulically or pneumatically operated or uses their combination. though these are not production machinery, they are auxiliary equipments which improve the flow of materials which in turn reduces stoppages in production and increase The productivity.

Hoisting equipments can be divided into following three categories:

1. **Lifting and Lowering Equipment.** These are used in the application of lifting a load, handling them in suspension, lowering and placing them at required locations. The loads are generally supported from overhead by a hook or travelling on a track. These devices are rugged, simple, dependable, easy to install versatile, and economical. These have limitation of handling lesser loads, slow, manual operation, fixed path and required overhead support.
2. **Elevating Equipment.** Vertical motion over fixed vertical or steepy inclined routes with a continuous or intermittent motion is followed by these equipment.
3. **Cranes.** Crane is a versatile and most useful machine and is only single machine capable of providing three dimensional movement to a load. It is able to perform its operations with speed, safety and precision. These have wide applications in construction projects, industries and shipping etc.

16.2 LIFTING AND LOWERING EQUIPMENTS

These equipments are generally known as "hoists". These are the devices used for lifting or lowering the loads, suspended from a hook on the end of retractable chains or cables. The loads are generally supported from overhead by a hook or travelling on a track.

Following types of hoisting equipments as discussed hereunder are generally used.

16.2.1. Block and Tackle (Pulley System)

Fig. 16.1. Block and tuckle.

Fig. 6.1 shows a block and tackle arrangement of lifting loads through a vertical distance. This is one of the oldest and simplest device. It is still used today for moving men and in hoisting machinery into position. It depends on manual power and gives only the mechanical advantages that is possible for the various rope formations. It is the most inexpensive in cost but the most wasteful of manpower.

A unit consisting of one or more sheaves (pulleys) mounted in a frame is known as 'block'. Block is mostly equipped with an eye, a hook or some similar device at one or both ends of the frame, by which objects can be attached. In pulley system fixed or movable pulleys are connected in different combinations.

16.2.2. Winches

Fig. 16.2 shows a winch. This is used to lift loads vertically by winding the rope or cable on a drum. Here, it is possible to use manpower, to get a much greater mechanical advantage than with a block and tackle. It is frequently used in loading heavy equipment into ships, construction of buildings and in similar jobs.

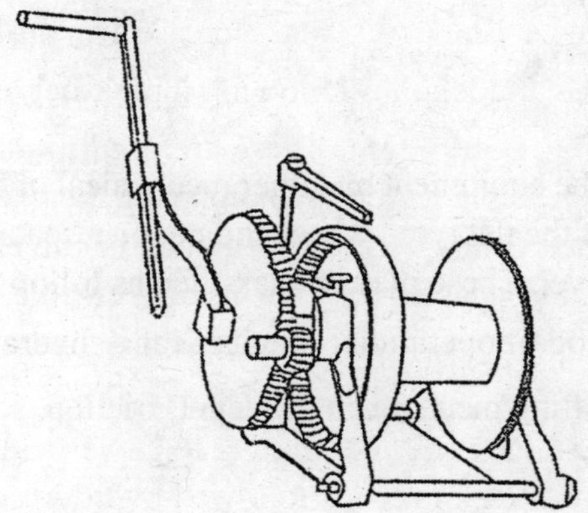

Fig. 16.2. Winch.

Capacity of a winch can be increased by increasing the number of gear trains. These are very useful, simple and how cost equipment for lifting heavy loads. Ratchet, brakes may be fitted on the winch to hold a suspended load and to prevent reverse travel of gears and thus provide safety against accidental dropping of load.

16.2.3. Power Hoists

Fig. 6.3 show a power hoist, which is often operated between fixed guide rails, for lifting things vertically. There is an infinite variety of hoists suitable for different purposes. The simplest is the chain hoist, which is operated by hand. There are hoists operated by compressed air, diesel engine or by electric power. The hoidss are smaller to elevators except that, a hoist does not carry the operator in it, but is operated from one or two other points.

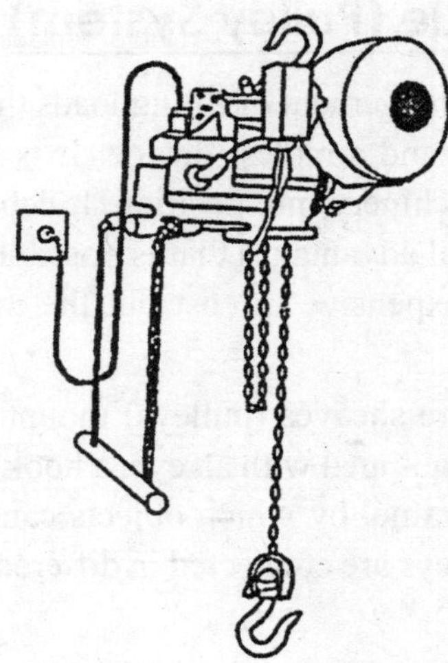

Fig. 16.3. Power hoist.

Hoists used for short distance horizontal hauls are suspended from trollies or carriers travelling over an overhead single rail track, commonly a rolled 'I–beam'.

Electrical hoists capable of hoisting speeds between 5 and 25 m/min are available in capacities ranging between 0.25 and 15 tons.

16.2.4. Jacks

Jacks are used to lift a part of the equipment by either mechanical or hydraulic means to a sufficient height to enable the removal of the flat tyred wheel and put on a space wheel or for servicing. These work on the principle of the lever. These may be classified as follows:

1. With respect to method of operation : Mechanical or hydraulic.
2. With respect to the lifting mechanism : Ratchet, friction, screw, scissor, geared.

(*a*) Mechanical Jacks

These may be ratched type, screw type, scissor type, or geared type. Geared types are generally used for construction equipments, whereas others mentioned above are used for lift works.

The geared mechanical jack is shown in Fig. 6.4. is operated by a lever having a pawl and ratchet and which is attached to the bevel gears as shown.

(*b*) Hydraulic Jacks

Hydraulic jacks work on the principle that, the pressure exerted by liquid on a surface, is proportional to the area of the surface. Due to this principle, jacks are capable of lifting great weight. Surface pressure exerted by a liquid depends upon the ratio of face areas of the ram and plunger.

Hydraulic jack works on the same principle as that of hydraulic press. These are used to lift loads, and are commonly used for lifting automobiles to facilitate cleaning and repair work. These are used to lift heavy loads through a short distance.

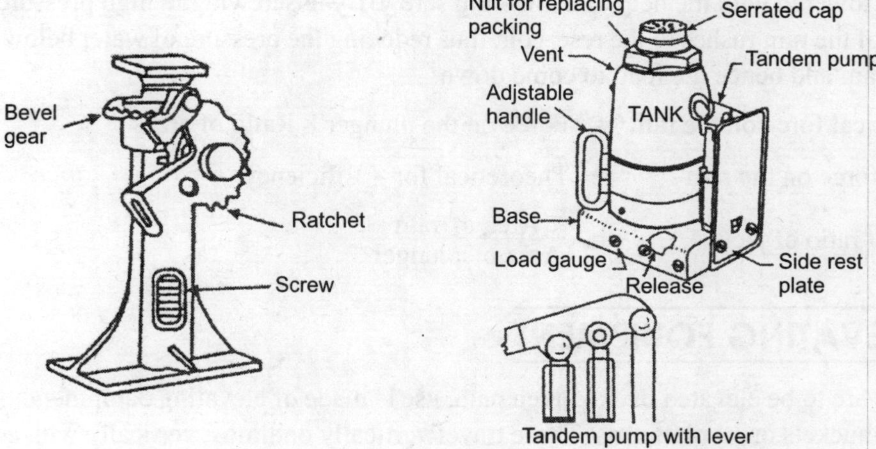

Fig. 16.4. Gear type mechanical jack.

Fig. 16.5. Heavy duty hydraulic jack.

Fig. 16.6 shows a simple hydraulic jack. It is provided with plunger on one side. This plunger is reciprocated by a handle attached to it. Another side of the jack is provided with a ram, by means of which loads are lifted.

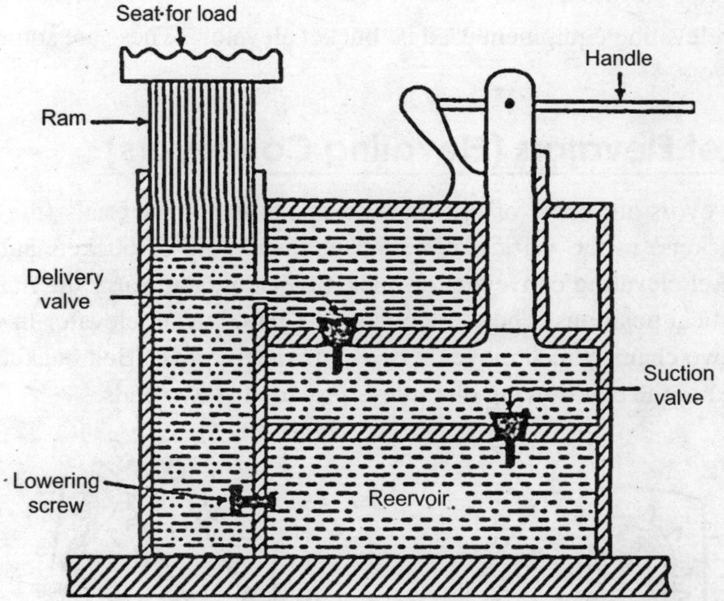

Fig. 16.6. Hydraulic jack principle.

Up and down motion of the handle reciprocates the plunger. This movement of the plunger draws the water (or oil) from the reservoir and delivers to the cylinder fitted with the lifting ram through the delivery valve. Both the suction and delivery valves are the non-return valves. Addition of high pressure water at the bottom of the ram causes it to move upward. Loads can be lifted if placed at the top of the ram.

Load is lowered with the help of a lowering screw. By unscrewing it, high pressure water from the bottom of the ram rushes to the reservoir, thus reducing the pressure of water below the ram and causes the ram and hence the load to come down.

Theoretical force on the ram = Force on the plunger × Ratio of areas

Actual force on the ram = Theoretical for × Efficiency

$$\text{Velocity ratio of jack} = \frac{\text{Area of ram}}{\text{Area of plunger}}$$

16.3 ELEVATING EQUIPMENT

When loads are to be elevated over a given path, use is made of elevating equipment to lift loads in cages, cars, buckets or on platforms. These travel vertically or almost vertically with continuous or intermittent motion over a rigidly built shaftway.

In construction industry commonly used elevating equipment are:

(*a*) Continuous motion vertical conveyors.
(*b*) Freight elevators/lifts.
(*c*) Vertical skips elevators

Most common elevating equipment used is 'bucket elevator'. These are sometimes also known as elevating conveyors.

16.3.1. Bucket Elevators (Elevating Conveyors)

The elevating conveyors are used for transporting dry granular materials (the material that will not stick to the buckets) in the vertical direction with the help of buckets and trays. These are also known as bucket elevating conveyors or bucket elevators and carry the material in bucket to vertical or near vertical positions. These are either (*i*) Chain bucket elevator in which buckets are attached to one or two chains which move on two end wheels, or (*ii*) Belt bucket elevator in which buckets are attached to the belt moving on pulleys provided at two ends.

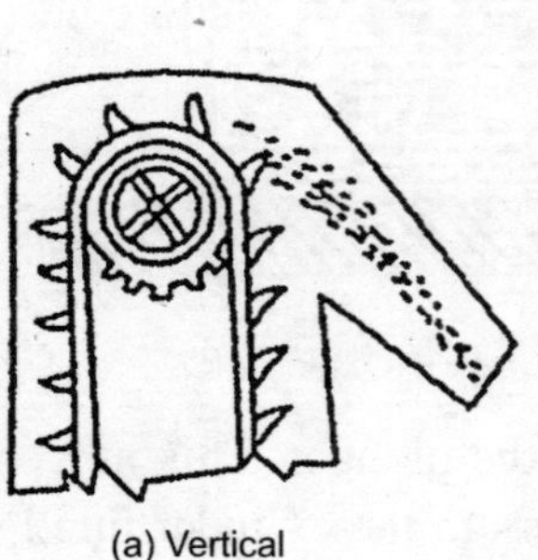
(a) Vertical

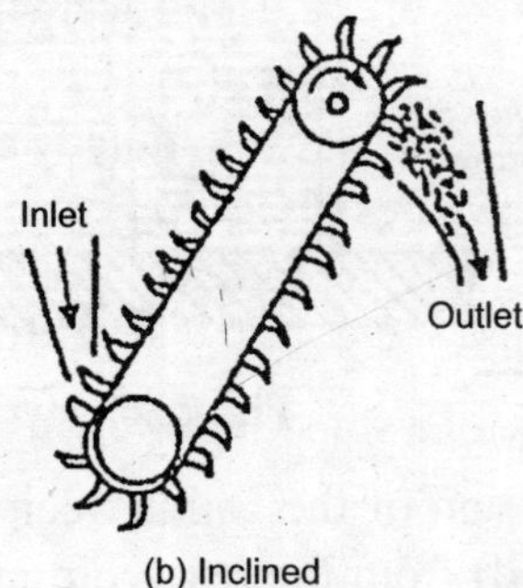

(b) Inclined

Fig. 16.7. Bucket elevator.

Bucket elevators, also known as elevating conveyors, are operated between head and foot terminal machinery in a supporting fame. Buckets are loaded by scoop action at foot end or by a chute above.

These consists of an endless chain of buckets which elevates the minerals. The elevators are usually enclosed in a sheet of metal casing with necessary inlet and outlet openings and chutes.

Normal operating speeds of these conveyors is between 0.15 and 0.5 m/sec. These conveyors are usually variable upto 400 tons/hr and normal lifts upto 30 meters.

The belt type elevators have higher speed and capacity, smoother and quieter, while chain types are heavier duty and suitable for abrasive materials.

Such elevators are sometimes used as portable elevators, which are known as stackers, consists of a base on wheels, a vertical frame or mast, a lifting or lowering mechanism and a load carrying platform. Lifting of such elevators is accomplished by hydraulic or mechanical action. The mast may be telescopic or non telescopic.

The bucket elevators consist of a line of pivoted buckets end-mounted between two strands of rivetless chains so that the centre of gravity is below the pivot in every bucket. As a result, buckets travel in stable equilibrium and automatically drop back into the horizontal position after being discharge by tilting.

Normal operating speeds of these conveyors is between 0.15 and 0.4 m/sec. These conveyors are available upto 400 tons/hr capacity. Normal lifts of these range up to 30 meters, but in special cases these may range upto 100 meters.

16.4 CRANES

Cranes are a combination of separate hoisting mechanisms with a frame for lifting or a combination of lifting and moving loads which can be freely suspended on or secured to them.

Cranes have wide application in construction projects, industries and in shipping etc. These are used for lifting the loads (may be construction materials, loose materials, packages, containers, finished and semifinished products in industries etc.) and placing them at desired place. For this purpose the cranes have three motions in general, namely hoisting, *i.e.*, lifting or lowering, derricking and slewing. The cranes are generally electrically operated, diesel operated or may have diesel-electric drive.

The cranes are very much useful to pick up a load at one point and be able to deposit it at another point within a restricted area. These are used for loads that are heavy for other types of equipment. Each type of crane is designed and manufactured to work economically in specific site situations : modern-day sites often employ more than one type of crane and more than one crane of the same type. Construction cranes are generally classified into two major families, namely mobile cranes and tower cranes.

The cranes can be classified in following classes :

1. Derrick cranes (Revolving or rotary cranes)
2. Mobile cranes.
3. Hydraulic cranes
4. Overhead cranes, also known as Gantry cranes.
5. Tower cranes

16.4.1. Derrick Cranes (Stationary Cranes)

Derrick cranes consist of a mast, a boom, bullwheel on which it rotates about a vertical axis, and supporting members (also known as guys). These cranes are very widely used in construction projects, industrial and multistoreyed building construction, plant erection, loading and unloading of cargoes at ports, in ship building etc. When used with grab it can handle loose materials like sand, ballast, coal etc.,,

As shown in the figure the most of the derrick cranes are supported by a number of guys. The boom can revolve through 360° and can pass below these guys. A bullwheel is attached to the mast and rotates it. These are operated by either a diesel engine or by an electric motor.

As mentioned above, mast is generally supported by guy ropes but when this is replaced by trussed structure it becomes stiffleg derrick. In this case the mast is supported at its upper end by two braced stiffleg, therefore the boom rotates through angle of 270° to 290°.

The derrick cranes are sometimes mounted on rail wagons for quick transportation and for working near rail lines, and are known as Whirler cranes. These cranes are available up capacities of 40 tonnes with 25 meters boom length and 15 meter work radius.

The Derrick cranes can be used as Tower cranes for the purpose of erection of high industrial and residential buildings of height 100 meter or above. The Derrick cranes are also designed to work as port tower cranes (for use in ports for container and general cargo handling), as shipyard cranes (for use in ship building and ship repairing works with longer capacities and larger working radii).

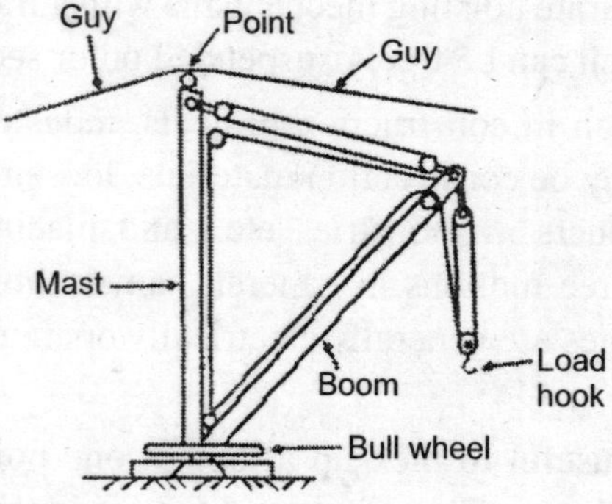

Fig. 16.8. Derrick crane.

16.4.2. Mobile Cranes

The mobile cranes have four motions viz., hoisting, derricking, slewing and travelling. Thus it has travelling motion as an extra feature as compared to Derrick cranes.

Mobile cranes are either crawler mounted or wheel mounted.

1. **Crawler mounted cranes** are highly manoeuverable and have the ability to operate on unmade ground. Long wide crawler mountings and wide crawler shoes give low ground bearing pressures and they can travel with their maximum load suspended from the hook. This improves crane stability. Since these move slowly, they are used in rough terrain

where truck mounted cranes can not travel and hence are economically used where ground conditions are poor and required to operate in a limited area.

The boom of these cranes comes in sections which are joined by pin connections. The straight boom thus formed can lift loads over a radius of 30 to 40 metres.

The full revolving super structure of crawler crane is mounted on a pair of continuous, parallel crawler tracks. Many manufacturers have different option packages available that enable configuration of the crane to a particular application. The crawlers provide the crane with good travel capability around the job site. Shifting a crawler crane between projects requires that it be transported by the truck, rail or barge. As the size of the crane increases, the time and cost to dismantle and load the crane, investigate the haul routes, and assemble the crane will also increase.

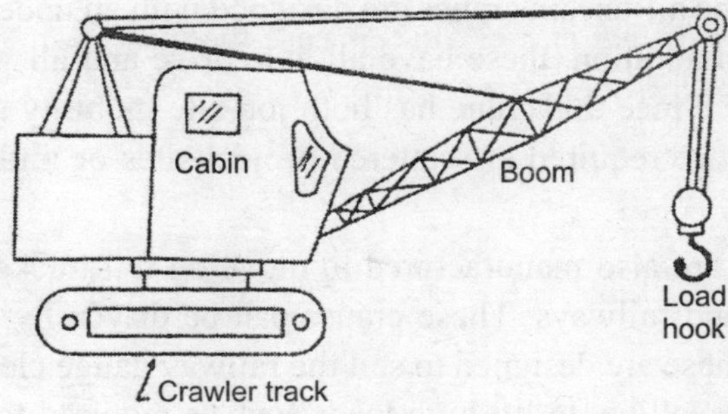

Fig. 16.9. Crawler mounted crane.

2. **Wheel mounted mobile cranes** are generally truck mounted or self-propelled and have high mobility, and can travel upto 70 to 75 km per hr on good roads. These are designed for long distance travel with high road speeds. The number of axles and drive depends upon load lifting capacity and the length of the boom. These cranes can operated on firm ground. For increasing base width and to improve stability during operations, out-riggers are used.

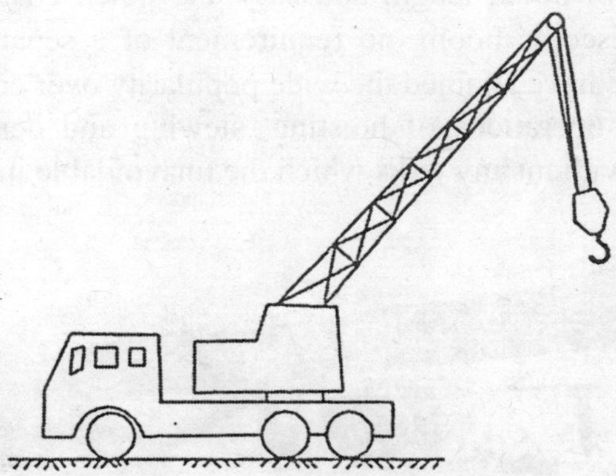

Fig. 16.10. Truck mounted crane.

Self-propelled rubber-tyred wheel cranes have greater mobility over hard surface and are widely used for shifting and transporting light loads over short distance, and for off-loading of medium to heavy loads.

Truck-mounted cranes are available with telescopic-jib and strut-jib. Telescopic-jib cranes provide flexibility in adjusting distances and heights of lifts. These have greater mobility on roads than other cranes. Strut-jib, cranes are suitable for shifting small loads at a distance, where ground obstruction restricts the utility of the crane.

Telescopic-boom mobile cranes can travel on public highways between projects under their own power.

Rough-terrain cranes are mounted on two axle carriers. These units are equipped with large wheels and closely spaced axles to improve manoeuverability at the job site. These have high ground clearance. To utilize their maximum lifting capacity the cranes, must be stabilized on outriggers while in operation.

All-terrain cranes : The all-terrain cranes are designed with an undercarriage capable of long-distance highway travel. In addition, these have all-axle drive and all wheel steering, large tyres and high ground clearance. Since this crane has both job-site mobility and transit capability, it is useful when multiple lifts are required at scattered project sites or multiple work locations on a single project.

Rail-mounted cranes are also manufactured to move on rail tracks. Such cranes are used in steel industry, dockyards and railways. These cranes can be driven by stream engines, by diesel engines or by electricity. These are designed to suit the railway gauge clearances available, so as to enable negotiating while travelling, to lift high loads such as required for salvage operations after accidents etc.

16.4.3. Hydraulic Cranes

Hydraulic cranes are usually truck-mounted cranes. All operations of these cranes are hydraulically powered including extension and retraction of out-riggers, extension and shortening of boom. The hydraulic power is supplied through oil pressure, generated by rotary oil pumps. Because of simplicity, quick change of boom length and easy and quick change of angle, mobility and manoeuverability of the telescopic boom, no requirement of a separate crew for extending or shortening of the boom, these have attained the wide popularity over conventional cable operated, latticed boom. Besides, the operations of hoisting, slewing and derricking can be performed simultaneously and that too without any jerks which the unavoidable in conventional cranes.

Fig. 16.11. Mobile hydraulic crane on outriggers.

Since ground contact of most cranes is longer than its width, they have greater stability in front and rear than off the sides. The balance of a crane can be improved by (*i*) adding counterweight (but not much, to avoid straining to its structure), (*ii*) by raising the boom to bring the load closer, or (*iii*) by using outriggers.

Hydraulic cranes consist of two main parts :

(*i*) Jigger, and (*ii*) Craning arrangement.

(*i*) **Hydraulic Jigger :** It is an arrangement of pulleys for increasing the velocity ratio. It consists of a fixed cylinder in which a ram moves (Refer Fig. 19.14). A number of pulleys (2 to 6 lifted side by side) are fixed on the top of a ram and similar number is fixed at the end of the cylinder. A wire rope, with one end fixed to a point on the cylinder, is taken round all the pulleys on ram and cylinder alternatively and finally to the other end is taken round a pulley fixed at the top of the mast (hollow vertical post) of the craning arrangement. It is then taken down after passing through a pulley fixed at the junction of jib and tie rods. At the free end, a hook is provided for hanging loads. Jigger may be kept either horizontally or vertically.

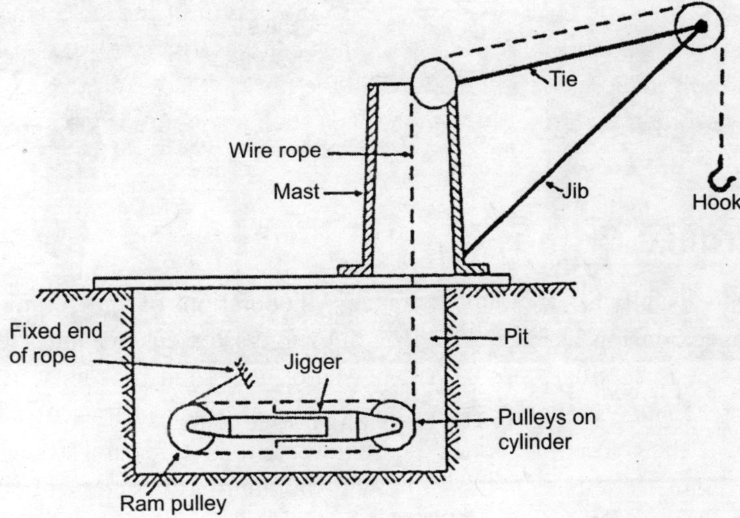

Fig. 16.12. Hydraulic crane.

(*ii*) **Craning Arrangement :** The crane consists of a hollow vertical post (known as mast) to which a jib and a tie is attached as shown in Fig. 19.14. The mast is supported on a bearing and can be revolved with its vertical axis, so that load from all directions can be handled. The jib can be raised or lowered together with the load.

Working : When the ram is at the end of the inward stroke, water under pressure is admitted in the cylinder. This water forces the ram to go in outward direction. While ram is making outward stroke, the wire rope is pulled and the heavy weight hanged with the hook will be lifted upward. The mast can be revolved in a desired direction, for placing load, water from the cylinder is then taken out and ram makes inward stroke. Thus the hook with load comes down on the ground.

Velocity Ratio : The velocity ratio of crane hook to the ram depends upon the number of sets of pulleys. If there be *n* pulleys on the ram and similar number of pulleys on the cylinder, them load will move through a distance which will be In times the travel of ram, *i.e.*, velocity ratio = $2n$.

16.4.4. Overhead (Gantry) Cranes

These are used for handling loads over a long rectangular area in factories, power stations, in shipyards and in workshops. The simplest form mainly consists of (*i*) structural bridge support on elevated tracks, and (*ii*) supporting a lifting block or hoist trolley. Control is from a cabin on the bridge or by pendant at floor level. Due to freedom from floor obstructions, these are widely used in erection, foundries, steel plants, power houses, shipyards etc.

The bridge consists of two main girders which are fixed at the end and supported on tram wheels and can more on gantry rails. The hoist trolley also known as crab is supported on these girders and is capable of travelling across main girders. Thus load hook has three motions viz. hoisting, *i.e.*, lifting and lowering; cross travelling with trolley in both directions; and longitudinal traversing with the bridge in both directions.

Fig. 16.13. Travelling bridge crane.

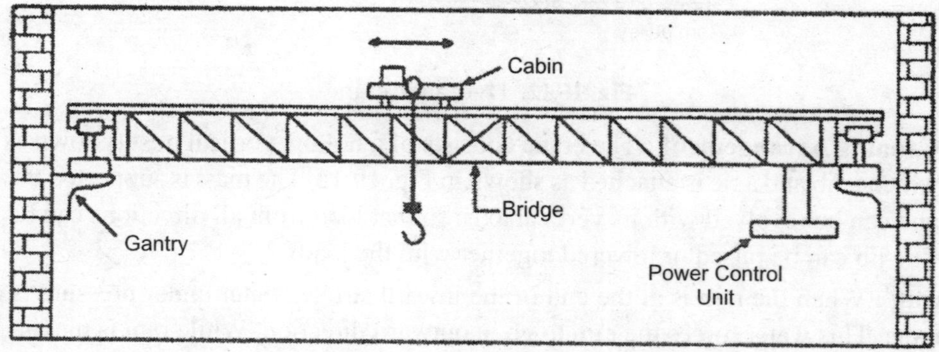

Fig. 16.14. Overhead crane.

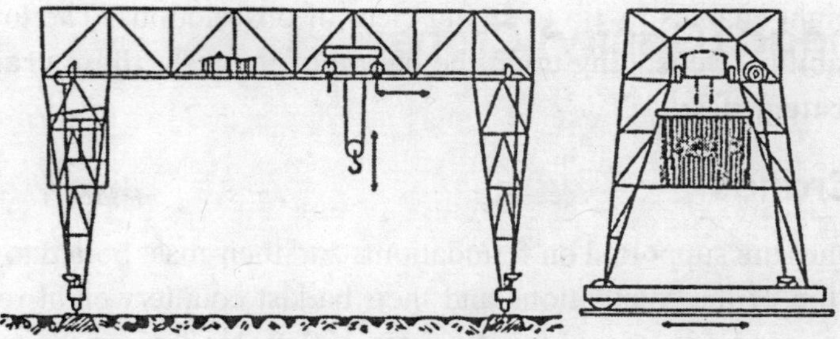

Fig. 16.15. Gantry crane.

16.4.5. Tower Cranes

Tower cranes are used in various construction projects of high buildings, bridges, cooling towers, television towers or power plants. The construction of the tower crane is such that it can work and then dismentled in restricted space. The tower has a truss structure welded from angle bars and channels. Tower extension pieces are available to increase the height as per requirement. Ladder are provided inside the whole height of the tower.

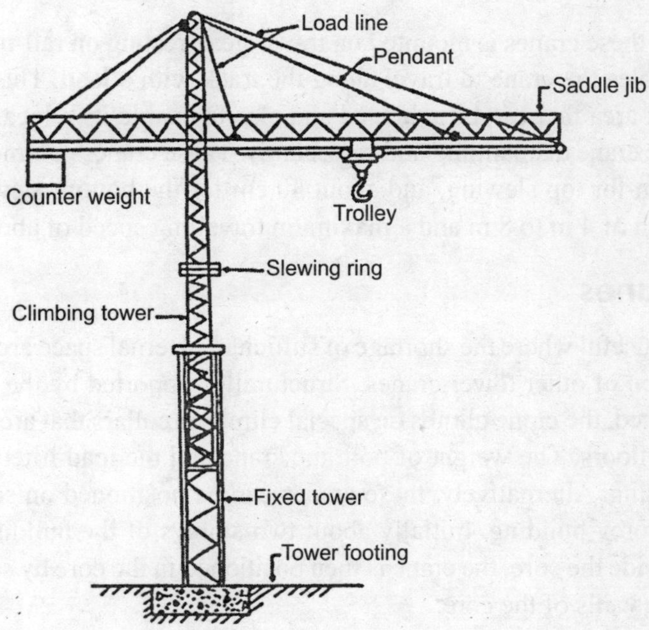

Fig. 16.16. Tower crane.

An operator's cabin is provided so as to enable full view. Jib is attached to the last highest section of the tower. Two sections of the jib, opposite to one another, are provided; one is saddle jib and another is counter weight jib. A saddle moves horizontally or rails provided on the saddle jib and is controlled from the hook block. The upper portion of the lower is fixed on the slowing head.

As mentioned above these cranes are used in high rise construction sites where concentration is in a limited area. The tower crane configuration enables the crane to be erected close to the high rise structure, and its height enables its jib to swing clear of obstructions. The tower cranes have 360 degree slewing capability. Depending upon the nature of the task, **these cranes can be grouped into the following categories:**

1. Stationary Cranes

These tower cranes are supported on foundations, and their mast bolted to a fixed position steel base placed on the top of the foundations and their ballast counterweight rests on the base. The height of these cranes can be increased further (upto 100 m) by bracing the mast with the building structure. Such cranes are of following two types :

(*a*) Top slewing tower cranes : These have a fixed tower and a swing circle mounted at the top, allowing only the jibs, tower top and operator cabin to rotate.

(*b*) Bottom slewing tower cranes : These tower cranes have the swing circle located under a slewing platform, and both the tower and jib assembly rotate relative to the base chassis. In this case the entire ballast is placed on the revolving base platform. In these cranes, the tower is essentially a telescopic mast for medium and large cranes. The smallest, hollow-section models is often of the folding type, and not telescopic type.

2. Travelling Cranes

The steel base of these cranes is mounted on travel-gear, resting on rail-tracks embedded to the foundation. This enables the crane to travel along the track with a load. This allows the increased coverage of the work area that can be achieved and also to enable its relocation with the progress of project without the crane dismantling and reassembly. These cranes commonly have a height not exceeding about 70 m for top slewing, and about 30 cm for the bottom-slewing type. The tracks, generally have ' width of 4 m to 8 m and a maximum travelling speed of about 20 to 30 m/min.

3. Climbing Cranes

These cranes are useful where the shortage of sufficient external space around the building does not permit the erection of other tower cranes. Structurally supported by the floors of the building that is being constructed, the crane climbs on special climbing collars that are fitted to the buildings completed structural floors. The weight of both the crane and the load lifted is transmitted be the structure of the building. Alternatively, these cranes can be positioned on solid cores like the lift shaft inside a multistorey building. Initially about two storeys of the building are constructed by placing the crane outside the core, the crane is then positioned in the core by securing it with special collars, resting on the walls of the core.

These cranes have only a relatively small mast because it moves upward with vertical construction progress. This climbing movement which causes work on the height of the mast employed. At the end of the construction, the tower crane is at the top of the structure. Its removal is then planned by using a derrirk, which is placed on the roof such that it can dismantle the tower crane, and then it is dismantled itself by using the hand tools, and its parts are then taken down in the elevator of of building.

From the above, it is clear that the selection and use of a tower crane requires considerable planning, because of its fixed position, its mast ability to cover all points from where the loads are to be lifted and to reach all the locations where the loads must be placed. The tower cranes are selected for use based on following factors:

(*i*) Weight, dimensions and lift and of the heavy loads.
(*ii*) Maximum free-standing height of the crane,
(*iii*) Maximum braced height of the crane.
(*iv*) Crane climbing arrangement.
(*v*) Weight of the crane to be supported by the structure.
(*vi*) Area that must be reached.

QUESTIONS

Q.1. Describe different types of hoisting equipment.

Q.2. Explain different types of lifting devices.

Q.3. Write short notes on:

(*a*) Gantry crane (*b*) Power winch
(*c*) Power hoist (*d*) Elevators

Q.4. Explain following in brief:

(*a*) Pillar crane (*b*) Overhead crane
(*c*) Screw jack (*d*) Hydraulic jack

Q.5. Describe construction, working and application of tower crane.

Q.6. Classify crane on the basis of broad construction features. Explain the application of each type and its method of operation.

Q.7. Explain a gantry crane. Explain its applications.

Q.8. Discuss different types of tower cranes.

Q.9. Explain bucket elevators in detail.

Q.10. What do you understand by "Hoisting equipment"? How will you classify them?

APPENDICES

Types of Structures

Building

Any structure for what so ever purpose and of what so ever materials constructed and every part thereof whether used as human habitation or not and includes foundation, plinth, walls, floors, chimneys, plumbing and building services, fixed platforms, verandah, balcony, cornice or projection, part of a building or any thing affixed there to or any wall enclosing or intended to enclose any land or space and signs and outdoor display structures is called as 'Building' Tants, turpentine shelters are not considered as building.

A building consists of three main parts :

(*i*) Sub-structure

(*ii*) Plinth

(*iii*) Super structure.

The part which is below the ground level and in contact with the sub-strata below and which transmits all the load to the ground below, is known as sub- structure. The part of budding, from the ground level to the ground floor level which is higher than the surrouding ground is called plinth and the portion of the building above the ground floor level is called superstructure. The plinth keeps the floor level above the normal ground level, protecting the inside area from rain water, unsanitary conditions outside and reptiles. The superstructure does the main function of enclosing the space and provides a barrier for protection of inside from the natural elements The components of superstructure are Walls. Floors, Roofs, Door and Windows, Stairs, Finishes and Fittings. Some important aspects of these components are discussed in this chapter.

There are four type of structures generally used in building construction depending upon method of transmitting load to sub-soil, and method of construction. These are

1. Load bearing structure used for low rise buildings and Bungalows.
2. Framed structure used for multistory buildings and for low rise buildings also.
3. Composit - a combination of both.
4. Industrial Structures generally constructed of steel work, steel trusses resting on steel or concrete columns and used for industrial buildings mainly but, now a days framed

structure or load bearing structures are also used for industrial purposes and hence may be called as industrial structure. The height of Building is generally more.

5. Timber structures

Structures are further classified as sub structure and super structure based on their position, with reference to ground level. A portion of any structure below ground level is called substructure and portion above ground level is called super-structure. In case of load bearing structure the load of super structure is transferred to subsoil through load bearing walls. In case of framed structure, the load of super-structure through beams is transferred to column footing to the sub-soil. The various components of a building are shown in Fig.

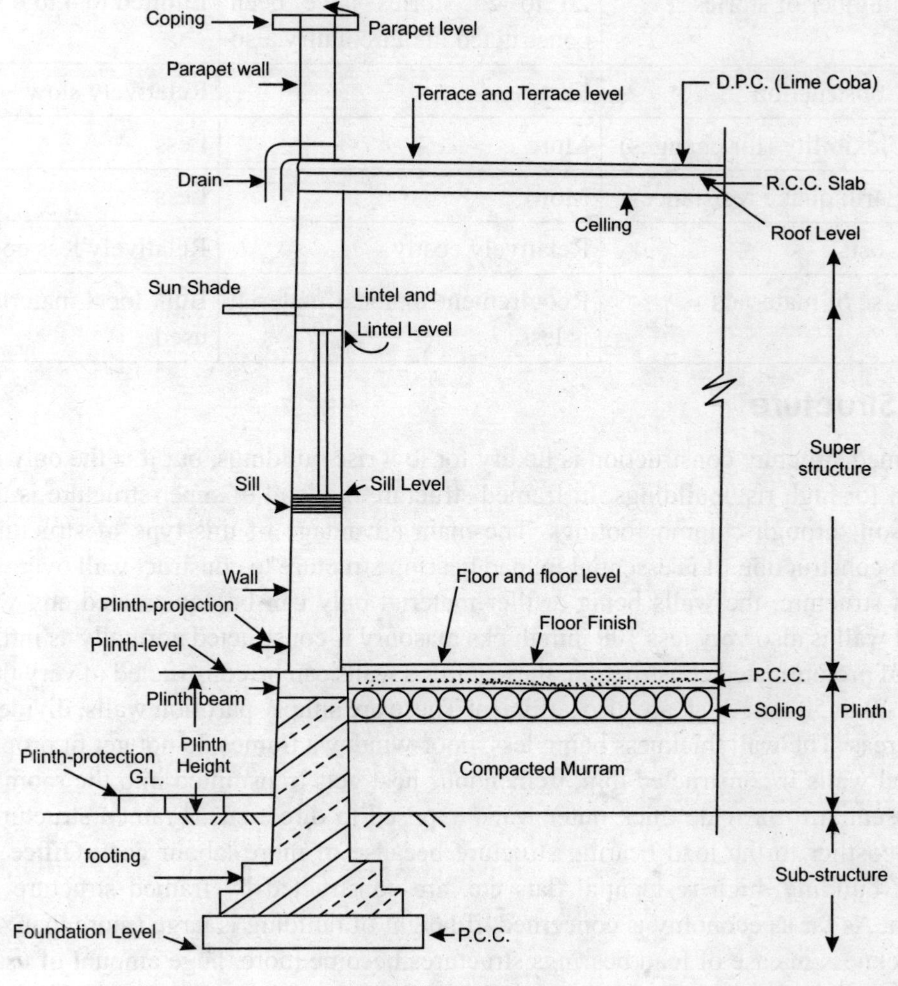

Fig. A-1

Load Bearing Structure

Load bearing structures are those in which load of super-structure is transferred to the sub soil through sub-structure *i.e.* wall foundation of that structure. The load of roof slab is transferred through walls to the foundation/footing. However load bearing structures are becoming obsolete in urban areas. In rural areas, people do construct load bearing structures.

Comparison of Framed Structure and Load Bearing Structure

Sr. No.	*Item of comparison*	*Framed structure*	*Load bearing structure*
1.	Floor space	More space because of thin walls	Less space because of thick walls
2.	Foundation condition	can be constructed even in difficult soils	Requires strong foundation soil continuously
3.	Dead load	Less	More
4.	Number of stories	20 to 25 stories have been constructed in our country also	Limited to 4 to 6 stories
5.	Construction	Fast	Relatively slow
6.	Flexibility (for changes)	More	Less
7.	Earth quake resistance	More	Less
8.	Cost	Relatively costly	Relatively less costly
9.	Use of materials	Requirement of local material is less.	Bulk local material can be used.

Framed Structure

Though framed structure construction is luxury for low rise buildings, but it is the only method of construction for high rise buildings. In framed structure the load of super-structure is transferred to the sub-soil through column footings. The main advantage of this type of structure is wide flexibility in construction. It is essential in load bearing structure to construct wall over wall where a in framed structure, the walls being a filler material only can be constructed any where. The thickness of wall is also very less 100 mm thicks masonry is constructed normally as infilled walls. Due to use of prefabricated construction, these infilled walls can be constructed of very thin precast wall panels 40 to 50 mm thick are also sufficient These are simply partition walls, divided big area into small areas. The wall thickness being less, door-windows frames do not get fit properly, more over external walls if constructed thin, tremendous heat gets transmitted into the rooms, hence it becomes essential to provide thick outer walls exposed to direct sun. Framed structure is about 15 to 20% costlier to the load bearing structure because of more labour cost. Office buildings, institutional building, high residential flats etc. are constructed by framed structure system of construction. As far as economy is concerned, if height of building is large (more than 4 storyed), the wall thickness in case of load bearings structures become more, huge amount of usable space is lost and this no larger remains economical. Framed structure become economical beyond 4 story building height

Steel Structure

It is a common method of construction for industrial steel sheds. Rolled steel joist are used as column, connected by steel beams. Grillage footing is provided as foundation and trusses are provided at top over which G.I., asbestos, plastic sheets are provided as covering material. The height of such buildings is always more, walls are generally constructed either of brick masonry or G.I. sheets or other light weight water proof materials are used for walling purpose These structure are very economical over to framed and load bearing structures but only single story structure is possible, but when long span structures are required to be constructed steel structure is only the alternative method of construction. Now-a-days cable structures are mostly used for very long column free constructions.

Composit Structure

Now a days due to very high structures composite structures are also becoming common in which steel column are constructed and other members are of concrete. A composite structure is one in which either load is distributed partly by column and partly by load bearing wall Industrial structures are generally composit one in which load of trusses is distributed on concrete columns or steel columns with infilled load bearing masonry walls.

The composite construction also consists of external load bearing walls with columns inside as some more space can be made available because of omitting internal thick walls.

The framed structures may be of (*i*) timber (*ii*) R.C.C. cast in situ or precast prestressed (*iii*) steel. Precasting is preferable for faster construction, giving more space, better quality control. steel given less dead load. Many times combination of steel frames above and precast prestressed concrete frames below may be used reducing the dead load. For steel frame skilled labour is required. For concrete semi skilled labour would do. Concrete givens better scope for creativity of the designer. In case of steel he has to resort to standard steel sections. However complexity of size and shape increases cost of form work transferred to the walls and then sub-soil through foundations. Wall footing may be called as strip footing as provided through out below the wall. The essential requirement of this type of structure is that as the load is transferred through the walls, walls are required to be constructed over the wall in case multistoryed buildings. Bricks used for masonry construction (walls) must be strong enough to resist the load and then to transfer it. As the walls resist the load, wall thickness becomes very large in case of multistories buildings and therefore a good amount of usable area is lost in walls, moreover brick being weaker material of construction relatively, high rise building can not be constructed using load bearing structure. For low rise building load bearing structures are very common and is economical by about 15 percent to framed structures. The minimum wall thickness is generally 200 mm in case of metric bricks and 230 mm in case of conventional bricks. Wall thickness being more, no sound is transmitted from one room to another Door-windows frames are best filted due to thick walls and electrical conduits or casing can be put over the walls easily. This type of structures is used mostly for low rise buildings (G + 1) or Bungalows. These buildings are more susceptible to vibrations than framed structures. These buildings can be constructed if hard strata is available at a reasonable depth i.e. about 1 m to 1.5 meter, else foundation cost increases and it no more remain economical one.

Timber Structures

These structures are used where timber is available in abundance at a reasonable cost. These structures are also used where earthquake occur frequently. Timber structures mostly have sloping roof. The loading frame is beams and columns are constructed of heavy wood and walls are mostly partition walls of plywood. The load is transferred through columns to foundations which are generally of concrete or brick. Such structures are very common in cold countries like Europe and America as wood is a bad conductor of heat and cold. These structures are also common in Japan.

Traditional Construction : traditional construction is a process of construction of a building using load bearing structure in which load of superstructure is transmitted to subsoil through plinth beam to wall footings that is wall Foundation or spread Foundation constructed of stone masonry or brick masonry or block work construction. In case of framed structure, the load of superstructure is transferred through slab to beam and then beam to column and then to sub soil through column footings or raft or village or through shell foundation or by deep foundation i.e. pile, pier or caisson/ well foundation.

Slabs supported on walls or on beams are classified as conventional slab. Conventional slab are generally rectangular in shape, but if can be of any shape such as triangular, circular, trapezoidal etc.

Traditional Construction

Traditional construction is a process of construction of a building using load bearing structure in which load of superstructure is transmitted to subsoil through plinth beam to wall footings that is wall foundation or spread foundation constructed of stone masonry or brick masonry or block work construction. In case of framed structure, the load of superstructure is transferred through slab to beam and then beam to column and then to sub soil through column footings or raft or grillage or through shell foundation or by deep foundation i.e. pile, pier or caisson/ well foundation.

Slabs supported on walls or on beams are classified as conventional slab. Conventional slab are generally rectangular in shape, but it can be of any shape such as triangular, circular, trapezoidal etc.

FOUNDATION

Importance - Type - Purposes

The foundation is the most critical part of a structure and most the failures are due to failure of foundation than any other causes. The stability of life of a structure depends on a good foundation. Therefore, it is important to construct a good foundation.

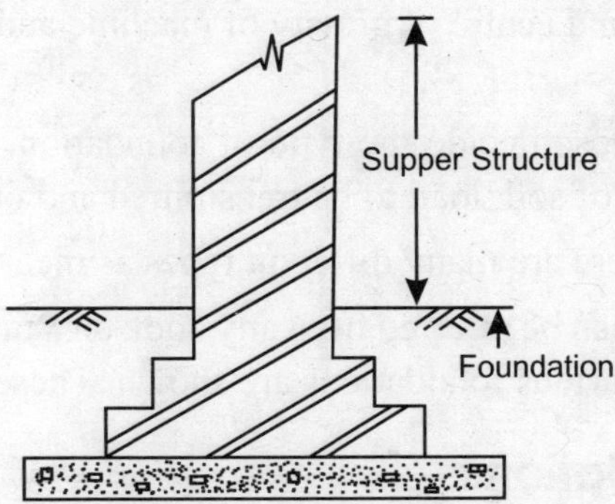

Every structure consists of two parts-one below the ground level and another above the ground level The portion which is below the ground level is called foundation" and the other which is above the ground level is called "super-structure".

The foundation work consists of excavation, preparation of bed, placing of P.C.C. and stone and brick masonry footings. The term "foundation" is defined as the artificially prepared lowest part of the structure, resting over the ground and where transmits the load of the structure over the subsoil below it.

The superstructure is the portion above the ground level. It consists of walls or columns. The load of furniture, living persons and slab etc. will be distributed on the walls or columns, then from walls or columns to the foundations and lastly, from foundation to the sub-soil below. In this way, the whole weight of the structure is transmitted to subsoi through the foundation.

Purpose of Providing Foundation

An ordinary common man is under the impression that the foundation supports the load of the structure, but it is not the object of the foundation. The purpose of a foundation are listed below (*a*) To distribute the load of the structure over the larger area so as to prevent from an settlement.

(*b*) To give stability to the structure to prevent it from over turning.

(*c*) To provide level surface for concrete bed and masonry courses.

Types of Foundations

They are classified as

1. Shallow foundation or open foundation
2. Deep foundation.
3. Machine foundation

Shallow foundation are those where depth of foundation is less than or equal to its width and constructed by open excavation.

Deep foundation are laid down at a considerable depth below ground level and usually involves special methods of construction. They are costly and difficult in construction.

Machine Foundation is also a type of shallow foundation, however used to support machineries, those vibrate during their working. The basic requirement is that the weight of foundation should be more than that of machine and centre of gravity of machine and that of foundation should be in same vertical Axis.

There are many factors those decide the type of foundation to be constructed for particular structure like bearing capacity of soil, load to be transmitted and others.

Under these categories, there are many different types as mentioned in figure.

Foundation and its details can be referred from any book on Building construction or foundation designs . Figures of all these various foundations are shown. These figures are self-explanatory.

Construction of Foundations

The construction methods of all these foundations depends upon the type of foundation, its design based on load and bearing capacity of soil over which Foundation rests. Now a days, machineries and equipments are used to construct these foundation. Power shovel; pile boring machines; excavators; breakers; concreting equipments are used to construct these foundations.

In case of Shallow Foundation like footing either of wall or of column; pits are excavated by excavators, power shovels, reinforcement cages are laid and concreting is done. In case of grillage foundation, Steel sections are laid over concrete base. These steel sections may be in one, two or more rows as per design, welded to each other by plates; steel columns is placed over these sections and welded. The sections i.e. grillage is then encased i.e. covered by concrete to avoid rusting of steel sections.

Shell Foundation is also a type of shallow foundation constructed to transfer load over a large surrounding areas. However this type of foundation is not used nowadays.

In case of pile Foundation, boring rigs, drilling machine rigs are used to bore a hole in soil to a required depth and of required diameter. Reinforcing cage is then put in excavated/ bored hole and concreting is done as usual. Sometimes is steel sections like H section, box section etc. are

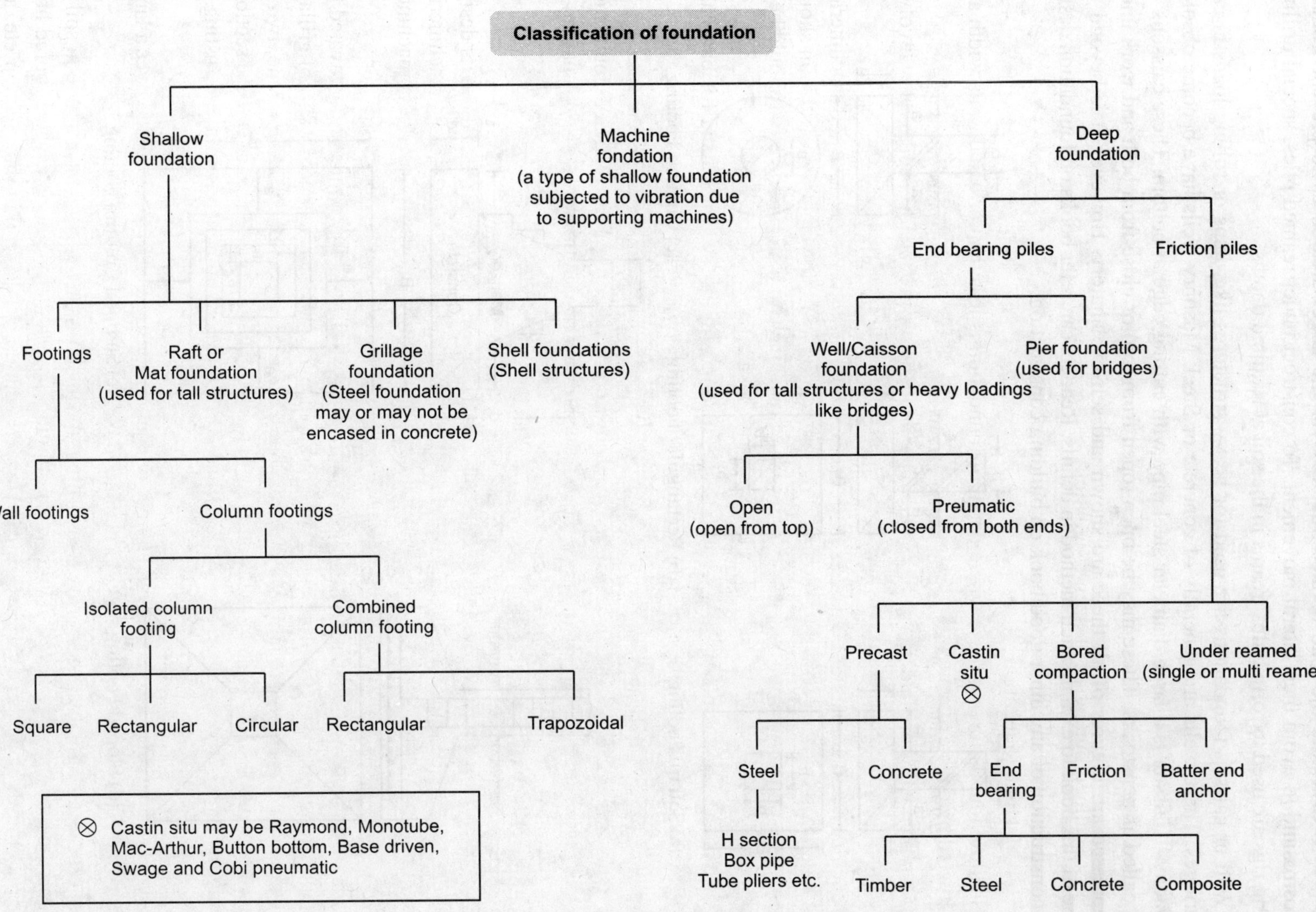

Classification of foundation
Shallow foundation
Machine fondation (a type of shallow foundation subjected to vibration due to supporting machines)
Deep foundation
Footings
Raft or Mat foundation (used for tall structures)
Grillage foundation (Steel foundation may or may not be encased in concrete)
Shell foundations (Shell structures)
Wall footings
Column footings
Isolated column footing
Combined column footing
Square
Rectangular
Circular
Rectangular
Trapozoidal
End bearing piles
Friction piles
Well/Caisson foundation (used for tall structures or heavy loadings like bridges)
Pier foundation (used for bridges)
Open (open from top)
Preumatic (closed from both ends)
Precast
Castin situ ⊗
Bored compaction
Under reamed (single or multi reamed)
Steel
Concrete
H section
Box pipe
Tube pliers etc.
End bearing
Friction
Batter end anchor
Timber
Steel
Concrete
Composite
⊗ Castin situ may be Raymond, Monotube, Mac-Arthur, Button bottom, Base driven, Swage and Cobi pneumatic

used. The sections are put in drilled hole and concreting is done around the sections to avoid their corrosion and to avoid their lateral movement. To construct under reamed piles, special drilling/ boring rigs are used to form bulbs/ reams in the soil at required depths.

Well or caisson foundations are used for heavy structures like dams or rolling loads in case of bridges. These wells are generally of concrete or Steel. Masonry wells have become obsolete nowadays. Caisson is a large diameter steel pipe with cutting edge at bottom. These caissons are also called as penstocks. These may be open (open from top) or close from both and work under compressed air. Figures of all these are shown and self-explanatory. However, it is beyond the scope of this book to explain each method in details. Readers can refer books on foundation design and construction or any other good book on building construction.

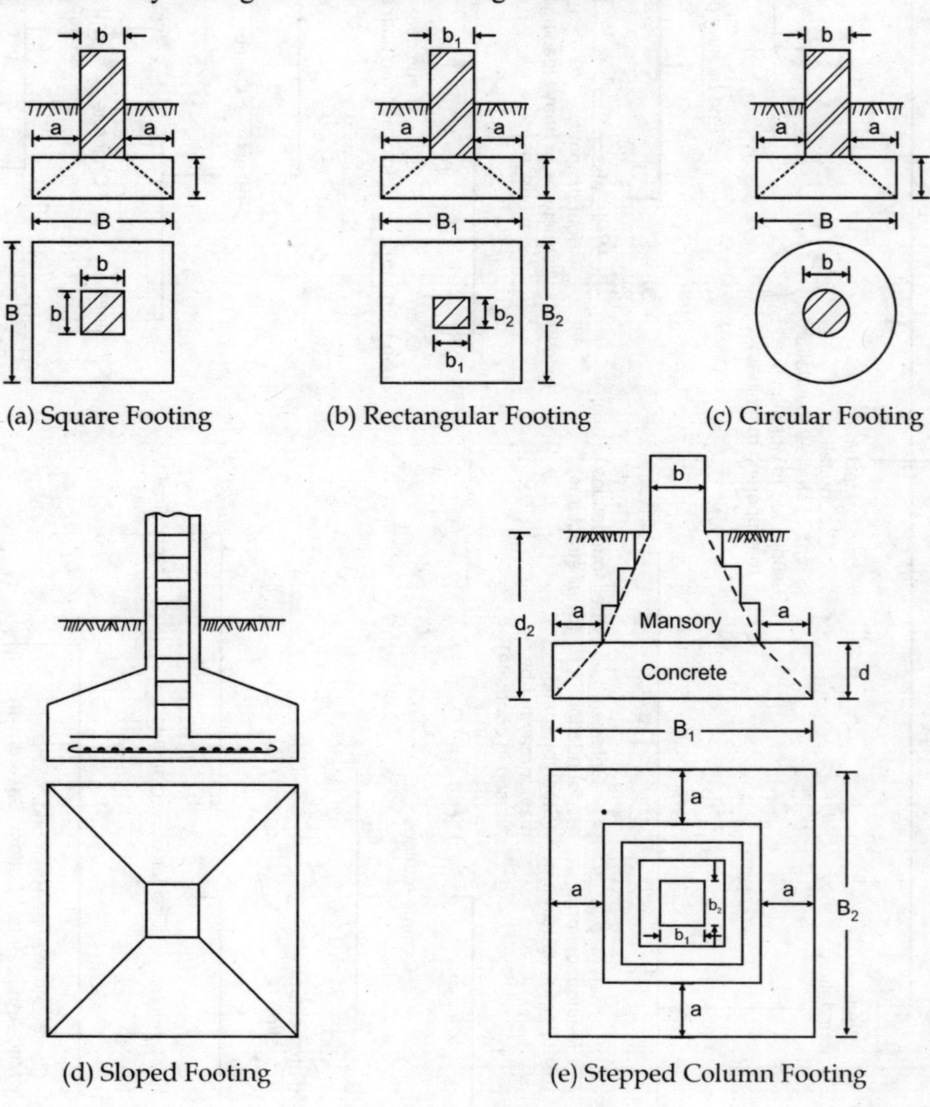

(a) Square Footing (b) Rectangular Footing (c) Circular Footing

(d) Sloped Footing (e) Stepped Column Footing

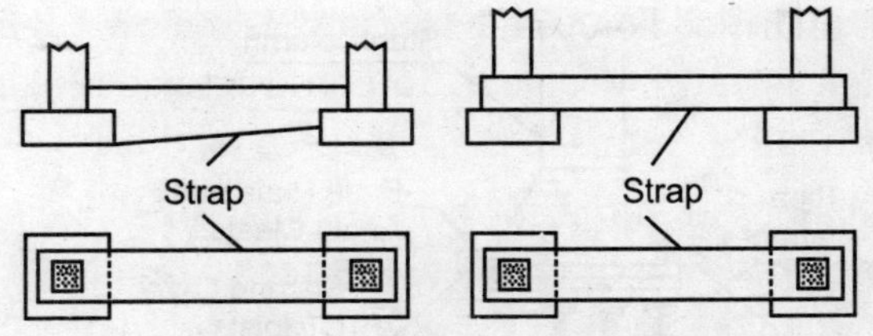

(f) Cantilever Footing or Strap Footing

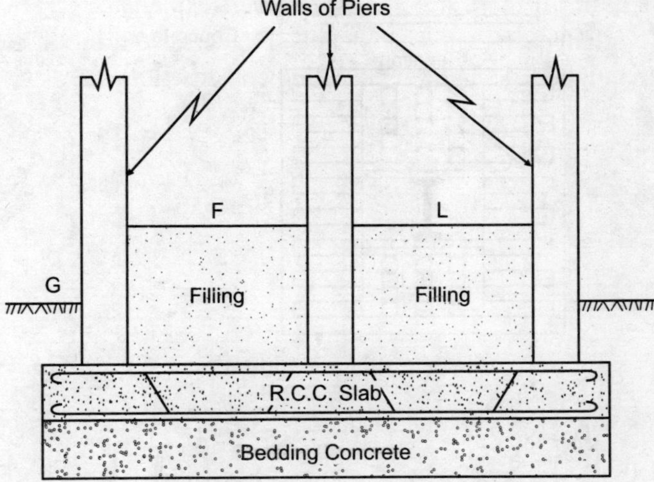

(g) Raft/Mat Foundation

Fig.

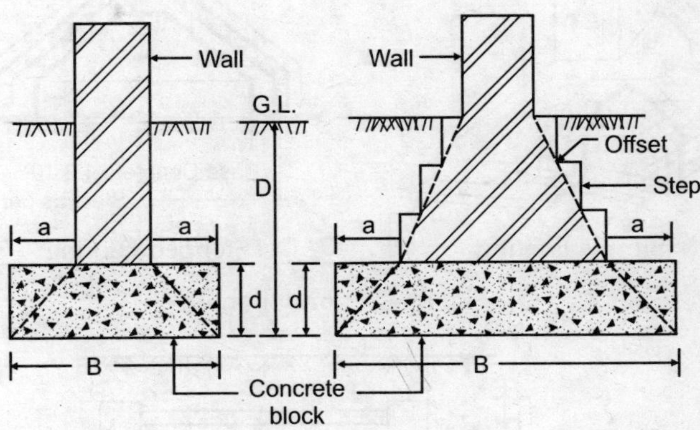

Fig. Simple Footing **Fig.** Stepped Footing

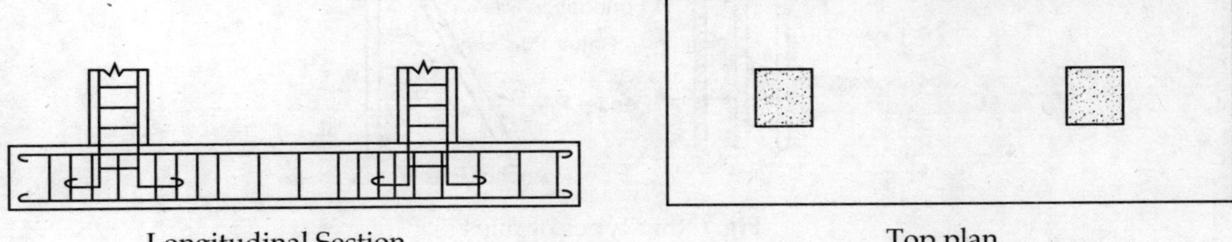

Longitudinal Section Top plan

Fig. Combined Footing

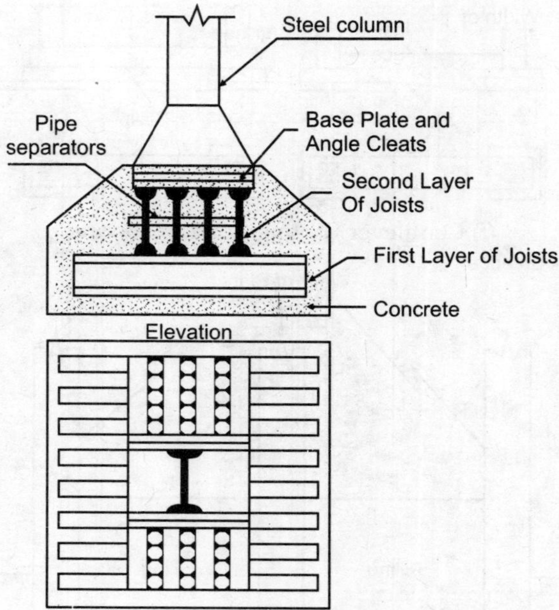

Fig. Grillage Foundation

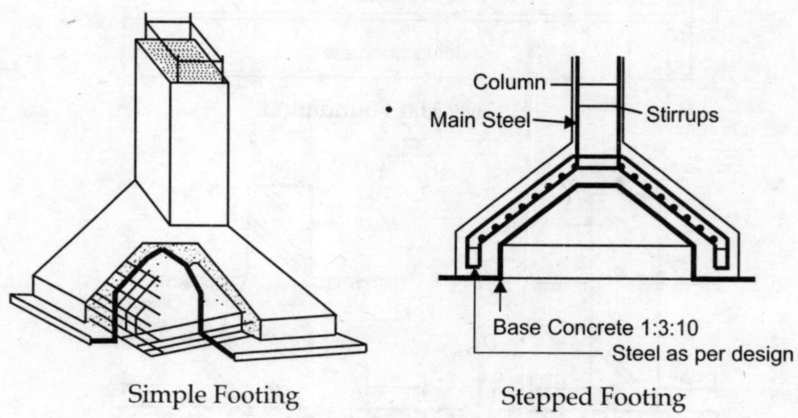

Fig. Shell Foundation

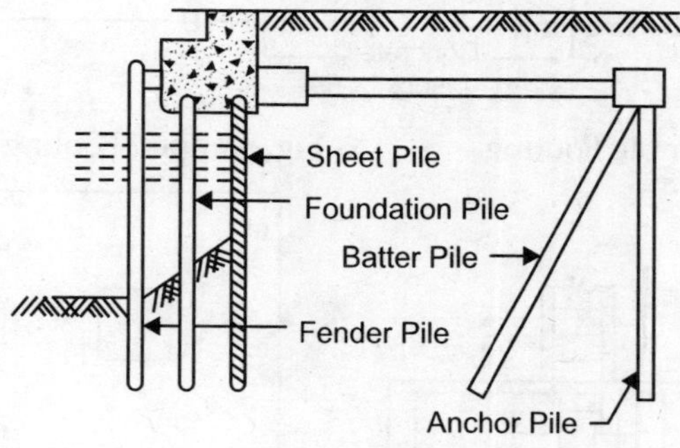

Fig. Other types of piles

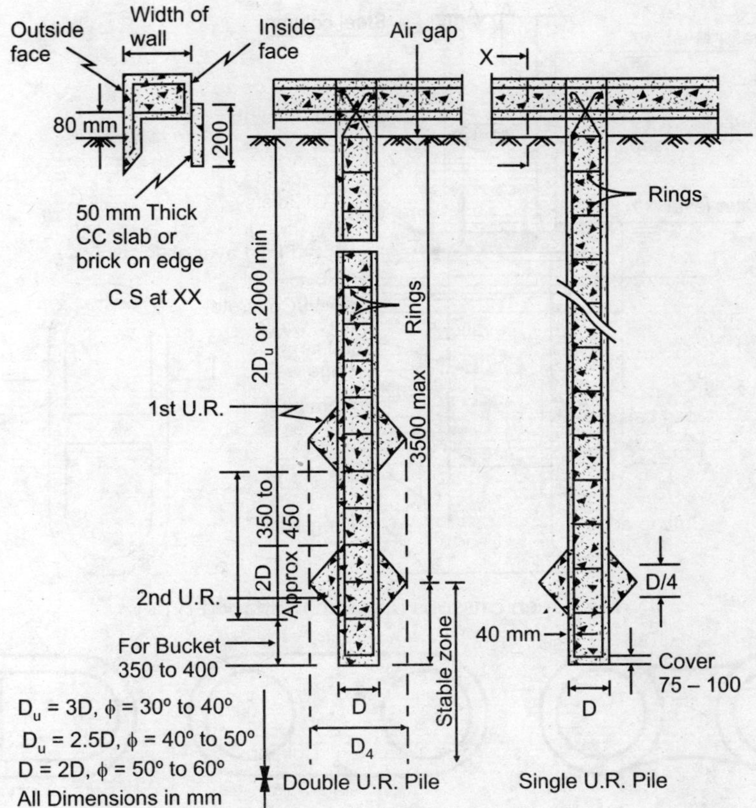

Fig. Under reamed pile foundation

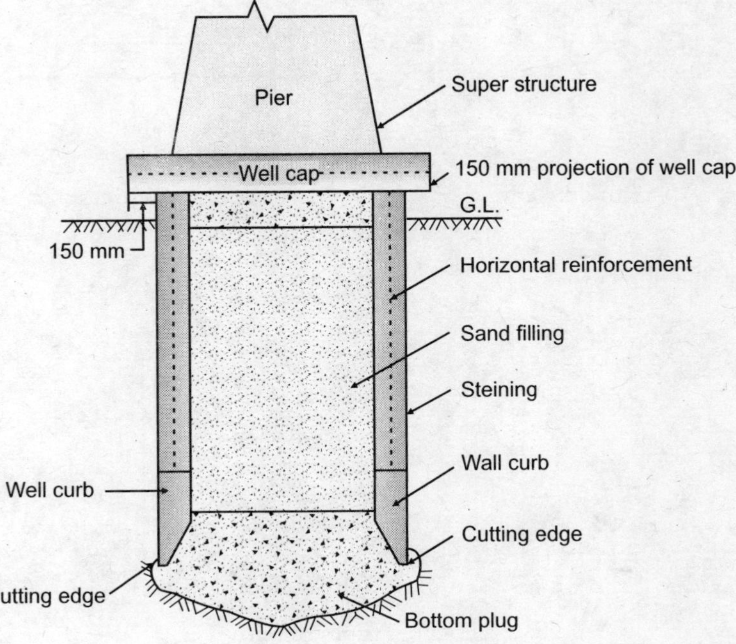

Fig. Components of Well foundation

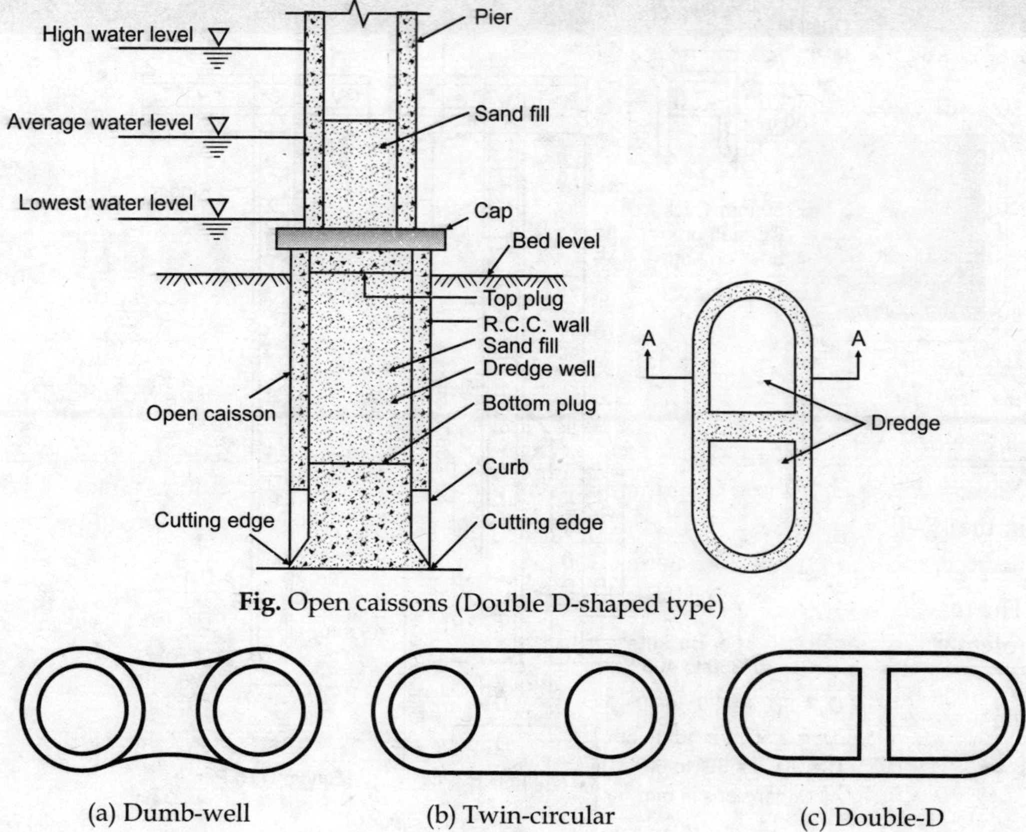

Fig. Open caissons (Double D-shaped type)

(a) Dumb-well (b) Twin-circular (c) Double-D

Fig.

HOURLY WORKING RATE OF EQUIPMENT

Note:– Various construction equipments as per AICTE model curriculum, although covered in the book in the chapters from 11 to 16, the other common construction equipments have been discussed in these "Appendices"
The readers interested in in-depth knowledge on entire range of construction equipment may refer the book titled as "Construction Equipment and Management" by the author S.C. Sharma.

Hourly working rate of construction equipment comprises of the following components :

1. Owning Cost : It is made up of following costs :

(*a*) Investment cost, (*b*) Depreciation cost,
(*c*) Major repair cost

2. Operation Costs : It includes the following costs :

(*a*) Cost of fuel (or power), (*b*) Cost of lubricants,
(*c*) Servicing and maintenance cost, (*d*) Labour cost,
(*e*) Cost of field repairs, (*d*) Various other overheads.

1. OWNING COSTS

1.1. Investment cost

This is a kind of fixed cost and continues to be incurred whether the equipment is used or not. Investment cost comprises of the following :

(*a*) Interest on the money invested in the procurement of the equipment.
(*b*) Various taxes on the equipment.
(*c*) Insurance expenses.
(*d*) Storage costs.

If the money spent on the purchase of the equipment, was invested in the bank or similar other agencies, it would have earned an interest. Now this amount, equivalent to the interest, should be recovered from this equipment. The rate of interest should be equal to the prevailing bank rates.

Taxes pertaining to ownership of the equipment are those paid by the owner for the equipment. Similarly owner is also required to pay the insurance premium for the equipment, in order to cover the risk of the equipment and any other loss caused due to any accident by the equipment. Storage cost is the cost of space required by the equipment and other expenses to be incurred by the owner for keeping the equipment.

Generally these investment costs are taken as about 10 to 15% of the total cost of the equipment. This total cost comprises of the following:

(*a*) F.O.B. price of the equipment with all attachments and accessories.

(*b*) Insurance and freight charges.

(*c*) Expenses on unloading, clearance and custom duty.

(*d*) Cost of transportation to the project site including loading and unloading.

(*e*) Errection and commissioning charges.

1.2. Depreciation Cost

Due to use and obsolescence every equipment loses its value. This loss is accounted for by depreciating the equipment every year. The depreciation, and obsolescence are described in detail hereunder.

1. Depreciation

Whenever any machine or equipment performs useful work its wear and tear is bound to occur. This can be minimised upto some extent by proper care and maintenance but can't be totally prevented. Its efficiency also reduces with the lapse of time and at one time it becomes uneconomical to be used further and needs replacement by new unit.

Therefore, we can say that efficiency and value of machine or asset constantly reduces with the lapse of time during use, which is known as "Depreciation". Some money must be set aside yearly from the profits, so that when an equipment becomes uneconomical, it can be replaced by the new one. Therefore, the initial cost of machine plus installation charges + repair charges – scrap value is charged against overheads and spread over the machine's useful life.

For this purpose depreciation account for an individual equipment is opened in the Company's Books and is knows as "Depreciation Fund", or "Sinking Fund". This amount is deducted yearly from the profits and kept separately to have sufficient money for replacement at the end of useful life.

2. Obsolescence

Suppose a contractor purchases a machine for his production but after some duration a better machine comes in the market, whose production rate is very high and is more economical, although the old machine is efficient but becomes out of fashion and uneconomical due to new better machine which has come in the market. This is known as "Obsolescence". Consideration of this factor is of much importance and some money should also be set aside from the profits for this cause.

Hence, "Obsolescence" is the depreciation of existing machinery or asset due to new and better invention, design of equipment or processes etc.

It is very difficult problem for the estimator to provide for the on-cost on obsolescence, because no body can predict when a revolutionary change in the machine is coming in the market. But it is general practice to reduce the life of a machine so as to account for the effect of obsolescence. The depreciation and obsolescence charges are then calculated for the reduced life.

Suppose an estimator expects the life of machine as 10 years when the depreciation rate will be 100/10 = 10%. By considering obsolescence also, its life may be taken as 8 years. Then the combined depreciation and obsolescence charge will be 100/8 = 12.50% instead of 10%. Therefore, the difference 12.50 – 10.00 = 2.50% will be obsolescence charges.

The other causes of depreciation are physical decay, accident, deferred maintenance and neglect, inadequacy etc.

1.3. Major Repair Cost

While minor or field repairs are carried out during the day to day working of the equipment, the major repairs are carried out after the substantial use of the equipment. Major repairs and overhauls are the replacement of major parts of the equipment because of the excessive wear through a long period of use. Since these repairs require a heavy amount of expenditure, they are met from the major repair fund. The major repair cost is spread out during the entrie life span of the equipment and a flat rate is levied per working hour in order to have a uniform rate. The amount thus collected in the pool is known as major repairs fund, and major repairs and overhauls are carried out from this pool.

Usual practice is to consider this major repair repair cost as a percentage of straight line depreciation cost and is generally taken as 80 to 200% of the cost of depreciation depending on the type of equipment.

2. OPERATING COSTS

(A) Cost of Fuel (or Power)

Prime-movers for construction equipments are generally diesel engines or electric motors. The actual consumption of fuel or electricity in these construction equipment depends upon:

(*i*) Engine B.H.P.

(*ii*) The load factor, which means the extent to which the engine will operate at full power. Load factor generally vary from 30% to 70%.

(*iii*) The condition of the engine.

Optimum fuel consumption in litres per hour per H.P. may be calculated using following formula.

Fuel consumption = 0.27 × Load factor

Equipment used in the construction industry seldom operate at a rated output, except for a short period of time.

During the balance part of its cycle the demand on the engine will be reduced substantially, resulting in decreased consumption of fuel. To allow for this the value of load factor (*C*) may be taken as follows as recommended by the "construction plant and machinery committee-1972 (central water and power commission)", taking into consideration the job condition also.

S. No.	Types of equipment	Excellent	Average	Severe
1.	Wheel type equipment (Highway)	0.25	0.30	0.40
2.	Wheel type equipment (off-Highways)	0.50	0.55	0.60
3.	Track type tractors	0.50	0.63	0.75
4.	Excavators	0.50	0.55	0.60

Construction equipment is seldom operated for 60 min. in an hour and therefore, the actual fuel consumption is calculated by multiplying utilisation factor to the fuel consumption arrived from above formula.

(B) Cost of Lubricants

(*i*) Engine oil

(*ii*) Air filter oil

(*iii*) Transmission oil

(*iv*) Hydraulic oil

(*v*) Greases etc.

The quantity of lubricating oil depends upon various factors. Some of these factors are :

(*i*) Capacity of the crank case.

(*ii*) Condition of piston rings.

(*iii*) Number of hours between oil changes. This depends on the operating conditions and the manufacture's recommendation. In extreme dusty conditions oil is required to be changed after every 50 hours of working, while in normal conditions it may vary from 100 hours to 200 hours.

Lubricating oil Consumed : Requirement of lubricants including engine oil, air cleaner oil, transinission oil, greases and hydraulic oil, can be assessed on the basis of informations supplied by the equipment manufacturer. Although actual consumption of lubricants can be taken as 30% of the fuel cost in diesel engine and about 20% of the fuel in petrol engines.

Following formula is used for estimating the quantity of oil required per hour:

$$Q = \frac{H.P. \times f \times 0.006 \times 4.5}{7.4} + \frac{C}{t} \text{ lit/hr}$$

Where, Q = Quantity of oil consumed in lit/hr

$H.P.$ = Rated horse power of the engine

f = Operating factor

C = Capacity of crankcase, in lit.

t = Number of hours between oil changes.

(C) Servicing and Maintenance Cost

Regular servicing and maintenance of each equipment is very essential in order to have the equipment in reliable and perfect condition of working. This cost includes :

(*i*) Change of lubricants.

(*ii*) Checking and servicing of fuel lubricating systems, including change of filter elements.

(*iii*) Care of tyres and tubes.

(*iv*) Care of battery and electrical system.

(*v*) Cleaning of the equipment and other similar works.

(*vi*) Salaries of staff engaged on maintenance and servicing.

(D) Labour Cost

Labour cost is an important element of operating cost and contributing a major part of it. Labour cost element includes the salaries of operators, and helpers engaged on the equipment. Provision for leave reserve (may vary from 10 to 12%) is also made under the head of labour cost. A percentage of the wages of supervisory staff, depending upon the attention required on that particular equipment, is also to be added.

(E) Cost of Field Repairs

This is the cost incurred on minor repairs which are carried out on the site or in field workshops. These repairs include replacement of minor parts, such as fan belts, filter, bearings, wire ropes etc. and other adjustments required to be carried out during normal working of the equipment.

(F) Other Overheads

These are the charges which are incurred on complete fleet of the equipment. Therefore, proportional expenditure is booked under this head on each equipment. These charges includes the pay on watchman, light and water charges in the storage yards, uniform to the operating and maintenance staff etc.

3. REPAIR RESERVE FUND

A provision for the repair of an equipment for its entire life is made by charging a percentage of the book value (excluding the value to tyre). The equipment earns this fund during the working of the equipment in order to carry out the repairs subsequently during its course of operation. Percentages for some of the major construction equipments mentioned hereunder:

- Crawler tractor/Dozer 240
- Wheeled tractor/Dozer 200
- Shovels/Draglines, hydraulic excavators, vibratory compactors, wheeled loaders motorised scrapers 150
- Dumpers, tractor trailors 140
- Crawler and mobile cranes, transit mixers 120
- Air compressors and water pumps with diesel engine 100
- Air compressors with electric motors, wagon drills, concrete mixers 80
- Batching and mixing plants, towed scrapers. 75

4. DETERMINATION OF OWNING AND OPERATING COSTS

To determine the cost of owning and operating, following procedure is adopted:

I. Ownership Charges

1. Depreciation

(A) Total cost of equipment less cost of tyres ₹ ______

(B) Deduct salvage value (generally 10% of A) ₹ ______

(C) Total investment to be depreciated (A – B) ₹ ______

(D) Economic life in hours Hrs. ______

(E) Depreciation per hours (C/D) ₹ ______

2. Interest, Insurance, Taxes and Storage

(F) Annual expenditure on these items ₹ ______

(G) Interest, Insurance, Taxes and Storage charges/hr ₹ ______

$= \left(\frac{F}{\text{Annual use in hrs}}\right)$

Total Owning Costs = E + G

II. Operating Charges

3. Fuel charge

(H) Consumption in lit. per hour ₹ ______

(I) Cost of fuel per lit. ₹ ______

(J) Fuel charges/hr. (H × I) ₹ ______

4. Lubricating and filters charges

(K) Engine oil (Consumption/hr × rate) ₹ ______

(*L*) Transmission (Consumption/hr × rate) ₹ ______

(*M*) Final drive oil (Consumption/hr × rate) ₹ ______

(*N*) Hydraulic oil (Consumption/hr × rate) ₹ ______

(*O*) Grease (Consumption/hr × rate) ₹ ______

(*P*) Filters (Consumption/hr × rate) ₹ ______

(*Q*) Total lub. and filter charges (total of K to P) ₹ ______

5. Cost of tyres

(*R*) Price of all the tyres ₹ ______

(*S*) Estimated tyre life Hrs. ______

(*T*) Cost of tyers $\left(\frac{R}{S}\right)$ per hr. ₹ ______

6. Repair charges

(*U*) Repair factor × Depreciation per hr. ₹ ______

7. (*V*) Operator's and helper's hourly wages ₹ ______

8. (*W*) Special items (like cutting edge, rippers,

drill rods, bits etc.) ₹ ______

Total operating cost = *J* + *Q* + *T* + *U* + *V* + *W* ₹ ______

Grand total of Hourly Owning and Operating Costs ₹ ______

Crushing Equipment

1. CRUSHERS

Aggregates of various sizes are required for most of the construction activities e.g., construction of roads, building, bridges, canals, tunnels, damas, air fields and other concrete structures. Production of aggregate involves drilling, blasting, loading; transporting, crushing, screening, handling and stocking.

Crushers are used to reduce, rock-mass or boulder to smaller and uniform sizes acceptable for concrete mix. The crusher production is screened to give various sizes of aggregates. The range of these sizes may vary as per requirements. These aggregates of various sizes are mixed in a prescribed proportion to give a desired quality of aggregate mix. The plants used for mixing the aggregates are known as "aggregate mixing plants".

As mentioned above crushers are used to produce aggregates. The crushers are fed with the stones broken from the rocks. The sizes of the stones fed depend upon the capacity of the crushing system adopted. Crushing may be accomplished by pressure, impact, attrition or by a combination of these operations. Pressure crushing is most common for large sized rock (or as primary crushers), whilst impact crushing is used for soft and medium sized rock or for secondary crushing. Attrition crushing is resorted for soft, small sized material or for making sand.

Pressure crusher include the Jaw crusher and gyratory crusher, though some amount of grinding or abrasion is always involved in the working of these machines. Impact crusher include hammer mills and disintegrators. Roll crushes constituted the grinding type machine.

Broadly speaking crushers are classified as primary, secondary and tertiary crushers. The primary crusher handle the raw quarry material, and relatively require larger hopper area. The final product of primary crusher is relative coarser and non-uniform. This is fed to secondary crusher which crush it to smaller and uniform product. Tertiary crusher are then used to produce fine aggregate.

(*a*) Primary Crushers : 1 Jaw crusher 2. Gyratory crusher
3. Hammer mill

(*b*) Secondary Crushers : 1. Cone crusher 2. Roll crushers
3. Hammer mills

(*c*) Tertiary Crushers : 1. Roll crushers 2. Rod mill
3. Ball mill

Crushers are heavy machines, surface of which are made of special alloys and are renewable. The selection of crushing equipment is done on the basis of :

1. Kind of stone to be crushed.
2. Size of aggregate required i.e, range, c.g. 15 to 20 mm aggregate, size.
3. Method of feeding stones to the crushers.
4. Required capacity of plant.

Crushing operation involves feeding of aggregate to crusher, crushing, screeming to screen different sizes of aggregates and transportation of material. There different operation are dealt herewith in detail. Sequence of these operations is shown in Fig. Appdx 2.1. (*a*) and (*b*)

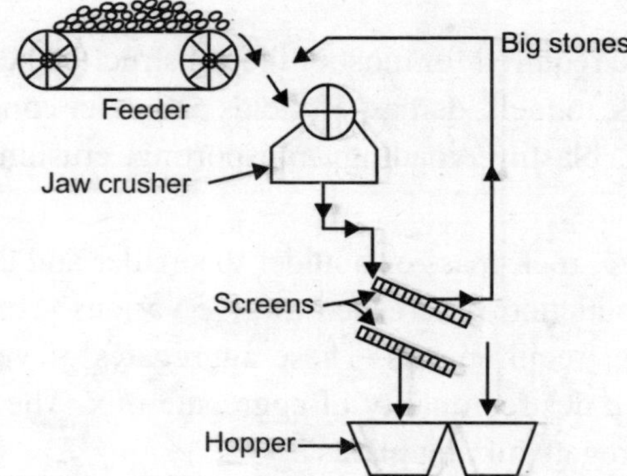

Fig. Appdx 4.1. (*a*) Material flow diagram for single stage crushing (open circuit diagram)

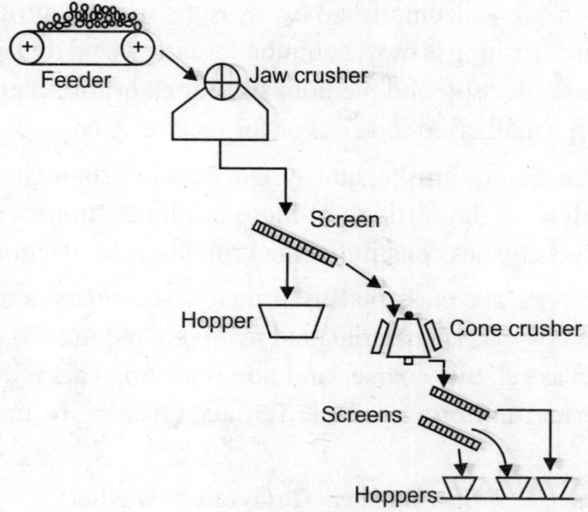

Fig. Appdx 4.1. (*b*) Material flow diagram for two stage crushing (close circuit diagram)

2. CRUSHING PLANTS

Crushing plants are being manufactured by large number of manufacturers all over the world with capacities upto 1000 tons per hour or so. These plants are either stationary or portable and changes in the design and construction from manufacturer to manufacturer and as per job requirement. The crushing plants employ 2 stages. Generally the primary crushing units are either jaw crusher or impact crusher and reduces the rocks to size between, say 75 mm and 250 mm. whereas the secondary units use jaw, impact or cone crushers, in which primary crushed rock is further reduced to sizes say 5 to 50 mm. One flow diagram is shown in Fig. Appdx 2.3. with jaw in primary and impact in secondary unit.

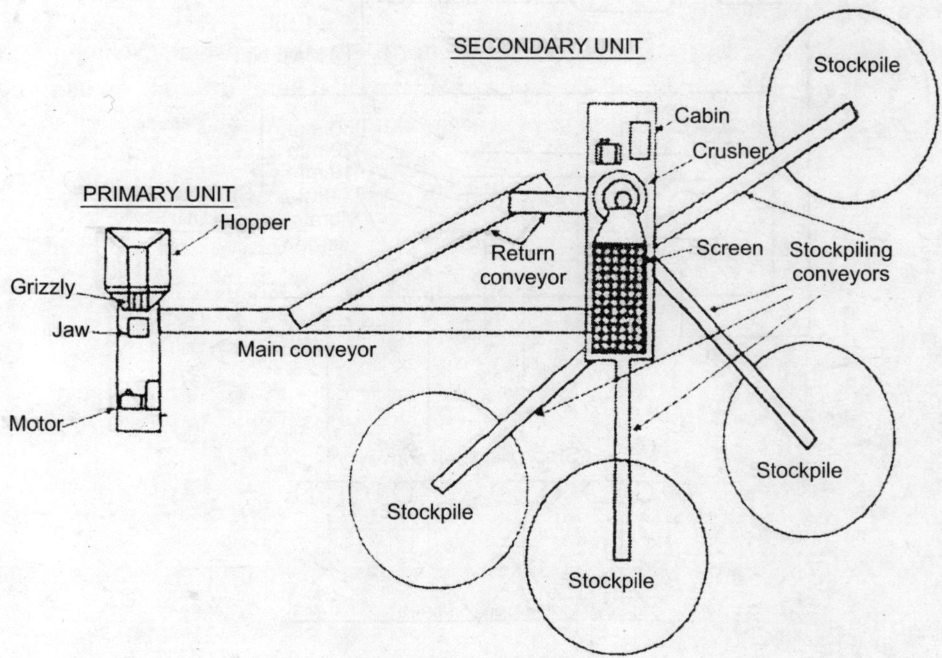

Fig. Appdx. 4.2. General layout of crushing plant.

General layout of this crushing plant is shown in Fig. Appdx 2.2, which illustrates the relative position of various units constituting the crushing plant. This also indicates the cycle of operation starting from feeding stones, boulders, gravel etc. to the crusher products of tiny and uniform granules of sizes 0–40 mm (or 0–50 mm) staggered into 4 sizes viz. 0–5 mm, 5–10 mm, 10–20 mm and 20–40 mm (or 20–50 mm) or 0–5 mm, 5–13 mm, 13–20 mm, 20–40 mm (or 20–50 mm).

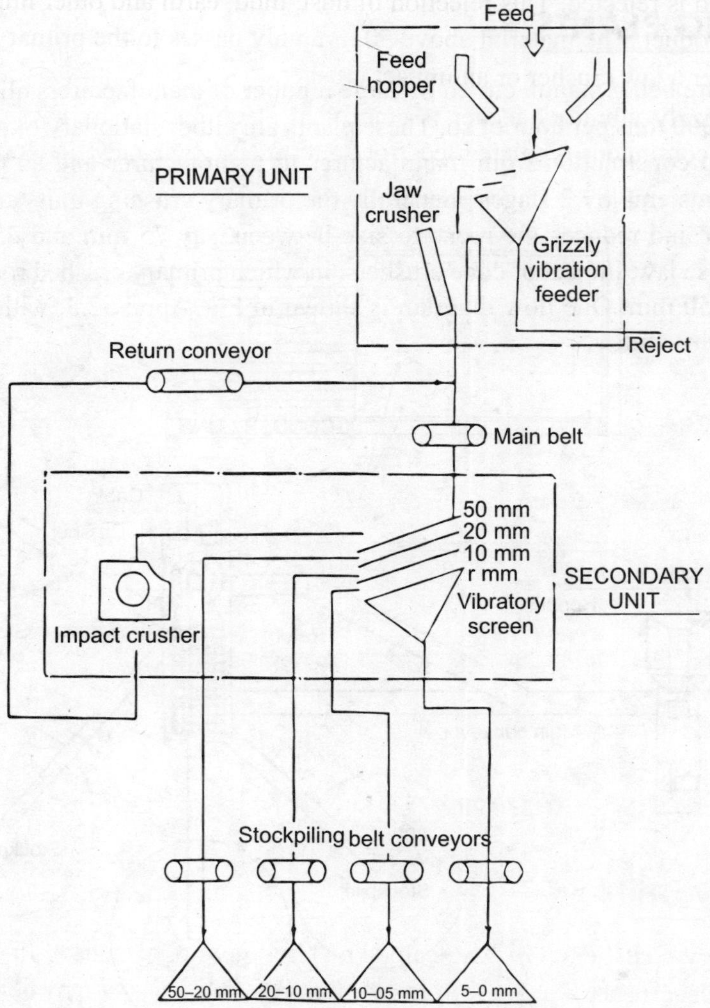

Fig. Appdx. 4.3. Flow diagram for a crushing plant.

3 CONSTRUCTION AND WORKING

3.1. Feeding Unit

The raw material (stones either from quarry, or river bed etc.) is dumped into feed hopper from a specially built platform and a ramp by means of a wheel loader or a dumper (in big size crushing plants).

(*a*) Feed Hopper : The feed hopper is a robust reinforced steel structure to withstand the severe impact shock. Big hopper volume assures constant regulated feed by grizzly vibrating feeder into the primary crusher.

(*b*) Grizzly Vibrating Feeder : The material fed to the feed hopper has dust, small pabbles, and stones of a various sizes. The material of sizes, below 50 mm passes through grizzly bars and is rejected. This rejection of dust, mud, earth and other impurities assures clean final product. The material above 50 mm only passes to the primary crusher, which may be either a jaw crusher or an impact crusher.

The grizzly vibratory feeder is controlled electronically by a variable motor, so as to ensure continuous regulated feed for better production and efficiency.

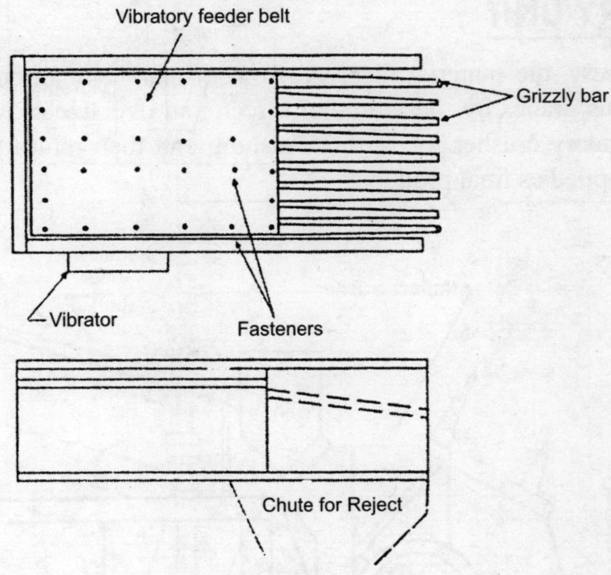

Fig. Appdx. 4.4. Grizzly vibrating feeder.

3.2. Primary Unit

(*a*) **Primary Crusher :** Grizzly scalped material with minimum of fines passes into the jaw or impact crusher fitted in the primary unit. The jaw crusher may be either single toggle or double toggle jaw crusher. It is supported on a heavy duty steel plate, annealed and stress relieved frame to withstand tough crushing operations. The jaw or impact crushers have construction features as described earlier in this chapter.

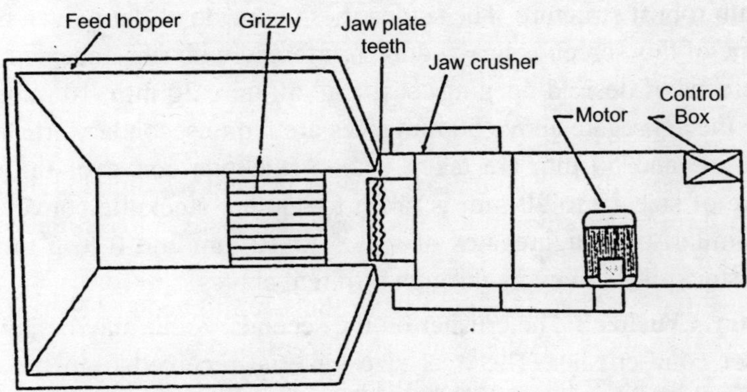

Fig. Appdx. 4.5. Primary crushing unit.

(*b*) **Main Belt Conveyor :** Material crushed by primary crusher are carried to the secondary unit by the main belt conveyor, one end of this conveyor is installed just under the discharge chute of primary crusher. The other end of this conveyor is kept just above the feed end of the vibrating screen.

3.3 SECONDARY UNIT

In the secondary process, the material received from primary unit through main belt conveyor is screened into various grades by the vibrating screen and oversize (+50 mm or 40 mm) is fed directly into the secondary crusher for further crushing and then return back to the screen. The various sizes are stockpiled as final products.

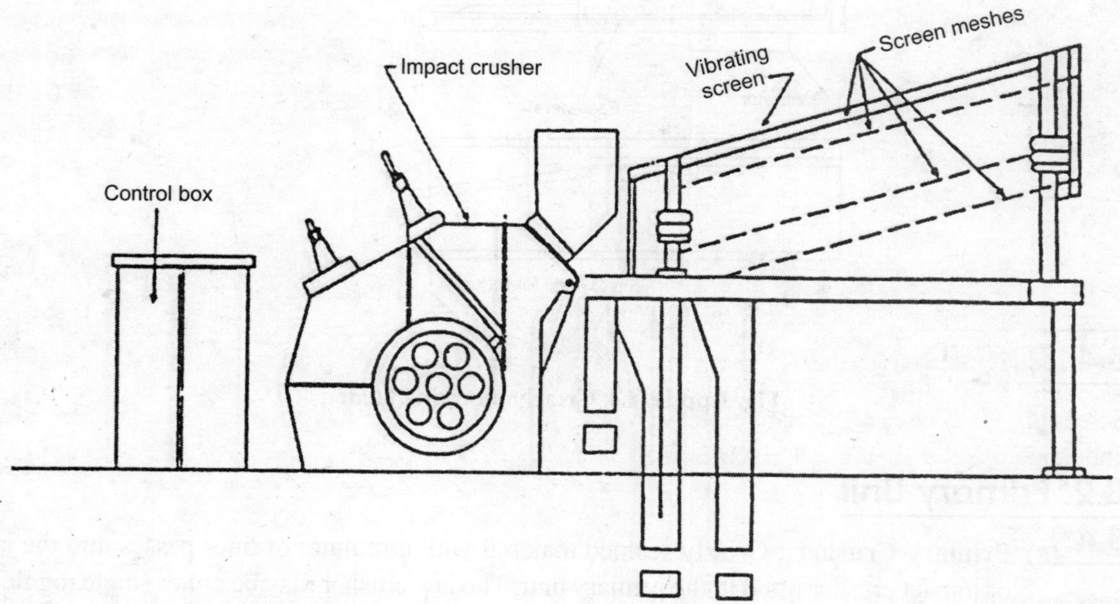

Fig. Appdx. 4.6. Secondary unit.

(*a*) **Vibrating Screen :** This is an inclined vibrating screen of multideck (generally triple deck) with robust structure. The screen mesh is made of flat woven hardened steel wire. The frame of this screen is heavy duty, reinforced with ribs annealed and stress relieved. It has screens of desired openings say, e.g., 50 mm; 20 mm; 10 mm; and 5 mm. In this example the aggregate above 50 mm sizes are fed to secondary crusher whereas of sizes between 20 and 50 mm are taken to one stockpile conveyor through a chute, while aggregate of size 10 to 20 mm is taken to another stockpile conveyor through another chute. Similarly, the aggregates of size 5 to 10 mm and 0 to 5 mm are taken to two separate stockpile conveyors through different chutes.

(*b*) **Secondary Crusher :** The crusher on the secondary unit may be jaw crusher or impact crusher or cone crusher. The over size material received from the vibrating screen is crushed in this secondary crusher and then allowed to drop in the return belt conveyor through a chute.

(*c*) **Return Belt Conveyor:** Products crushed by the secondary crusher are carried back to the vibrating screen by return belt conveyor.

(*d*) **Stockpiling Belt Conveyors :** The final product delivered from the vibrating screen, are stockpiled by stockpiling belt conveyors. Each size of the product is carried by separate conveyors to separate stockpiles located at convenient places.

Speed of the conveyors ranges between 1 m/sec to 4 m/sec depending on size of material, characteristics of material (like density, abrasiveness, and friability), width and inclination of the belt etc. The inclination may go upto 22° depending upon angle of repose, size and shape of material, method of loading, speed of belt, moisture content etc.

(*e*) **Chutes :** These are required to transfer material from one piece of equipment to another. They ensure flow of material without spillage, jamming, breakage or damage to the equipment. To increase the life of chute, lining suitable for the material is provided so as to withstand the impact of the material. Inclination of the chute should be sufficient to ensure smooth flow of material, but it should not be unnecessary steep, otherwise, the material may break and lining may wear faster. Flow of material from the chute should be in the same direction as that of the belt conveyor receiving it as to avoid spillage of material and damage to the belt.

(*f*) **Bucket Elevators :** These are used only when restriction of space rule out the use of conveyors.

3.4 SCREENING UNIT

Screening is a device of grading material in different sizes. Screening efficiency is an indication of undersize material going into the oversize.

3.4.1. Types of Screens

Screens are of following two main types:

1. **Rotary Screens :** These are of cylindrical shape with circular openings. Capacity of the rotary screens is limited. The initial cost of these rotary screens is less than that of a vibrating screen. These have lesser screening efficiency and hence are less suitable for quality work.
2. **Vibrating Screens :** These are versatile and are suitable for most of the screening requirements. These are mounted either horizontally or inclined over a frame, which must withstand stress imposed by high gravity forces derived from the vibration. Structure of a vibrating screen comprises two side frames held rigidly by the shaft housing and screen mesh support frame. The feed end is enclosed by rubber covers which prevent spillage and permit ease of access for screen mesh replacement.

Fig. Appdx 4.7. Vibrating screen

A feed box distributes the feed material evenly across the width of the screen without any reduction in the screening area. The whole assembly is supported on springs; to eliminate the vibrations to supporting structure. Rubber shock absorbers or a friction block reduce the tendency for excessive throw to occur during start-up and about-down.

3.4.2. Vibration Mechanisms

Vibrations are created by weights which rotate on shafts held on the screen frame. Different models of vibrating screens have their own method of generating the vibration, differentiated by shaft weight configuration.

(*a*) **Vibrating Single Shaft Mechanism :** Screen using this mechanism are of various sizes with 2, 3 or 4 decks so as to give upto 5 products (including oversize). In this mechanism, the shaft motion is circular, enabling rapid separation resulting in higher efficiency. The inclined screen (generally between 15 to 20°) is advantageous due to assistance of gravity in screening.

The vibration is generated by a single eccentric shafts, supported by roller bearings located at the centre of gravity of the screen. Counterweights are fixed to the shafts, and additions of extra weights enables adjustment of the amplitude. The whole assembly is encased in a protective housing, sealed against dust entry. Screen pulley design compensating for vibration in belt tension. The drive assembly can be mounted on either side of the screen.

(*b*) **Vibrating Double Shaft Mechanism :** These have a linear action, the motion deriving from contra-rotating shafts onto which counter weights are fixed. The linear action coveys the material across the screen without gravity assistance.

Linear motion is obtained by the shafts contra-rotating at the synchronous speed. The shafts are supported by bearings in housing which are bolted to the screen frame. Counter-weights are identical, and adjustable for different applications. The weights are keyed onto the shafts and separate drive motors are provided for each shaft.

The screens adopting this mechanism has 2 or 3 decks to provide 3 or 4 products (including oversize.)

3.4.3. Screenings Applications

Screening is done for following purpose :

(*i*) Grading (*ii*) Scalping (*iii*) Dust removal
(*iv*) Dewatering (*v*) Deslurrying

1. Grading

Mesh opening is kept fairly close to the required grading depending upon the feed material (*i.e*, its sieve analysis, shape, moisture, content etc.). Water content in the feed material is most important factor. The screenings are therefore classified as dry, moist or wet.

(*a*) **Dry Grading :** When water content is less than 2%, it hardly affects the screening characteristics. With dry material, large throughputs per square metre are obtained and screening is possible on small mesh openings down to about 0.5 mm.

(*b*) **Moist Grading :** The higher the water content in feed material the higher the cohesion between the particles and lower their free movements and the screening efficiency. If water content further increases, after a certain limit, the cohesion decreases and screening is then again possible. The smaller the mesh openings, the higher the adverse effect of the water content.

(*c*) **Wet Grading :** In this process water content is increased to a limit when cohesion is reduced to a minimum. The fines are rinsed with the water and then dewatered.

2. Scalping

Scalping is done on relatively large sieve openings. These are used to remove unwanted size of material, may be either too small or too big.

Relief screening is required before crushing to prevent smaller pieces going into the crusher. For example a crusher is operating at 150 mm setting and is being fed with 500 mm size boulders along with small piece. If a scalper having 150 mm openings is employed, then material below 150 mm will bypass the crusher and only boulder will pass through the crusher. This will avoid unnecessary overload on the crusher.

Control screenings are used for the removal of coarse lumps and larger size material then required. For example, in drum mix asphalt plants, where oversize lumps or stone metals are required to be removed and rejected, these scalping screens are used. The sieve opening is larger by many times, than that of the maximum allowable particle size, the facilitates large throughput.

3. Dust Removal

It is the screening of granular products with a view to remove unwanted very fine particles which were formed during the process (by size reduction or abrasion).

4. Dewatering

It is like straining unwanted water from the feed material. This water was added during wet gradation, wet sorting flotation etc. The dewatered material is then dried further either, mechanically, or thermically or both.

5. Deslurrying

It aims at removing certain particles range along with the water, e.g., in order to reduce ash content in coal or to remove dissolved clay from sand and gravel.

SHOVELS (CABLE OPERATED)

1. INTRODUCTION

Shovel family represents rope operated excavators and uses different types of front-end attachments. Following are the four types of front-end attachments, by changing them only, one type of machine is converted into another types, basic structure remains unchanged.

(*i*) The Dipper or Power Shovel

(*ii*) The Backhoe or Hoe.

(*iii*) The Dragline.

(*iv*) The Clamshell.

Some times the first type is called by the name 'shovel' only, which is used for excavating and is fitted with a short-length boom working in forward strokes. The distance, the material must be moved by the machine from the digging point to the dumping point, will determine which type of front-end machine is most advisable. The type of material to be excavated also determine the actual type of machine employed. Shovel and Hoe booms are somewhat short as compared with the longer booms on a dragline or clamshell. Shovels and Hoes can dig tougher material than lies within the possibilities of the longer boom machines.

The shovels are available in various sizes varying from a fraction of cu. m. to over hundred cu. m. in bucket size and can handle the material weighing from a few tonnes to hundreds of tonnes. The basic structure of all types of shovels, namely; Dipper, Dragline, Clamshell, Back Hoe and Mobile Shovels may be either crawler mounted or truck mounted. If the work is concentrated in one area and has a large job, mobility is of less importance, then crawler mounted shovels are more useful. Where the jobs are scattered at various places truck- mounted shovels are preferred. The crawler mounted shovel can operate on ground surfaces which are not strong enough to support the truck-mounted shovels.

Structure of the basic shovel provides basic operations, namely,

(*a*) Movement of the machine, known as 'walking',

(*b*) Revolving action of the superstructure for handling loads, known as 'Swinging', and (*c*) Actual working of the attachment like digging, hoisting etc.

The power for all these operations is provided by the prime-mover.

2. DIPPER SHOVEL OR POWER SHOVEL

These are used for digging hard rock in quarries or ore in mines. These are used for digging above the machine base level, and have faster and more accurate digging and dumping cycle as compared to that of draglines, hoes and clamshells.

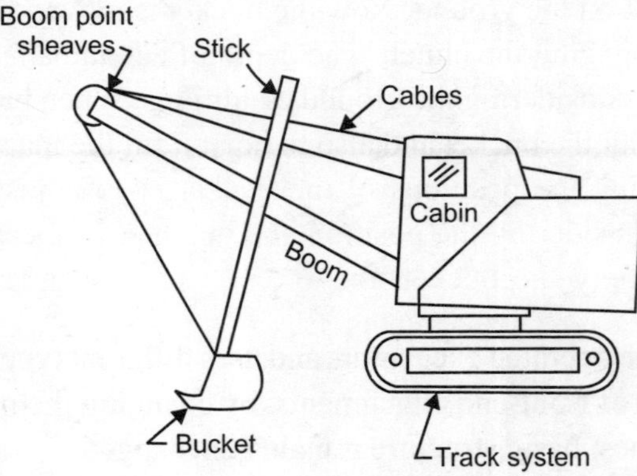

Fig. Appdx 5.1. Dipper shovel.

The power shovel consists of bóom, dipper stick, the bucket and the mechanism of operation. The boom and stick are connected as shown in the Fig. Appdx. 5.1. The lower end of the boom is hinged to the support bracket of the revolving deck and the upper end (also known as point), carries the bucket hoist sheaves. The dipper stick moves back and forth on a shipper drum through a guide formed by a saddle block attached to the boom. The shipper drum is mounted on a shipper shafts which is passing through the boom and supported on it at the ends. The shipper drum is rotated by pulling the cable and causing the stick to slide back and forth. The bucket is provided at the end of the dipper stick.

The buckets of the power shovels are strong built steel box which are open at the top and provided with a hinged door at the bottom. The buckets are made in the sizes varying from 1/2 cu m to hundred cu m or more depending upon the size and capacity of the power shovel. The design and shape of the bucket is such that, it can easily handle the material to be excavated. Front edge is given a lip formation so as to make cuts conveniently. Lip has tapered sockets to receive shanks made into the teeth. The teeth are made of manganese steel and are replaceable. Depending upon the job conditions the teeth designs are of sharp, rock and general purpose types.

So as to ensure optimum production, rebuilding of the bucket tooth points is necessary. The bucket teeth material austenitic manganese steel is characterized for high strength, high ductility and extreme toughness. Due to severe impact the surface becomes work hardened while the interior remains soft and tough. This prevents the teeth from breaking into pieces. Re pointer bars of austenitic manganese steel are available for reclaiming worn dipper teeth. The exterior surface of the worm tooth is cut and removed by oxyacetylene flame. The size of the re pointer bar is selected depending upon the extent of damage of the both. The bar is cut to the required size and then welded with a medium heavy coated, all-position, low hydrogen type austenitic stainless electrodes. In case austenitic manganese steel pointer bars are not available, carbon steel pointer bars may be used but they shall give a lesser life.

Operation of Power Shovels

Before actual operation, the power shovel faces the job with stick projecting forward horizontally. For operation, first step is to release hoist to bring the bucket down and move the sticks in vertical plane and hoist brake is applied again before it hits the ground down on the shipper drum till the bucket teeth rest on the ground. Now the bucket is moved forward and downward, by lifting the hoist brake and engaging the clutch. The depth of cut and angle of digging is controlled as per requirement of the job considering the ground conditions. When bucket is filled, the next step is to retract the stick to pull out the bucket and also to hoist it till the machine can be swung around for dumping. During swigging the positions of the bucket is so adjusted that it comes in most suitable dumping position considering the position and height of the carrier used for transporting the excavated earth. The empty the bucket, the door opened by unlatching it through the trip arrangement. The cycle is thus repeated again and again.

Power Shovel Applications

As mentioned earlier Power or Dipper shovel is most suitable for close range works. These are used for digging very hard material and can remove big boulders. For working, it needs to be kept on fairly solid footing on the ground. These can be used on various jobs of digging in rock, gravel, clay etc. Therefore, they find applications in road works for trimming and dressing slopes and banks, mining work (open cut) etc.

3. HOES OR DRAG SHOVELS

This equipment is also known as Drag-shovel or Pull-shovel. Since the digging mechanism resembles to an ordinary garden hoe, this equipment is named as Hoe. The equipment has an ability to penetrate even the toughest of materials. Due to this ability these are commonly used in quarries which have tough digging conditions and are prone to flooding, rendering the use of conventional shovel operations impractical.

The drag shovels are extensively used in trench digging applications, and can dig easily to greater depth below the level of the equipment. Like power shovels, hoe also has accurate spotting of the digging bucket. The dumping into the carrier units is difficult and not as efficient as a power shovel, and also has a slow working speed.

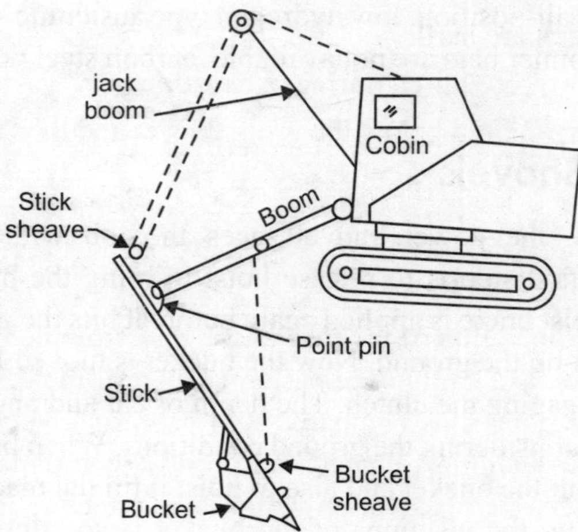

Fig. Appdx 5.2. Drag shovel or Hoe.

As the name Drag or Pull shovel suggests, digging is done by dragging or pulling of the bucket toward the machine, whereas Powers shovel makes outward strokes while digging. This shovel consists of a boom and stick; a jack boom and the bucket attached at the end of the stick. The stick in Hoe is hinged by a pin with a boom as shown in the Fig. Appdx. 5.2. and thus enables to turn over the pin to take any desired direction best suited for digging to dumping operations. The lower end of the stick carries the bucket, while the upper end carries a sheave having the hoist cable supported on a jack boom at the other end and passes on to the hoist drum. Therefore, when a pull is applied to the hoist cable, the reaction at the stick hinge enable to move the boom with stick, up or down. Thus the boom can take any position in vertical plane,

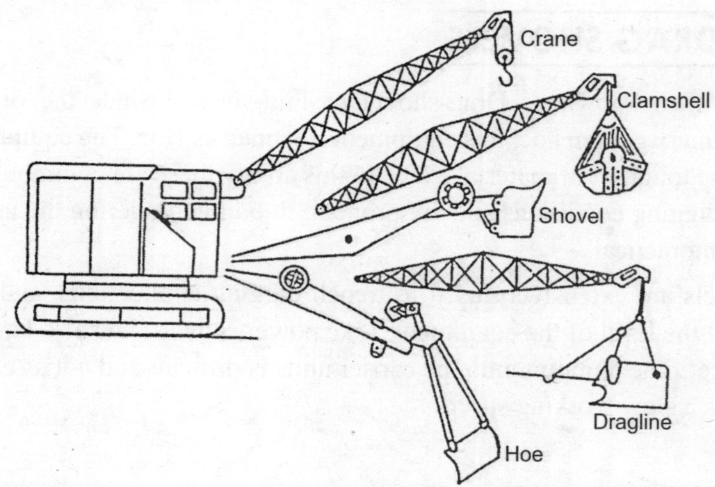

Fig. Appdx 5.3. Boom attachments

Applications :

Hoes or Drag Shovels are used mainly for following types of jobs :

(*a*) For digging trenches, footings or basements.
(*b*) To dig materials which are hard.
(*c*) Where close trimming is required during excavation
(*d*) Where excavation is required below the ground level and then dumping is done at a short range.

4. DRAGLINE

Since the prominent operation of the machine is dragging the bucket against the material to be dug, it is named as Dragline. Long light boom of the dragline allows it to dig and dump over longer distances as compared to a power shovel or a Hoe. Therefore, it is most suitable for excavation of channel and canals, can handle wet or foggy material, and suitable for underwater excavation while standing away from the pit, on a firm soil. Unlike power shovel it can dig materials below its track level and can handle only softer materials.

Capacity of a dragline is indicated by the bucket capacity measured in cubic meters and generally available in 1 to 15 cu m capacities. Front end of a dragline consists of a boom, a bucket, a fairlead, and hoist, dump and drag cables as shown in the Fig. Appdx. 5.4. The boom is like a crane boom, the top and bottom ends of which are known as the 'point' and the 'foot' respectively. To make the boom lighter the section can be made up of tubes. The boom is attached to the deck with pins at foot and supported at the point through the boom hoist cable passing over a pair of sheaves placed one on each side.

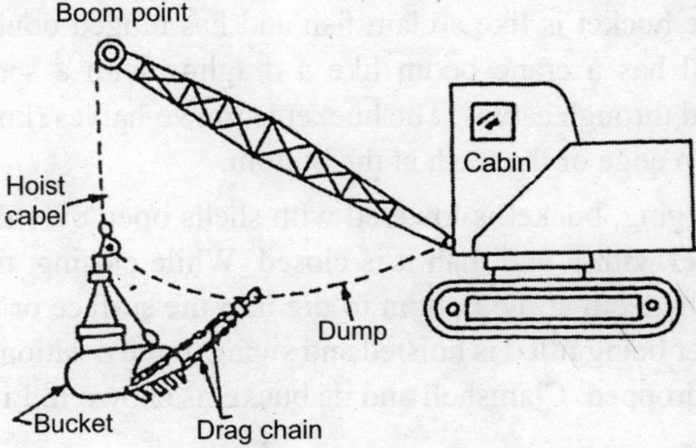

Fig. Appdx. 5.4. Dragline.

5. CLAMSHELL

Clamshell is a machine having most of the characteristics of dragline and crane in common. Digging is done like a dragline and once the bucket is filled, it works like a crane. It is very useful for accurate spot dumping of material in a confined space in vertical plane. It can be used for handling of loose or soft and medium hard material only. Although it can be used for all the jobs expected from a shovel family but is not very efficient due to its slow operation.

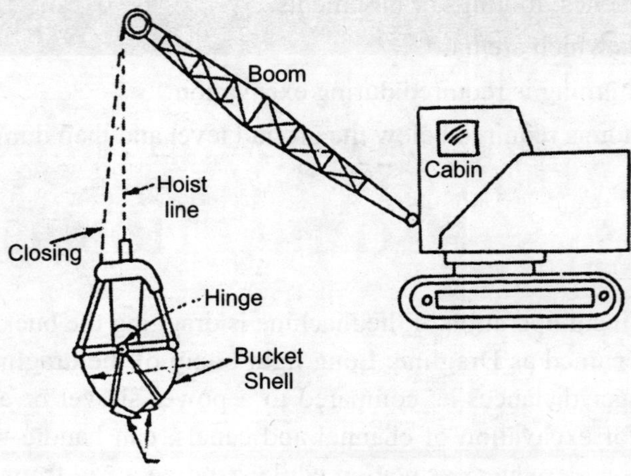

Fig. Appdx 5.5. Clamshell.

Clamshell can be used in following types of applications:

(*a*) Where digging or dumping in a vertical plane *i.e.*, below, at or above ground level is required.

(*b*) Where material is relatively soft or medium hard.

(*c*) For digging trenches.

(*d*) For charging the materials in a bin or a stock pile.

(*e*) Where accurate dumping is required.

Since the shape of the bucket is like a clam fish and has hinged double shell, it is named as 'clamshell'. The clamshell has a crane boom like a dragline with a specially designed bucket attached to it the upper end through cables. The bucket is in two halves (known as shells) hinges at the top and has either sharp edge or the teeth at the bottom.

For the purpose of digging, bucket is lowered with shells open over the surface to be dugged till it makes a good contact with it and than it is closed. While closing, the weight of the bucket helps the sharps edge or the teeth at the bottom to dig into the surface or material, thereby filling the bucket. The bucket after being filled is hoisted and swing to the position where dumping is done and then the contents are dropped. Clamshell and its bucket is shown in Fig. Appdx 5.5.

6 COMPACTION EQUIPMENT

1. INTRODUCTION

Effective compaction of material is an essential part of construction engineering operations concerned with foundations, construction and maintenance of roads, airports, dams etc. The equipment required for these work range from the smallest vibrating rammers to the largest rollers of vibrating type. Quality characteristics such as strength, wearing resistance, evenness and impermeability of a pavement depend on the degree of compaction of various layers of the pavement and the sub grade supporting it. The better the compaction, the better will be shear strength, density and bearing capacity of the individual layers and with that the lasting quality of the embankment or road construction.

Rollers are used for compaction since long olden days. Over a thousand years ago the Chinese used huge cylindrical shaped rollers for road works, and the word famous builders of Rome used towed stone rollers. The self propelled road rollers, powered by steam engines were built in England in nineteenth century. As the time passed, road roller design continually improved. Since the manufacture of the first diesel-engine-powered road roller in 1929, the development of compaction equipment has taken place rapidly.

Compaction

Compaction is the process whereby material particles are constrained to pack more closely together through a reduction of air void content, generally by mechanical means. Compaction can also be defined as the process of densifying or increasing the unit weight of a soil mass through the application of static or dynamic force, with the resulting expulsion of air and, in some cases, moisture, Static forces are produced by a roller or compactor which densifies soil primarily by its heavy weight, whereas dynamic forces utilise a combination of weight and energy to produce a vibratory or tamping effect on the soil. Compactors utilise both static and dynamic forces to achieve the required soil density.

Primary function of compaction is to produce the reasonable surface finish in accordance to the layer position within the whole embankment construction. Compaction is measured in terms of dry density. The amount of compaction applied can be controlled by the choice of compaction equipment, the thickness being compacted, the speed of the compaction equipment and the number of passes of compaction equipment.

2 TYPES OF COMPACTORS

Today's compaction equipment are represented by many large, highly responsive, and versatile self-propelled units. These compactors and rollers can be divided into following major classes.

2.1. Static Smooth Wheeled Rollers

These are rollers used with or without ballast and may be 3 wheeled or tandem types (two rolls of equal width). These are the conventional rollers used for almost all types of rolling. It is not effective on uniformly graded sand, gravel or silt on over-moist cohesive soil due to poor traction.

Fig. Appdx. 6.1. Static three wheeled roller.

Diesel powered rollers with modern technology of hydrostatic transmission rollers have replaced the traditional steam rollers which were first introduced 150 years ago. The dead weight rollers (static rollers are sometimes called by this name) rely on weight alone to compact the material over which they pass. The unit of 8 to 10 tonnes imparting pressure of 20 to 40 kg per linear cm are generally used. For light works, lighter rollers of weight up to 1 ton are also available.

Other Static Compaction Equipment

These can be divided into the following groups :

(*a*) Towed static smooth compactors.
(*b*) Static sheep foot or pad foot compactors.
(*c*) Static three wheel self-propelled compactors.
(*d*) Static tandem compactors.
(*e*) Three axle static compactors.

2.2. Sheep Foot or Pad Foot Rollers

These may be self driven or tractor driven and are suitable on cohesive especially when water content is on the higher side. The inherent dynamic beat developed by these tamping pads or feet can in many cases produce a high degree of compaction at a faster rate.

2.3. Pneumatic-Tired Rollers

These are used for compaction both in earthwork and bituminous road construction. Because of the oscillating axle layout, they produce more even compaction across the rolling width steel smooth rollers, bridge the material leaving uncompacted areas of fill. These rollers are available in weights ranging from 6 to 30 tonnes with 500 to 4000 kg per tire and tire pressure from 2 to 8 kg/sq. cm.

The research discovered that a certain amount of compaction was performed by heavy commercial vehicles during the first few days of the opening of the road to traffic. It was for this reason that the pneumatic tired rollers (P.T.R) came into being. The tire configuration is of the odd number arrangement–five, seven, nine or more, with certain degree of overlap. Usually even numbers, are provided on the rear axle. Rubber tired or pneumatic compactors have been used more and more in last few years for compaction in earth-works and in bituminous road construction. These are generally self-propelled and are built over a wide range of weights. They have proved to be very successful for the compaction of cohesive and non-cohesive soils and stabilized soils in airfield, embankment and road construction applications.

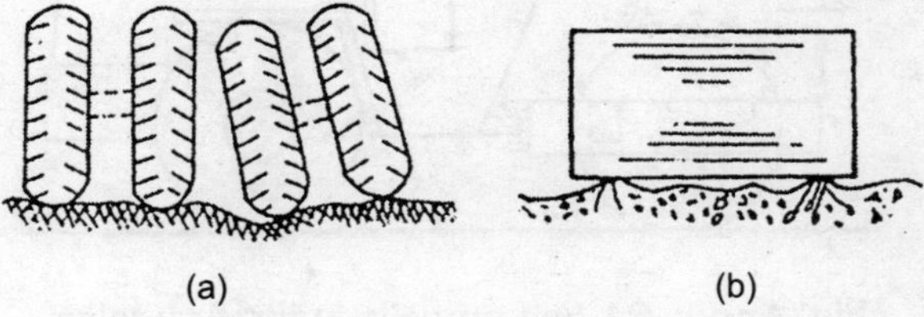

Fig. Appdx 6.2. (*a*) Oscillating wheels of rubber tire roller (*b*) Bridging with steel wheel roll.

These provide a more uniform compaction effect than steel wheel compactors. Compaction with wide steel rolls sometimes produces a bridging of the material being compacted. Rubber tired compactors with oscillation of the front and rear tires, and equal wheel load ensures that the compactor follows the cross-section of the ground contour. This assures uniform compaction throughout the layer.

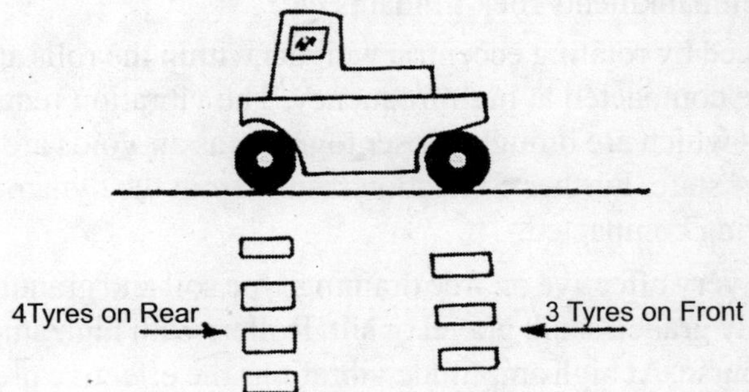

Fig. Appdx 6.3. Pneumatic tired roller

Main advantages of rubber tired rollers over steel wheel rollers are :

(*i*) Surface of the layer is not bridged.

(*ii*) Bituminous layers which are compacted with a rubber-tired roller are better sealed to keep out dirt and moisture.

(*iii*) When rubber-tired rollers are used, post compaction by traffic will be negligible.

(*iv*) The compaction effect of the force supplemented by the elasticity of the rubber tires provide a 'Kneading effect' resulting in a better surface compaction and surface sealing.

(*v*) Unlike a steel wheel compactor, the rubber tired compactor does not push the material ahead of it. Therefore, the compaction crack often produced by steel compactors are eliminated by using the rubber tired rollers.

2.4. Vibratory Rollers

Different types and sizes of vibratory rollers to suit specific conditions are available, like:

(*i*) Rollers with the large vibration drum in front with two steel or pneumatic tired rear driving wheels. The pneumatic tires offer better traction on gradient and in over moist soil. These may be articulated or non-articulated type.

Fig. Appdx 6.4. Self propelled vibratory roller.

(*ii*) Roller with tandem smooth wheels, either one wheel vibrating and driven or both wheels vibrating and drive.

(*iii*) The small duplex roller with twin rolls weighing 1 to 1.5 tons for trench reinstatement.

(*iv*) Single roll pedestrian vibratory rollers weighing nearly 400 kg are used for compacting small areas such as pavements, foundation etc.

(*v*) Towed vibrating rollers weighing 5000 to 11000 kg are used for mass earthworks, base construction, embankments rock fill dams etc.

Vibration are produced by rotating eccentric weights within the rolls at high speed, the roll thus hitting the material to be compacted at high frequency. The vibration reduced the friction between the particles of materials which are brought closer together as air voids are eliminated, thus binding the material into a denser state. Further comaction results from the dynamic forces by the vibrating roll onto the material being compacted.

Vibrating rollers are very effective on free draining type soil and granular base course. These are not effective on uniformly graded sand, gravel or silt. Rollers with high amplitude are preferable for soil and granular base course. At high amplitude vibrations the effective pressure on the compacting material is around twice the static weight of the roller.

2.5. Tampers and Vibrating Plates.

Tampers with small vibrating feet are used to tamp footing, trench bottoms, position posts, lay paving slabs or in any small area where high manoeuvrability is required. These can also be used on small road repair jobs, footpath constraints and compacting trench bottoms for pipe laying. The tampers can deliver from 800 to 4500 blow per min. and generally powered by engines of 2.5. to 12.5 H.P.

Design Features of Vibratory Compactors

(*a*) Hydrostatic drive to ensure smooth, jerk free, reversing and an infinitely variable speed control.

(*b*) Large roll diameters to reduce the rolling resistance and to reduce the horizontal material displacement.

(*c*) Hydrostatic steering to ensure smooth, jerk free steering.

(*d*) Good accessibility to all drive components providing easy serviceability.

(*e*) Hydrostatic vibrator drive with infinitely variable frequency and centrifugal force to match the various soils and asphalt material.

(*f*) Easy operation of the vibrator unit during reversing for its fast disengage-engage.

3. COMPACTION SPEED AND NUMBER OF PASSES

The number of passes and compaction speed required when using vibratory compactor are determined by the necessary "compaction time". Tests indicate that, a limited increase in the compaction speed can be an advantage. Over 5 km/hr speed is not recommended. Since with more speed the number of vibration cycle per cm are not sufficient for achieving desired result. Generally a speed around 4 km/hr. is preferred for getting better compaction results.

4. FACTORS INFLUENCING THE COMPACTION

Following parameter influence the degree of compaction achieved by a vibratory roller:

(*a*) Static weight.

(*b*) Number of vibratory drums.

(*b*) Roller speed.

(*c*) Drum diameter.

(*e*) Frequency and amplitude.

(*f*) Relationship between frame and drum weight.

(*g*) Driven or non-driven drum.

(*h*) Centrifugal force.

7 PAVER FINISHERS

1. INTRODUCTION

A paver finisher or simply known as 'paver' is a machine for laying and spreading sub-base courses or asphalt for road construction to give accurate depth, even levelling and good surface finish before rolling. Pavers are also used in various construction activities like construction of footpaths, parking areas, sports grounds, highways, airport runways and railways (for laying track ballast).

Earlier paver finishers were used only on asphalt works, but during last about thirty years the use of pavers for sub-base and base course has gained popularity. Use of paver finisher, compared to the traditional Motor-Grader method is found more suitable, because:

(*a*) Better and easier control of levels, both longitudinal and transverse.

(*b*) Improved homogenity of the plant prepared mix.

(*c*) Uniform pre-compactin.

(*d*) Complete elimination of final levelling, reduce the efforst and thus costs.

(*e*) Improved adherence of asphaltic courses.

The pavers can easily handle asphalt and all asphalt mixes, coated and uncoated materials, crushed stones and dry lean concrete, in fact all road making material other than wet concrete.

A paver finisher, during constructing bituminous pavements, are fed by the mix transporting vehicles into the hoppers at the front. Two conveyor belts lake up the mix which has been fed into hopper, to an auger which distributes it in the prescribed thickness over the operating width. The mix in then laid in required width and depth and levelled and compacted partially by the screed. The screed is heated either by propane gas burners or diesel oil burners, evently distributed over the entire operating width. Although the degree of compaction achieved by the screed has, in course of time, been steadily increased, final compaction by roller is still necessary. Compatibility of a bituminous mix is considerably reduced as the mix cools and then requires more compacting energy to achieve a degree of compaction at a lower temperature.

To enable pavers to deal with changes in road width, without interruption operation, hydraulically extendible screeds for both sides of the finishers were developed during last few years. With such units it is possible to infinitely vary the basic operating width of the paver finishers. With these screeds, the width can be increased to 6.0 m. For pavers with rigid extensions widths upto 12.0m can be reached. Single layer upto 30 cm thickness can be placed. The operating speeds which depend upon the size of the object and the type of work range from 1 to 60 m m/min and can be infinitely selected. Maximum extendable width to the screed, may be hydraulically or mechanically, determines the size of the paver.

The screeds used in paves are of thee types :

(*i*) having only tamping.

(*ii*) having only vibrating, and

(*iii*) having a combined tamping and vibrating.

The pavers are either mounted on tyres or on crawler tracks. The power is provided by a suitable engine.

2. CONSTRUCTION OF PAVER FINISHERS

Main constructional features of the paver finisher are described hereunder :

2.1. Chassis

Chassis is robust and composed of the under carriage and super structure. The under carriage carries the material conveyor system and front hopper, crawler system or pneumatic wheels and rubber tyres duly fitted. It also accommodates the power unit, transmission system and augers. The hydraulic system and various control units are also mounted on the undercarriage. Push roller bar is provided at the correct height for standard material transport vehicles. Driver's seat is located at a place so that he sits directly above the auger and can see the overall distribution of material in front of the screed, and thus he can observe and control all the functions of the paver finishers.

2.2. Transmission

The prime mover for the paver finishers are diesel engines. The power generated by the diesel engine is supplied to the gear box, traction conveyor drives, auger and screed vibrations. The gear box has very wide range of speeds. The speed selected depends upon various factors such as :

(*i*) If the paver is shifted, high spced is required.

(*ii*) When the supply of material is high the speed is also high and vice-versa.

(*iii*) When laying for lesser width and lesser thickness the speed should be more whereas, while laying more width and higher thickness the speed of paver should be less. Thus the operator choses the working speed according to the paving material, the relevant mixing capacity, the feeding capacity, the thickness and width of layer to be placed. Speed of the paver while working is generally kept upto 3 km/hr (or 50 meters per min).

2.3. Hopper

Self-dumping hopper consists of two wings and an apron, all independently operated with the aid of hydraulic rams. This operation is carried out by the driver from his seat by activating the control valve with the help of a connecting rod. It can be held in any position.

The material is fed into the hopper by the trucks which back up against the guide rollers of the paver finisher. During this dumping process, the truck is pushed forward by the self propelled paver finisher. Generally the capacity of the hoppers are from 12 tons to 14 tons.

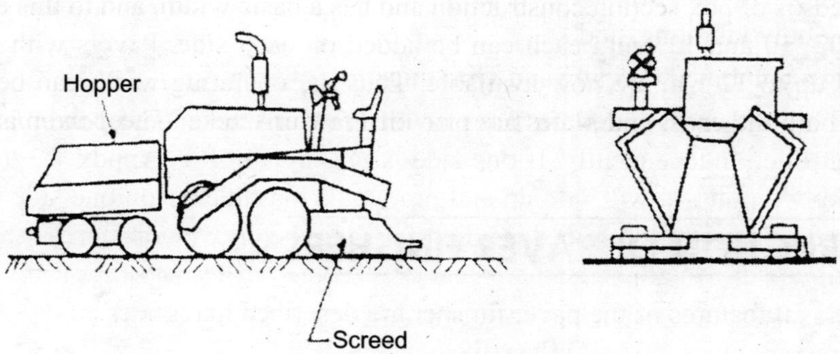

Fig. Appdx. 7.1. Paver finisher.

2.4. Conveyors

Conveyors, two in numbers are driven independently from the engine power generally by roller chains and sprockets. Speed of the conveyors can be controlled by the driver as per requirements. The conveyors are self cleaning, *i.e.*, they are designed in such a way that material does not settle on them. The material poured into the hopper by the tipping trucks is conveyed to the augers by these two conveyor belts. Each belt with its corresponding auger has its own drive (mechanical or hydraulic) so that the material is independently conveyed on either side. This helps to suit the varying material requirements of each side.

2.5. Auger

The auger can rotate to both left and right, therefore they can convey material both inward and outwards. Thus it spreads the material uniformly in the auger box so that the material can be layered by the screed in the desired thickness throughout the width. The augers have super hardened cast faces bolted to the flights to give long life.

The amount of material delivered to the auger box is maintained at a level which will ensure a steady supply by the augers across the leading edge of the screed. To achieve this flow control gates situated on the rear wall of the hopper are set, so as to get the material as per requirement only. As excessive build-up of material will create a bulldozing effect by the screed and also material will become cold and the quality of mat will result in loss of level and holes will appear in the mat.

The height of the auger is required to be adjusted depending upon the thickness of the material to be laid. Although, no hard and fast rules can be laid down for auger height settings, but since the augers are used to gauge the level of the material, an incorrect height setting could result in an over or underfilled auger box. As a general rule, for bituminous materials, the lowest point of the auger should be not more than 50 to 70 mm above the level of the finished mat.

Auger extensions are available with each paver finisher so as to extend or reduce the paving widths and contains the material within the required width limits.

2.6. Screeds

Each of the screeds is of box section construction and has a basic width, and to this extension parts of widths of 500, 250 and 125 mm each can be added on each side. Pavers with screeds which can be extended up to 12.5 m are now available. Thus the operating width can be varied as per requirement. At both the ends, end plates are provided to form the edges of the mat. Each side of the screed is controlled independently. If one side is raised as in Fig. Appdx 7.2. (*a*); by vertical adjustment of the two point, it will ride up and produce a mat thicker on one side then the other. To form a crown the centre of the screed is adjusted using the crown adjuster, as shown in Appdx 5.2 (*b*). The maximum crown adjustment are generally done upto 4% (positive) and 2% (negative). Negative crossfall makes it possible to use for constructing farm roads with centre drainage, or for removing rutting in the wheel paths of the pavements.

Fig. Appdx 7.2. Screed adjustments.

The material received from conveyors is distributed by the auger in front of the screed upto the screed end plates. As soon as enough material is accumulated the paver finisher can start paving.

Both screed halves are connected at the bottom centre by union bearings and spring loaded tie bolts and at the top by 2 turn buckles, These turnbuckles constitute the crown control for both screed halves.

As mentioned earlier the screed are of 3 types having tamping only; having vibrations only : and tamping and vibration combined. Now-a-days screeds with tamping and vibrating combined are preferred since it gives better initial compaction and good mat surface.

(*a*) **Tamping System :** Generally good screeds are composed of two tampers driven by same eccentric shaft which are arranged in front of the vibrating screed plate, *i.e*, in the material flow area of the spreading augers. Due to its chamfered surface the first tamper has a proportioning, pre-compacting effect whereas the main tamper with its differently shaped bar reaches the final compaction. The amplitude to the first tamper is adjustable between 0 and 12 mm and that of the main tamper between 3 and 9 mm independent of each other. These adjustments of tamper amplitudes together with the variation of tamper speed and vibration frequency controls the compaction requirements of different types of mixes and mat thicknesses.

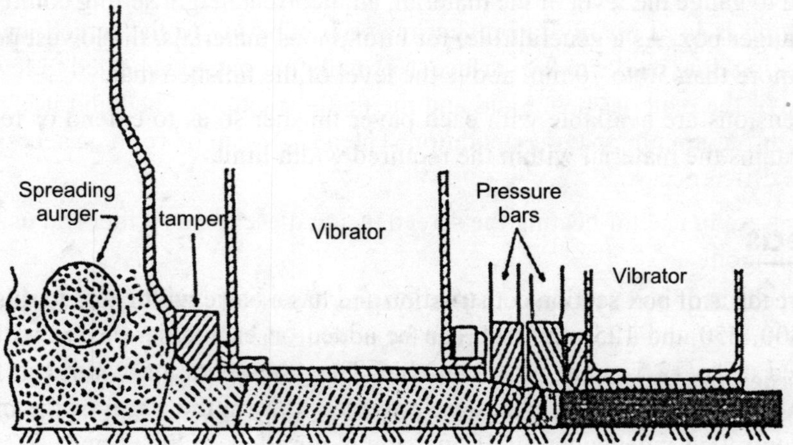

Fig. Appdx 7.3. Tamping and vibrating arrangement of screed.

The speed of tamper shaft is controlled from the control panel where tamper blows per min are indicated. The speed of the shaft can be varied as per requirement varying from 0 to nearly 1500 r.p.m. (25 Hz).

(*b*) **Vibrating System :** Vibrating screed plate behind the tamper plate portion provides additional compaction and also improves the surface texture, especially when laying asphalt wearing courses. The vibrations shaft speed is variable between 0 and maximum 4200 r.p.m. (70 Hz).

(*c*) **Tamping and Vibrations Combined :** The amplitude adjustment of both tampers, independently of each other, and the infinite variation of the tamper speeds and vibration frequency make it possible to adapt the screeds compacting effect to different types of mix and varying layer thicknesses.

Experience shows that Marshall and Procter densities reached with paver finishers equipped with high density screeds having dual tamper and vibrating arrangement, are about 7–10% higher than those obtained with single tamper/vibrating speeds :

(*a*) Vibrating screed — approx. 80–82% Proctor density.

(*b*) Tamping and vibrating screed — approx. 88–92% Proctor density.

(*c*) Double tamping and vibrating screed — approx. 95–100% Proctor density.

Economic Advantages of High Density Screeds

1. A substantial reduction of rolling work. In many cases elimination of one or even more rollers, lower labour cost.
2. Uniform densities and excellent surface evenness, surface irregularities caused by rolling work are virtually eliminated.
3. Considerable lower risk of disputes over penalty/payment deduction for not meeting specification requirements.
4. Further extension of the pavers' field of use, e.g. laying roller compacted concrete in thick layers.

Screed Heating

Screed heating is used mainly to prevent material adhering to the screed and to prevent the screed from making drag marks to the surface. Normally the screed is heated to few degrees above the temperature of the material being laid and the flame is extinguished and it is not necessary to heat the screed throughout the paving operation. The heaters can be re-lit as necessary to maintain the temperature difference.

Two systems are in use for heating the screed, using diesel oil as a fuel and using propane gas, later is more common.

(*a*) **Diesel Oil Burner System :** In this system diesel oil is fed from fuel tank by a burner pump through the filter to the burner for heating the screed.

(*b*) **Propane Gas Burner System :** Propane gas for the burner system is supplied from a bottle. As propane gas burns very cleanly no carbon deposits will be found on the burners.

Screed Maintenance

After the use, all traces of paving material are removed using the diesel spray hose for bituminous materials and water for dry materials, special attention should be paid to the tamping machines, to the tampers and tamper shields cleaning. Keep all nuts and bolts tights.

Requisites for a Good Paver

A good paver finisher must fulfill following requirements:

(*i*) Must be able to handle all types of mixes.

(*ii*) Must evenly spread and initially compact, leaving a finished surface ready for final compaction by the steel and pneumatic tyred rollers.

(*iii*) Must be able to compensate for minor grade irregularities.

(*iv*) Must be able to form proper crowns and super-elevations.

(*v*) Must be able to push the large trucks while unloading the bituminous mix into the hopper of the paver.

3. TRACKED V/S WHEELED PAVERS

Earlier, tracked pavers were associated with new construction paving, while rubber tired with resurfacing. Rubber tired pavers are now-a-days predominate, and are used on laying of all kinds of sub-base and base materials, mixes as well as wet-mix aggregate and lean concrete. This eliminate the need for a motor grader for spreading base courses, and also reduces the number of compactor passes required, as paver screed itself does some work of pre-compaction due to its tamping and vibrating actions.

The wheeled paver has greater mobility. It can travel on its own. With the textendable screed, the latest pavers can travel narrow and lay wide.

Wheeled pavers have following **advantages** over tracked pavers.

(*i*) Wheeled pavers can be transported under their own power easily and quickly on site either from one end of the site to another or even from the end of a laid mat to commence a matching joint on its adjacent run.

(*ii*) Wheeled paver lay a truer mat, due to its free rolling movement which does not inherently places into the mat a series of ripples produced by its tracked counterpart.

(*iii*) A paving machine, because of its duties, has inherent vibration from its operating machinery, its tampers and/or vibrators. On wheeled pavers, this inherent vibration is, to a large extent, absorbed due to pneumatic tyres and, therefore, the amount of vibrations affecting the operator is cut to the minimum leading to less fatigue. In tracked pavers the absorption of vibration is very less.

Tracked pavers have following advantages:

(*i*) These have advantages when the ground conditions are very soft or loose, as the greater pressure area over the tracks as against the foot print of the rubber tyres, assist it in travelling over these conditions on which the wheel version could quite easily develop wheel spin.

(*ii*) It has an advantages in paving oddly shaped intersections as it can manoeuver faster than the wheel paver.

CONSTRUCTION OF BRIDGES

Construction of Bridges

A bridge is a structure built to span a physical obstacle, such as a body of water, Valley, or Road, without closing the way underneath. It is constructed for the purpose of providing passage over the obstacle usually something that can be detrimental to cross otherwise. There are many different designs that serve a particular purpose and apply to different situations. Designs of bridges vary depending on the function of bridge, the nature of the Terrain where the bridge is constructed and anchored, the material used to make it, and the funds available to build it.

Bridges are mostly made for crossing rivers, valleys or roads. Nowadays, most big bridges are made to carry vehicles but people have also walked across bridges for thousands of years. Bridges called highway over passes carrier Road over another Road.

Types of Bridges

Bridges are classified as Road/ Highway bridges or Railway bridges. Highway bridges are mostly of concrete and Railway bridges are of Steel. There are five common types of bridges, and these are, Girder, arch, cable, rigid framed and truss. Each has its unique features and uses. Truss bridges are extremely effective because they have a high strength. Two of the most used truss bridges are of the Pratt and Howe design.

Beam bridges are the simplest Bridge type normally consists of one or more spans, supported by abutment or pier at each end. Beam bridges are usually constructed of RCC or steel or a combination of both RCC and steel. Types of beam bridges are girder Bridge, box girder Bridge, and plate girder Bridge etc. The most common types of modern bridges include : Beam, truss, arch, cantilever, cable stay and suspension. Beam bridges are horizontal beam supported at each end by substructure units and can be either simply supported when the beams only connect across a single span, or continuous when the beams are connected across two or more spans. When there are multiple spans, the intermediate supports are known as piers. Double decked (double decker) bridges have two levels.

A Truss Bridge is a bridge whose load bearing superstructure is composed of a truss. This truss is a structure of connected elements forming triangular units. The connected elements may be stressed form tension, compression, are sometimes both in response to dynamic loads. Truss bridges are one of the oldest types of modern bridges.

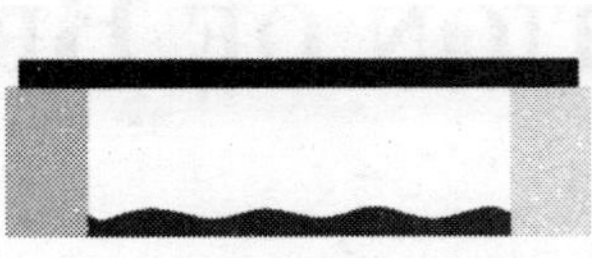

Beam bridge

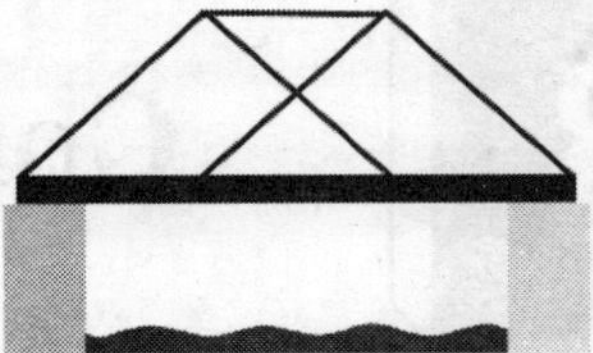

Truss bridge

In suspension bridges, piers those are built over a body of water, foundations are made by sinking caissons into the river bed and filling them with concrete. Next, the anchorages are built on both ends, usually of reinforced concrete with embedded eye bars to which the cables will be fastened. For permanent bridges in sea water the most common approach is to sink a pier down into the earth below the water allowing for shorter bridge spans. If the water is relatively shallow, and the loads not too great, individual piles or a precast concrete pile group can be driven through the water and into the river bed.

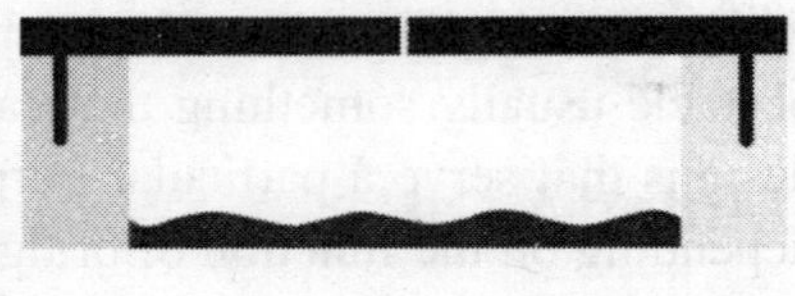

Cantilever bridge

Cantilever bridges are built using cantilevers horizontal beams supported on only one end. Most cantilever bridges use a pair of continuous spans that extend from opposite sides of the supporting piers to meet at the centre of the obstacle the bridge crosses. Cantilever bridges are constructed using much the same materials and techniques as beam bridges. The difference comes in the action of the forces through the bridge.

Arch bridges have abutments at each and. The weight of the bridge is thrust into the abutments at either side. The earliest known arch bridges were built by the Greeks. Tied arch bridges have an arch shaped superstructure, but differ from conventional arch bridges. Instead of transferring the weight of the bridge and traffic loads in to thrust forces into the abutments, the ends of the arches are restrained by tension in the bottom chord of the structure. They are also called bowstring arches.

Arch bridge

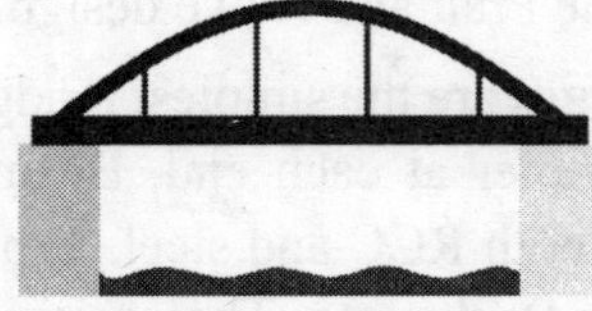

Tied arch bridge

Suspension bridges are suspended from cables. The earliest suspension bridges were made of ropes or wines covered with pieces of bamboo. In modern bridges, the cables hang from Towers that are attached to caissons or cofferdams.

Cable stayed bridges, like suspension bridges, are held up by cables. However, in a cable stayed Bridge less cable is required and the towers holding the cables are proportionately higher.

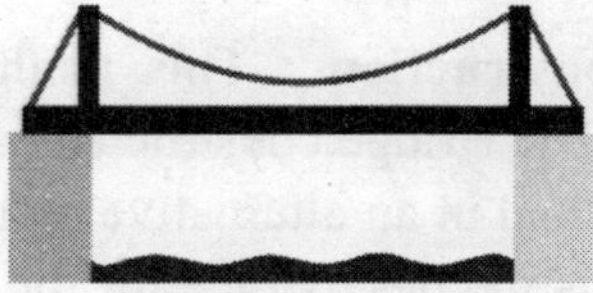
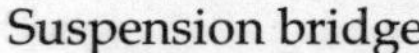

Suspension bridge

Cable-stayed bridge

Some engineer subdivide 'beam' bridges in two slab, beam and slab and box girder on the basis of their cross-section. A slab can be solid or voided reasons while beam and slab consists of concrete or steel girders connected by a concrete slab. A box girder cross section consists of a single cell or multicellular box. In recent years, integral bridge construction has also become popular.

Fixed or movable bridges : Most bridges are fixed bridges, meaning they have no moving parts and stay in one place until they fail or are demolished. Temporary bridges are designed to be assembled and taken apart, transported to a different site and reused. They are important in military engineering and are also used to carry traffic while and old Bridge is being rebuilt. Movable bridges are designed to move out of the way of boats or other kinds of traffic, which would otherwise be too tall to fit.

Materials used in bridge construction

Generally bridges are constructed or manufactured either of concrete, steel or cable depending upon span, type and load. Small footbridges are a small cantilevers maybe simple beams; however, large cantilever bridges designed to handle road or rail traffic, use of trusses built from structural Steel, or box girders built from pre-stressed concrete are used. The cables are usually made of steel cables galvanized with zinc are used for most of the suspension bridges, but some bridges are still made with steel-concrete. Stone, brick and other such materials that are strong in compression and somewhat so in shear are used for arch bridges. Beam bridges can use pre-stressed concrete, an inexpensive building material, which is then embedded with rebar. The resulting bridge can resist both compression and tension forces. The triangular pieces of truss bridges are manufactured from straight or steel sections.

Methods of Bridge Construction

- Cast-in-situ method of bridge construction
- Balanced cantilever method of bridge construction
- Precast method of bridge construction
- Span by span casting method of bridge construction
- Incremental launching method of bridge construction
- Cable stayed method of bridge construction
- Arch method for bridge construction

1. **Cast-in-situ Method of Bridge Construction :** This method is a flexible method of bridge construction where complex and unusual geometrical shapes can be constructed easily. A situation when it is hard to transport prefabricated elements either due to size or unreachability, this method is a good choice
2. **Balanced Cantilever Method of Bridge Construction :** This method is used for constructing bridges with the span 50 to 250m. The Bridge constructed can either be cast in place or precast. Here, the segments are attached in an alternative manner at opposite ends of the cantilever supported by piers. This is the best choice for the construction of long span bridges and cable-stayed bridges.
3. **Precast Method of Bridge Construction :** In this method, the bridge is constructed with the help of precast concrete elements. The prefabrication is performed in different methods. The precast elements include : precast beams, precast decks and precast segmental decks.
4. **Span by Span Casting Method of Bridge Construction :** This method is associated with cantilever construction method but with much advancement in the technique. It is considered as most economic and rapid method in construction. For long bridges and viaducts within individual span up to 60m, the method is feasible. Decks are begun at one abutment and constructed continuously by placing segments to the other end of the bridge. Segments can be positioned by either a temporary staying mass system through more commonly using an assembly truss.
5. **Incremental Launching Method of Bridge Construction :** The incremental launching method (ILM) is employed mainly for the construction of continuous concrete bridges or steel girder bridges. The ILM method performs the procedure and increments. With this method of construction, the bridge deck is built in sections by pushing the structure outwards from an abutment towards the pier. The method can be used for Bridge decks with the length greater than 250m.
6. **Cable Stayed Method of Bridge Construction :** In the cable stayed method of construction, cables are used to carry the bridge deck from one or both sides of the supporting tower. The cables carry and transfer all the loads to the foundation. Cable-stayed method of construction is used for constructing bridges that span more than 300m.
7. **Arch Method for Bridge Construction :** Arch shaped bridge construction is one of the most economical choices when the bridge under consideration is required to cross over landscape that are inaccessible. Many modern arch construction methods have made the arch construction more economical. The arch construction can be built with concrete or precast concrete. The cast in situ free cantilever method and slip formed sections are two main construction types.

Factors affecting selection of bridge construction methods :

Before a bridge can be built an appropriate method of construction must be chosen. The decision is made by the design team. The principal factors affect the selection of methods are :

1. The type of bridge and its span.
2. The obstacles to be crossed and nature of soil strata.
3. The horizontal and vertical profiles of the bridge decks.
4. The accessibility of the site and time allowed for construction.

INDEX

D

E

S

T

U

V

W

CONSTRUCTION ENGINEERING AND MANAGEMENT

DR. S.V. DEODHAR • S.C. SHARMA

FEEDBACK FORM

1. Which topics from your examination point of view are inadequately or not covered in this book?

____

____

2. Is there any misprints/ mistakes/ factual inaccuracy in this book? If so, please specify.

____

____

3. What is your assessment of this book as regards the presentation of the subject-matter, expression, precision and price in relation to the other recommended books available on this subject?

____

____

4. Which competing books you regard as better than this? Please mention their author and publishers.

____

____

5. Any further suggestions/comments you would like to make for the improvement of the book?

____

____

PS: If needed, please attach a separate sheet.

Name : ____________________

College : ____________________

Address : ____________________

Email : ____________________

Mobile : ____________________

Please mail the feedback form at:

KHANNA BOOK PUBLISHING CO. (P) LTD.

4C/4344, Ansari Road, Daryaganj, New Delhi-110002

You can also email this feedback form at contact@khannabooks.com *or* Whatsapp at +91 – 99109 09320

BEST FEEDBACK WILL BE REWARDED SUITABLY